Shape

ALSO BY JORDAN ELLENBERG

How Not to Be Wrong
The Grasshopper King

Shape

THE HIDDEN GEOMETRY OF INFORMATION, BIOLOGY, STRATEGY, DEMOCRACY, AND EVERYTHING ELSE

Jordan Ellenberg

PENGUIN PRESS
NEW YORK
2021

PENGUIN PRESS
An imprint of Penguin Random House LLC
penguinrandomhouse.com

Maps on page 361 by Caitlin Bourbeau

Portions of this book appeared in different form as "How Computers Turned Gerrymandering Into a Science" and "Five People. One Test. This is How You Get There" in the *New York Times*; as "The Supreme Court's Math Problem" and "Building a Better World Series" on Slate.com; and as "A Fellow of Infinite Jest" in the *Wall Street Journal*.

LIBRARY OF CONGRESS CATALOGING-IN-PUBLICATION DATA

Names: Ellenberg, Jordan, 1971– author.
Title: Shape : the hidden geometry of information, biology, strategy, democracy, and everything else / Jordan Ellenberg.
Description: First. | New York : Penguin Press, [2021] |
Includes bibliographical references and index.
Identifiers: LCCN 2020054440 (print) |
LCCN 2020054441 (ebook) | ISBN 9781984879059 (hardcover) |
ISBN 9781984879066 (ebook) |
ISBN 9780593299739 (export edition)
Subjects: LCSH: Geometry. | Shapes.
Classification: LCC QA446 .E45 2021 (print) |
LCC QA446 (ebook) | DDC 516—dc23
LC record available at https://lccn.loc.gov/2020054440
LC ebook record available at https://lccn.loc.gov/2020054441

Printed in the United States of America
1st Printing

Designed by Amanda Dewey

To the inhabitants of space in general
And CJ and AB in particular

CONTENTS

Shape

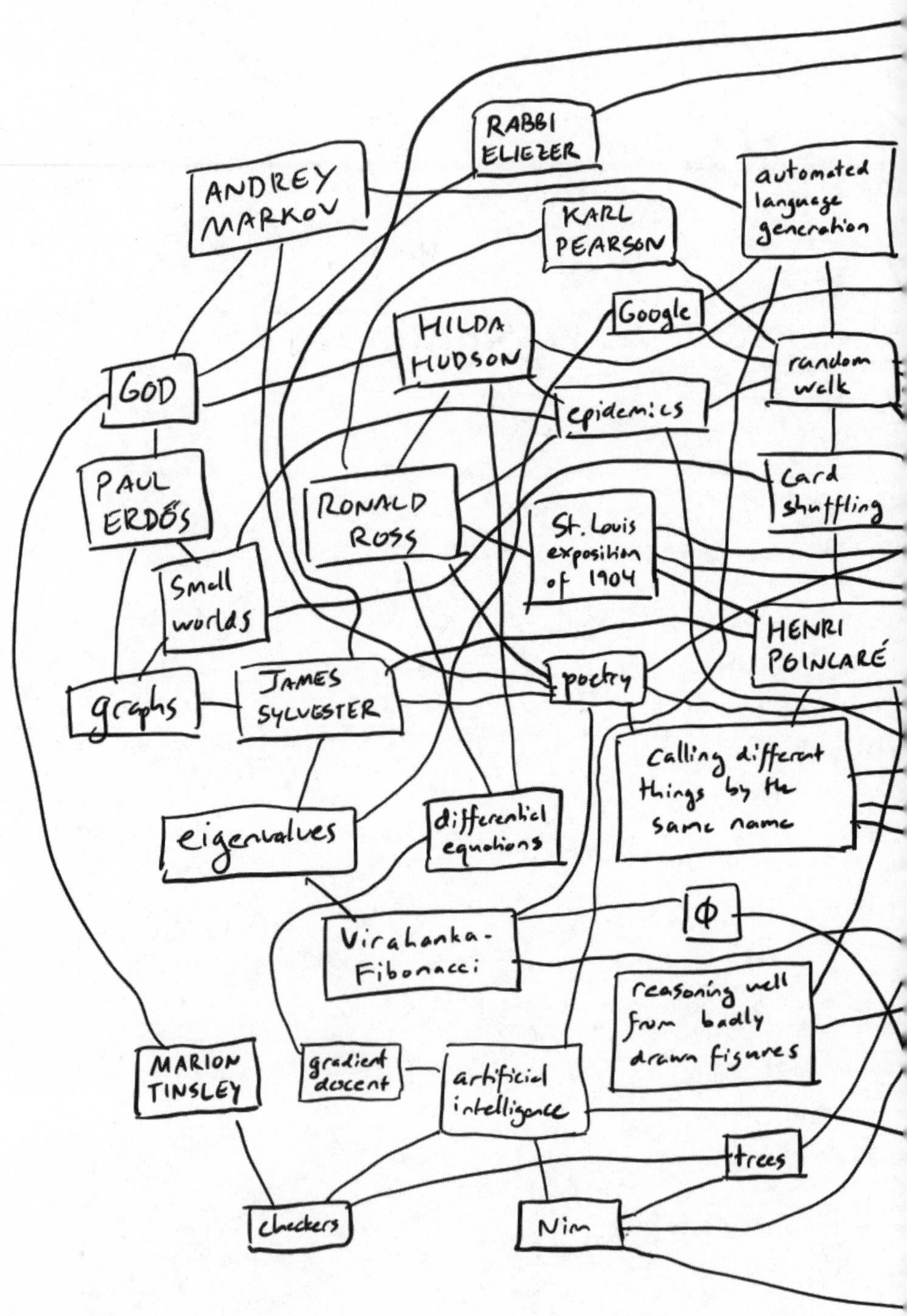
RABBI ELIEZER
ANDREY MARKOV
KARL PEARSON
automated language generation
Google
HILDA HUDSON
random walk
GOD
epidemics
PAUL ERDŐS
card shuffling
RONALD ROSS
St. Louis exposition of 1904
Small worlds
HENRI POINCARÉ
JAMES SYLVESTER
poetry
graphs
calling different things by the same name
eigenvalues
differential equations
ϕ
Virahanka-Fibonacci
reasoning well from badly drawn figures
MARION TINSLEY
gradient descent
artificial intelligence
trees
checkers
Nim

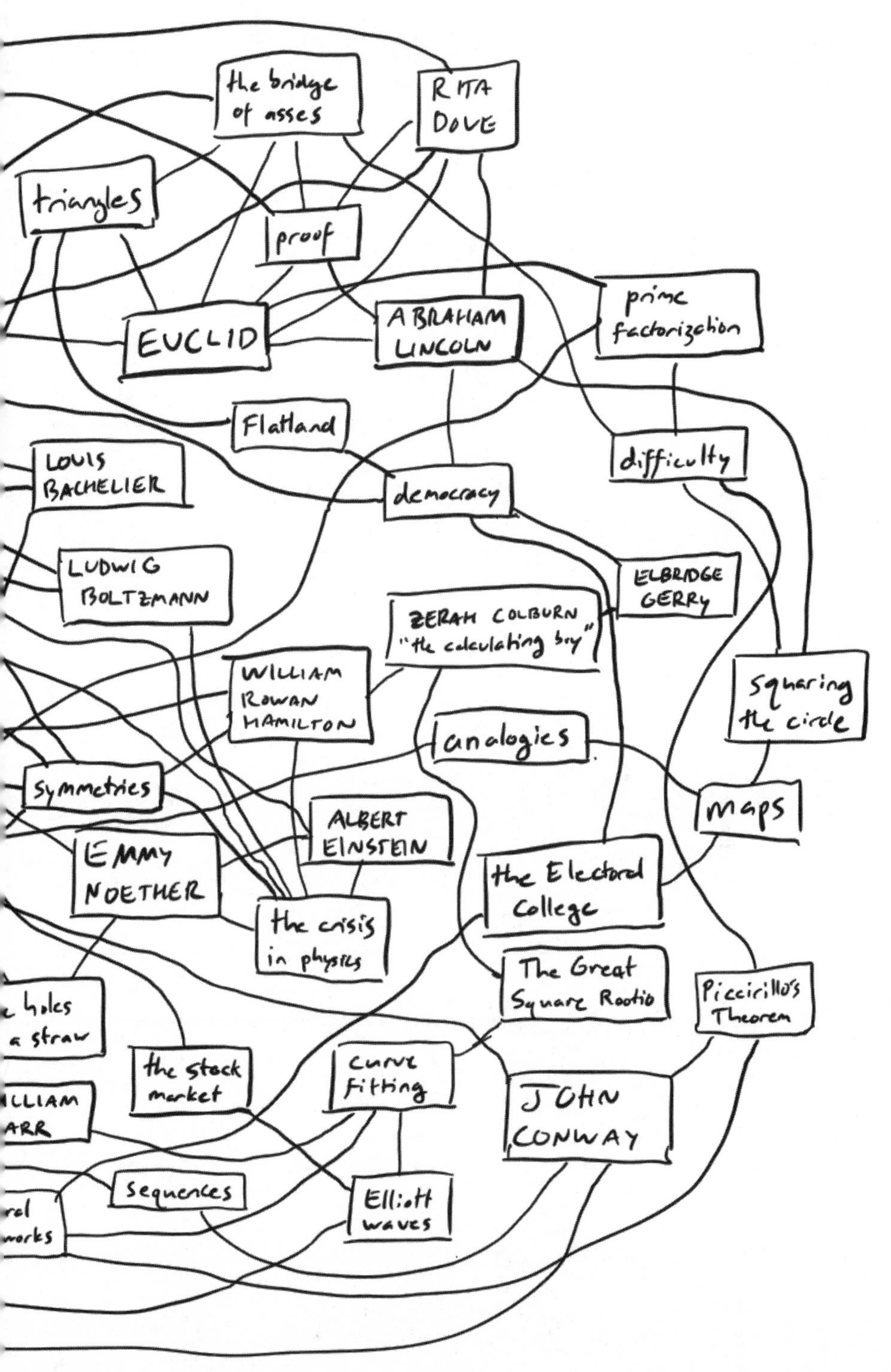

the bridge of asses
RITA DOVE
triangles
proof
EUCLID
ABRAHAM LINCOLN
prime factorization
Flatland
LOUIS BACHELIER
democracy
difficulty
LUDWIG BOLTZMANN
ELBRIDGE GERRY
ZERAH COLBURN "the calculating boy"
WILLIAM ROWAN HAMILTON
squaring the circle
analogies
symmetries
maps
ALBERT EINSTEIN
EMMY NOETHER
the Electoral College
the crisis in physics
The Great Square Root
Piccirillo's Theorem
holes
a straw
the stock market
curve fitting
JOHN CONWAY
LLIAM
ARR
sequences
Elliott waves
ral
works

INTRODUCTION

Where Things Are and What They Look Like

I am a mathematician who talks about math in public, and this seems to unlock something in people. They tell me things. They tell me stories I sense they haven't told anyone in a long time, maybe ever. Stories about math. Sometimes sad stories: a math teacher rubbing a kid's ego in the mud for no reason but meanness. Sometimes the story is happier: an experience of abrupt illumination that burst open a child's mind, an experience the grown-up wanted to find a path back to, but never quite could. (Actually, this one is kind of sad, too.)

Often these stories are about geometry. It seems to stand out in people's high school memories like a weird loud out-of-scale note in a chorus. There are people who hate it, who tell me geometry was the moment math stopped making sense to them. Others tell me it was the *only* part of math that made sense to them. Geometry is the cilantro of math. Few are neutral.

What makes geometry different? Somehow it's primal, built into our bodies. From the second we exit hollering from the womb we're reckoning where things are and what they look like. I'm not one of those people who will tell you everything important about our inner lives can be traced back to the needs of a shaggy band of savannah-dwelling hunter-gatherers, but it's hard to doubt that those folks had to develop knowledge of shapes, distances, and locations, probably before they had the words to talk about them. When South American mystics (and their non–South American imitators) drink ayahuasca, the sacred hallucinogenic tea, the first thing that happens—okay, the first thing that happens after the uncontrollable vomiting—is the perception of pure geometric form: repeating two-dimensional patterns like the latticework in a classical mosque, or full three-dimensional visions of hexahedral cells clustered into pulsating honeycombs. Geometry is still there when the rest of our reasoning mind is stripped away.

Reader, let me be straight with you about geometry: at first I didn't care for it. Which is weird, because I'm a mathematician now. Doing geometry is literally my job!

It was different when I was a kid on the math team circuit. Yes, there was a circuit. My high school's team was called the Hell's Angles and we filed into every meet in matching black T-shirts with a boom box playing "Hip to Be Square" by Huey Lewis and the News. And on that circuit I was well-known among my peers for balking whenever presented with "show angle APQ is congruent to angle CDF," or the like. Not that I didn't do those questions! But I did them in the most cumbersome possible way, which meant assigning numerical coordinates to each of the many points in the diagram, then grinding out pages of algebra and numerical computation in order to compute the areas of triangles and lengths of line segments. Anything to avoid actually doing geometry in the approved manner. Sometimes I got the problem right, sometimes I got the problem wrong. But it was ugly every time.

If there's such a thing as being geometric by nature, I'm the opposite. You can give a geometry test to a baby. You present a series of pairs of pictures; most of the time the two pictures are of the same shape, but every third time or so, the shape on the right-hand side is reversed. The babies spend more time looking at the reversed shapes. They know *something's* going on and their novelty-seeking minds strain toward it. And the babies that spend more time gaping at the mirrored shapes tend to score higher on math and spatial reasoning tests when they're preschoolers. They're quicker and more accurate at visualizing shapes and what they would look like if rotated or glued together. Me? I lack this ability almost completely. You know the little picture on the credit card machine at the gas station that shows you how to orient the card when you swipe it? That picture is useless to me. It's beyond my mental capabilities to translate that flat drawing into a three-dimensional action. Every single time, I have to run through each of the four possibilities—magnetic stripe up and to the right, magnetic stripe up and to the left, magnetic stripe down and to the right, magnetic stripe down and to the left—until the machine consents to read my card and sell me some gas.

And yet geometry is felt, generally, to be at the heart of what's required for real figuring in the world. Katherine Johnson, the NASA mathematician now well-known as the hero of the book and movie *Hidden Figures*, described her early success at the Flight Research Division: "The guys all had graduate degrees in mathematics; they had forgotten all the geometry they ever knew. . . . I still remembered mine."

MIGHTY IS THE CHARM

William Wordsworth, in the long, mostly autobiographical poem *The Prelude*, tells a somewhat implausible story about a shipwreck victim hurled ashore on an uninhabited island with nothing in his possession but a copy of Euclid's *Elements*, the book of geometric axioms and propositions that launched geometry as a formal subject about two and a half millennia ago. Good luck for shipwreck guy: depressed and hungry though he is, he consoles himself by working his way through Euclid's demonstrations one by one, tracing out the diagrams in the sand with a stick. That's just what it was like to be young, sensitive, poetic Wordsworth, writes middle-aged Wordsworth! Or to let the poet speak:

> Mighty is the charm
> of those abstractions to a mind beset
> With images, and haunted by itself.

(Ayahuasca drinkers have a similar take—the drug reboots the brain, and lifts the mind above the tortured labyrinth it thinks it's stuck in.)

The strangest thing about Wordsworth's shipwreck-geometry story is that it's basically true. Wordsworth borrowed it, with several phrases lifted intact, from the memoirs of John Newton, a young apprentice slave merchant who, in 1745, found himself, not shipwrecked per se, but left on Plantain Island off Sierra Leone by his boss, with little to do and less to eat. The island wasn't uninhabited; the enslaved Africans lived there with him, and his chief tormentor was an African woman who controlled the flow of food: "a person of some consequence in her own country," Newton describes her, then complains, in a truly astonishing failure to grasp the situation, "This woman (I know not for what reason) was strangely prejudiced against me from the first."

A few years later, Newton almost dies at sea, gets religion, becomes an Anglican priest, writes "Amazing Grace" (which has a very different prescription for what book you should study when you're depressed), and finally renounces the slave trade and becomes a major player in the movement to abolish slavery in the British Empire. But back on Plantain Island, yes—he had one book along, Isaac Barrow's edition of Euclid, and in his dark moments he retreated into its abstract comforts. "Thus I often beguiled my sorrows," he writes, "and almost forgot my feeling."

Wordsworth's appropriation of Newton's geometry-in-the-sand story wasn't his only flirtation with the subject. Thomas De Quincey, a contemporary of Wordsworth, wrote in his *Literary Reminiscences*: "Wordsworth was a profound admirer of the sublimer mathematics; at least of the higher geometry. The secret of this admiration for geometry lay in the antagonism between this world of bodiless abstraction and the world of passion." Wordsworth had done poorly in mathematics at school but formed a mutually admiring friendship with the young Irish mathematician William Rowan Hamilton, who some believe inspired Wordsworth to add to *The Prelude* the famous description of Newton (Isaac, not John): "A mind forever / Voyaging through strange seas of Thought, alone."

Hamilton was fascinated with all forms of scholastic knowledge—mathematics, ancient languages, poetry—from his earliest youth, but found his interest in math hyperactivated by a childhood encounter with Zerah Colburn, the "American Calculating Boy." Colburn, as a six-year-old boy from a Vermont farm family of modest means, was discovered by his father, Abia, sitting on the floor reciting multiplication tables he had never been taught. The boy proved to have immense powers of mental calculation, unlike anything before seen in New England. (He also, like all the men in his family, had six fingers on each hand and six toes on each foot.) Zerah's father brought him to meet with various local dignitaries, including the governor of Massachusetts, Elbridge Gerry (we'll come back to this guy later in a very different context), who advised Abia that only in Europe were there people capable of understanding and nurturing the boy's peculiar skills. They crossed the Atlantic in 1812, at which point Zerah was alternately educated and exhibited for money across Europe. In Dublin he appeared alongside a giant, an albino, and Miss Honeywell, an American woman who performed feats of dexterity with her toes. And in 1818, now aged fourteen, he engaged in a calculation competition with Hamilton, his Irish teen math counterpart, in which Hamilton "came off with honor, though his antagonist was generally the victor." But Colburn did not go on in mathematics; his interest was purely in mental computation. When Colburn studied Euclid, he found it easy, but "dry and devoid of interest." And when Hamilton met the Calculating Boy two years later and quizzed him about his methods ("He has lost every trace of his sixth finger," Hamilton recalls; Colburn had had it cut off by a London surgeon), he found that Colburn had little insight into the reasons his arithmetic methods worked. After abandoning his education, Colburn tried his hand on the English stage, didn't succeed there, moved back to Vermont, and lived out his life as a preacher.

When Hamilton met Wordsworth in 1827, he was just twenty-two, and had already been appointed a professor at the University of Dublin and the royal astronomer of Ireland. Wordsworth was fifty-seven. Hamilton wrote a letter to his sister describing their meeting: the young mathematician and the old poet took "a *midnight walk* together for a long, long time, *without any companion* except the stars and our own burning thoughts and words." As his style here suggests, Hamilton had not wholly given up his poetic ambitions. He immediately began to send his poems to Wordsworth, who responded warmly but critically. Shortly afterward, Hamilton renounced poetry; in fact, he did so in verse, directly addressing the Muse in a poem called "To Poetry," which he sent on to Wordsworth. Then, in 1831, he changed his mind, marking his decision by writing *another* poem called "To Poetry." He sent that one to Wordsworth, too. Wordsworth's response is one of the all-time classic gentle letdowns: "You send me showers of verses, which I receive with much pleasure, as do we all; yet have we fears that this employment may seduce you from the path of Science which you seem destined to tread with so much honour to yourself and profit to others."

Not everyone in Wordsworth's circle appreciated the interplay between passion and cold strange lonely reason as much as he and Hamilton did. At a dinner party at the painter Benjamin Robert Haydon's house at the end of 1817, Wordsworth's friend Charles Lamb got drunk and began to tease Wordsworth by abusing Newton, calling the scientist "a fellow who believed nothing unless it was as clear as the three sides of the triangle." John Keats joined in to accuse Newton of having stripped the rainbow of all its romance by showing that a prism exhibits the same optical effect. Wordsworth laughed along, his jaw set shut, one imagines, to avoid a quarrel.

De Quincey's portrait of Wordsworth goes on to advertise yet another math scene in *The Prelude,* still unpublished at the time. Poems had trailers in those days! In this scene, which De Quincey excitedly promises "reaches the very *ne plus ultra* of sublimity in my opinion," Wordsworth falls asleep while reading *Don Quixote* and dreams of meeting a Bedouin riding a camel across the empty desert. The Arab has two books in his hand, except one of the books, in the way of dreams, is not just a book but a heavy stone, too, and the other book is also a glowing seashell. (A few pages later, the Bedouin himself turns out to be Don Quixote.) The seashell-book issues apocalyptic prophecies when you hold it up to your ear. And the stone-book? That's Euclid's *Elements* again, here appearing not as a humble instrument of self-help but as a means of connection with the uncaring and unchanging cosmos: the book "wedded soul to soul in purest bond / Of reason, undisturbed by space or time." It makes sense De Quincey would be pretty into this psychedelic stuff; he was a former child prodigy who picked up a tenacious laudanum habit and wrote up his dizzying visions in *Confessions of an English Opium-Eater,* a sensational bestseller of the early nineteenth century.

Wordsworth's take is typical of geometry as viewed from a distance. Admiration, yes, but the way we admire an Olympic gymnast, executing flips and contortions that seem impossible for ordinary humans. It's what you get, too, in the most famous geometry poem of all, Edna St. Vincent Millay's sonnet "Euclid alone has looked on Beauty bare."* Millay's Euclid is a singular, unearthly figure, blasted with enlightenment by a shaft of insight on a "holy, terrible day." Not like the rest of us, who Millay says might, *if we're lucky,* get to hear Beauty's footsteps hurrying off down a faraway hallway.

That's not the geometry this book is about. Don't get me wrong—as a mathematician, I get a lot of benefit from geometry's prestige. It feels good when people think the work you do is mysterious, eternal, elevated above the common plane. "How was your day?" "Oh, holy and terrible, the usual."

But the harder you push that point of view, the more you incline people to see the study of geometry as an obligation. It acquires the slightly musty smell of something admired because it is good for one. Like opera. And that kind of admiration isn't enough to sustain the enterprise. There are plenty of new operas—but can you name them? No: you hear the word "opera" and you think of a mezzo-soprano in furs bellowing Puccini, probably in black and white.

There's plenty of new geometry, too, and, like new opera, it's not as well publicized as it could be. Geometry isn't Euclid, and it hasn't been for a long time. It's not a cultural relic, trailing an odor of the schoolroom, but a living subject, moving faster now than it ever has before. In the chapters to come we'll encounter the new geometry of pandemic spread, of the messy U.S. political process, of professional-level checkers, of artificial intelligence, of the English language, of finance, of

* By 1922, when Millay wrote this, Euclid was in fact no longer alone; non-Euclidean geometries, in their own way just as beautiful unclad, had not only been discovered but were understood, thanks to Einstein, to be the true underlying geometry of space, as we'll see in chapter 3. I wondered whether Millay knew about this and was adopting an intentionally anachronistic persona here, but my poetry scholar friends tell me she was likely not up on the latest in mathematical physics.

physics, even of poetry. (A *lot* of geometers secretly dreamed, like William Rowan Hamilton, of being poets.)

We are living in a wild geometric boomtown, global in scope. Geometry isn't out there beyond space and time, it's right here with us, mixed in with the reasoning of everyday life. Is it beautiful? Yes, but not bare. Geometers see Beauty with its work clothes on.

Chapter 1

"I Vote for Euclid"

In 1864, the Reverend J. P. Gulliver, of Norwich, Connecticut, recalled a conversation with Abraham Lincoln about how the president had acquired his famously persuasive rhetorical skill. The source, Lincoln said, was geometry.

> In the course of my law-reading I constantly came upon the word *demonstrate*. I thought, at first, that I understood its meaning, but soon became satisfied that I did not. . . . I consulted Webster's Dictionary. That told of "certain proof," "proof beyond the possibility of doubt;" but I could form no idea what sort of proof that was. I thought a great many things were proved beyond a possibility of doubt, without recourse to any such extraordinary process of reasoning as I understood "demonstration" to be. I consulted all the dictionaries and books of reference I could find, but with no better results. You might as well have defined *blue* to a blind man. At last I said, "Lincoln, you can never make a lawyer if you do not understand what *demonstrate* means;" and I left my situation in Springfield, went home to my father's house, and staid there till I could give any propositions in the six books of Euclid at sight. I then found out what "demonstrate" means, and went back to my law studies.

Gulliver was fully on board, replying, "No man can talk well unless he is able first of all to define to himself what he is talking about. Euclid, well studied, would free the world of half its calamities, by banishing half the nonsense which now deludes and curses it. I have often thought that Euclid would be one of the best books to put on the catalogue of the Tract Society, if they could only get people to read it. It would be a means of grace." Lincoln, Gulliver tells us, laughed and agreed: "I vote for Euclid."

Lincoln, like the shipwrecked John Newton, had taken up Euclid as a source of solace at a rough time in his life; in the 1850s, after a single term in the House of Representatives, he seemed finished in politics and was trying to make a living as an ordinary traveling lawyer. He had learned the rudiments of geometry in his earlier job as a surveyor and now aimed to fill the gaps. His law partner William Herndon, who often had to share a bed with Lincoln at small country inns in their sojourns around the circuit, recalls Lincoln's method of study; Herndon would fall asleep, while Lincoln, his long legs hanging over the edge of the bed, would stay up late into the night with a candle lit, deep in Euclid.

One morning, Herndon came upon Lincoln in their offices in a state of mental disarray:

> He was sitting at the table and spread out before him lay a quantity of blank paper, large heavy sheets, a compass, a rule, numerous pencils, several bottles of ink of various colors, and a profusion of stationery and writing appliances generally. He had evidently been struggling with a calculation of some magnitude, for scattered about were sheet after sheet of paper covered with an unusual array of figures. He was so deeply absorbed in study he scarcely looked up when I entered.

Only later in the day did Lincoln finally get up from his desk and tell Herndon that he had been trying to square the circle. That is, he was trying to construct a square with the same area as a given circle, where to "construct" something, in proper Euclidean style, is to draw it on the page using just two tools, a straightedge and a compass. He worked at the problem for two straight days, Herndon remembers, "almost to the point of exhaustion."

> I have been told that the so-called squaring of the circle is a practical impossibility, but I was not aware of it then, and I doubt if Lincoln was. His attempt to establish the proposition having ended in failure, we, in the office, suspected that he was more or less sensitive about it and were therefore discreet enough to avoid referring to it.

Squaring the circle is a very old problem, whose fearsome reputation I suspect Lincoln might actually have known; "squaring the circle" has been a metaphor for a difficult or impossible task for a long time. Dante name-checks it in the *Paradiso*: "Like the geometer who gives his all trying to square the circle, and still can't find the idea he needs, *that's* how I was." In Greece, where it all started, a standard exasperated comment when someone is making a task harder than necessary is to say, "I wasn't asking you to square the circle!"

There is no *reason* one needs to square a circle—the problem's difficulty and fame is its own motivation. People with a conquering mentality tried to square circles from antiquity until 1882, when Ferdinand von Lindemann proved it couldn't be done (and even then a few diehards persisted; okay, even *now*). The seventeenth-century political philosopher Thomas Hobbes, a man whose confidence in his own mental powers is not fully captured by the prefix "over," thought he'd cracked it. Per his biographer John Aubrey, Hobbes discovered geometry in middle age and quite by accident:

> Being in a Gentleman's Library, Euclid's *Elements* lay open, and 'twas the 47 *El. Libri* 1. He read the Proposition. By G_, saydd he (he would now and then sweare an emphaticall Oath by way of emphasis) *this is impossible!* So he reads the Demonstration of it, which referred him back to such a Proposition; which proposition he read. That referred him back to another, which he also read. *Et sic deinceps* that at last he was demonstratively convinced of that trueth. This made him in love with Geometry.

Hobbes was constantly publishing new attempts and getting in petty feuds with the major British mathematicians of the time. At one point, a correspondent pointed out that one of his constructions was not quite correct because two points P and Q he claimed to be equal were actually at very slightly different distances from a third point R; 41 and about 41.012 respectively. Hobbes retorted that his points were big enough in extent to cover such a minor difference. He went to his grave still telling people he'd squared the circle.*

An anonymous commentator in 1833, reviewing a geometry textbook, described the typical circle-squarer in a way that quite precisely depicts both Hobbes, two centuries prior, and intellectual pathologies still hanging around here in the twenty-first:

> [A]ll they know of geometry is, that there are in it some things which those who have studied it most have long confessed themselves unable to do. Hearing that the authority of knowledge bears too great a sway over the minds of men, they propose to counterbalance it by that of ignorance: and if it should chance that any person acquainted with the subject has better employment than hearing them unfold hidden truths, he is a bigot, a smotherer of the light of truth, and so forth.

In Lincoln, we find a more appealing character: enough ambition to try, enough humility to accept that he hadn't succeeded.

What Lincoln took from Euclid was the idea that, if you were careful, you could erect a tall, rock-solid building of belief and agreement by rigorous deductive steps, story by story, on a foundation of axioms no one could doubt: or, if you like, truths one holds to be self-evident. Whoever *doesn't* hold those truths to be self-evident is excluded from discussion. I hear the echoes of Euclid in Lincoln's most famous speech, the Gettysburg Address, where he characterizes the United States as "dedicated to the proposition that all men are created equal." A "propo-

* The long and frankly hilarious story of Hobbes's war against his patient mathematical critics is told in chapter 7 of *Infinitesimal*, by Amir Alexander.

sition" is the term Euclid uses for a fact that follows logically from the self-evident axioms, one you simply cannot rationally deny.

Lincoln wasn't the first American to look for a basis of democratic politics in Euclidean terms; that was the math-loving Thomas Jefferson. Lincoln wrote, in a letter read at an 1859 Jefferson commemoration in Boston he was unable to attend:

> One would start with great confidence that he could convince any sane child that the simpler propositions of Euclid are true; but, nevertheless, he would fail, utterly, with one who should deny the definitions and axioms. The principles of Jefferson are the definitions and axioms of free society.

Jefferson had studied Euclid at William and Mary as a young man, and esteemed geometry highly ever afterward.* While vice president, Jefferson took the time to answer a letter from a Virginia student about his proposed plan of academic study, saying: "Trigonometry, so far as this, is most valuable to every man, there is scarcely a day in which he will not resort to it for some of the purposes of common life" (though he describes much of higher mathematics as "but a luxury; a delicious luxury indeed; but not to be indulged in by one who is to have a profession to follow for his subsistence").

In 1812, retired from politics, Jefferson wrote to his predecessor in the presidency, John Adams:

> I have given up newspapers in exchange for Tacitus and Thucydides, for Newton and Euclid; and I find myself much the happier.

Here we see a real difference between the two geometer-presidents. For Jefferson, Euclid was part of the classical education required of a cultivated patrician, of a piece with the Greek and Roman historians and the scientists of the Enlightenment. Not so for Lincoln, the self-educated

* Though "we hold these truths to be self-evident" wasn't Jefferson's line; his first draft of the Declaration has "we hold these truths to be sacred & undeniable." It was Ben Franklin who scratched out those words and wrote "self-evident" instead, making the document a little less biblical, a little more Euclidean.

rustic. Here's the Reverend Gulliver again, recalling Lincoln recalling his childhood:

> I can remember going to my little bedroom, after hearing the neighbors talk of an evening with my father, and spending no small part of the night walking up and down, and trying to make out what was the exact meaning of some of their, to me, dark sayings. I could not sleep, though I often tried to, when I got on such a hunt after an idea, until I had caught it; and when I thought I had got it, I was not satisfied until I had repeated it over and over, until I had put it in language plain enough, as I thought, for any boy I knew to comprehend. This was a kind of passion with me, and it has stuck by me, for I am never easy now, when I am handling a thought, till I have bounded it north and bounded it south, and bounded it east and bounded it west. Perhaps that accounts for the characteristic you observe in my speeches.

This is not geometry, but it's the mental habit of the geometer. You don't settle for leaving things half-understood; you boil down your thoughts and trace back their steps of reason, just as Hobbes had amazedly watched Euclid do. This kind of systematic self-perception, Lincoln thought, was the only way out of confusion and darkness.

For Lincoln, unlike Jefferson, the Euclidean style isn't something belonging to the gentleman or the possessor of a formal education, because Lincoln was neither. It's a hand-hewn log cabin of the mind. Built properly, it can withstand any challenge. And anybody, in the country Lincoln conceived, can have one.

FROZEN FORMALITY

The Lincolnian vision of geometry for the American masses, like a lot of his good ideas, was only incompletely realized. By the middle of the nineteenth century, geometry had moved from college to the public high school; but the typical course used Euclid as a kind of museum piece, whose proofs were to be memorized, recited, and to some extent appreciated. How anyone might have *come up* with those proofs was not to be spoken of. The proof-maker himself almost disappeared: one writer of the time remarked that "many a youth reads six books of the *Elements* before he happens to be informed that Euclid is not the name of a science, but of a man who wrote upon it." The paradox of education: what we most admire we put in a box and make dull.

To be fair, there is not much to say about the historical Euclid, because there is not much we know about the historical Euclid. He lived and worked in the great city of Alexandria, in North Africa, sometime around 300 BCE. That's it—that's what we know. His *Elements* collects the knowledge of geometry possessed by Greek mathematics at the time, and lays the foundations of number theory for dessert. Much of the material was known to mathematicians prior to Euclid's time, but what's radically new, and was instantly revolutionary, is the *organization* of that huge body of knowledge. From a small set of axioms, which were almost impossible to doubt,* one derives step by step the whole apparatus of theorems about triangles, lines, angles, and circles. Before Euclid—if there actually was a Euclid, and not a shadowy collective of geometry-minded Alexandrians writing under that name—such a structure would have been unimaginable. Afterward, it was a model for everything admirable about knowledge and thought.

There is, of course, another way to teach geometry, which emphasizes invention and tries to put the student in the Euclidean cockpit, with the power to make their own definitions and see what comes of them. One such textbook, *Inventional Geometry,* starts from the premise that "the only true education is self-education." Don't look at other people's constructions, the book counsels, "at least until you have discovered a construction of your own," and avoid anxiety and comparing yourself with other students, because everyone learns at their own pace and you're more likely to master the material if you're enjoying yourself. The book itself is no more than a series of puzzles and problems, 446 in all. Some of these are straightforward: "Can you make three angles with

* Except one; but the vexed question of the "parallel postulate," and the two-thousand-year journey toward non-Euclidean geometry it launched, is well-told elsewhere and will only be glanced at here.

two lines? Can you make four angles with two lines? Can you make more than four angles with two lines?" Some of them, the author warns, are not actually solvable, the better to put yourself in the position of a *true* scientist. And some of them, like the very first one, have no clear "right answer" at all: "Place a cube with one face flat on a table, and with another face towards you, and say which dimension you consider to be the thickness, which the breadth, and which the length." Altogether, it is just the kind of "child-centered," exploratory approach that traditionalists deride as what's wrong with education nowadays. It came out in 1860.

A few years ago, the mathematics library at the University of Wisconsin came into possession of a huge trove of old math textbooks, books that had actually been used by Wisconsin schoolchildren over the last hundred years or so* and eventually discarded in favor of newer models. Looking at the weathered books, you see that every controversy in education has been waged before, multiple times, and everything we think of as new and strange—math books like *Inventional Geometry* that ask students to come up with proofs on their own, math books that make problems "relevant" by relating them to students' everyday lives, math books designed to advance social causes, progressive or otherwise—is also old, and was thought of as strange at the time, and no doubt will be new and strange again in the future.

A note in passing: the introduction to *Inventional Geometry* mentions that geometry has "a place in the education of all, not excepting that of women"—the book's author, William George Spencer, was an early advocate of coeducation. A more common nineteenth-century attitude toward women and geometry is conveyed in (but not endorsed by) *The Mill on the Floss*, by George Eliot†, published the same year as Spencer's textbook: "Girls can't do Euclid, can they, sir?" one character asks the schoolmaster Mr. Stelling, who responds, "They've a great deal of superficial cleverness; but they couldn't go far into anything." Stelling

* In one of the books of basic arithmetic, last used around 1930, I found a small penciled notation in the margin: "turn to p. 170"—on p. 170 there was another instruction, "turn to p. 36," where I got a new command, and so on and so on, until I came to the last page, where I found written, "You're a fool!" Pranked by a ten-year-old from beyond the grave.

† In this context it's relevant that "George Eliot" was a pen name for Mary Ann Evans.

represents, in satirically exaggerated form, the traditional mode of British pedagogy Spencer was rebelling against: a long march through memorization of the masters, in which the slow messy process of building understanding is not just neglected but actively guarded against. "Mr. Stelling was not the man to enfeeble and emasculate his pupil's mind by simplifying and explaining." Euclid, a kind of tonic of manliness, was to be suffered straight, like a strong drink or an ice-cold shower.

Even in the highest reaches of mathematical research, dissatisfaction with Stellingism had begun to build. The British mathematician James Joseph Sylvester, whose geometry and algebra (and distaste for the stultified deadness of British academia) we'll be talking about later, thought Euclid should be hidden "far out of the schoolboy's reach," and geometry taught through its relation to physical science, with an emphasis on the geometry of *motion* supplementing Euclid's static forms. "It is this living interest in the subject," Sylvester wrote, "which is so wanting in our traditional and mediaeval modes of teaching. In France, Germany, and Italy, everywhere where I have been on the Continent, mind acts direct on mind in a manner unknown to the frozen formality of our academic institutions."

BEHOLD!

We don't make students memorize and recite Euclid anymore. In the late nineteenth century, textbooks started including exercises, asking students to construct their own proofs of geometric propositions. In 1893, the Committee of Ten, an educational plenum convened by Harvard president Charles Eliot and charged with rationalizing and standardizing American high school education, codified this shift. The point of geometry in high school, they said, was to train up the student's mind in the habits of strict deductive reasoning. This idea has stuck. A survey conducted in 1950 asked five hundred American high school teachers about their objectives in teaching geometry: the most popular answer by far was "To develop the habit of clear thinking and precise expression," which got almost twice as many votes as "To give a knowledge of the facts and principles of geometry." In other words, we are not here to stuff our students with every known fact about triangles, but to develop in them the mental discipline to build up those facts from first principles. A school for little Lincolns.

And what is that mental discipline for? Is it because, at some point in the student's later life, they will be called upon to demonstrate, finally and incontrovertibly, that the sum of the exterior angles of a polygon is 360 degrees?

I keep waiting for that to happen to me and it never has.

The ultimate reason for teaching kids to write a proof is not that the world is full of proofs. It's that the world is full of *non-proofs*, and grown-ups need to know the difference. It's hard to settle for a non-proof once you've really familiarized yourself with the genuine article.

Lincoln knew the difference. His friend and fellow lawyer Henry Clay Whitney recalled: "[M]any a time have I seen him tear the mask off from a fallacy and shame both the fallacy and its author." We encounter non-proofs in proofy clothing all the time, and unless we've made ourselves especially attentive, they often get by our defenses. There are tells you can look for. In math, when an author starts a sentence with "Clearly," what they are really saying is "This seems clear to me and I probably should have checked it, but I got a little confused, so I settled for just asserting that it was clear." The newspaper pundit's analogue is the sentence starting "Surely, we can all agree." Whenever you see this, you should at all costs *not* be sure that all agree on what follows. You are being asked to treat something as an axiom, and if there's one thing we can learn from the history of geometry, it's that you shouldn't admit a new axiom into your book until it really proves its worth.

Always be skeptical when someone tells you they're "just being logical." If they are talking about an economic policy or a culture figure whose behavior they deplore or a relationship concession they want you to make, and not a congruence of triangles, they are not "just being logical," because they're operating in a context where logical deduction—if it applies at all—can't be untangled from everything else. They want you to mistake an assertively expressed chain of opinions as the proof of a theorem. But once you've experienced the sharp *click* of an honest-to-goodness proof, you'll never fall for this again. Tell your "logical" opponent to go square a circle.

What was distinctive about Lincoln, Whitney says, wasn't that he possessed a superpowered intellect. Lots of people in public life, Whitney writes ruefully, are very smart, and among them one finds both the good and the bad. No: what made Lincoln special was that "it was morally impossible for Lincoln to argue dishonestly; he could no more do it than he could steal; it was the same thing to him in essence, to despoil a man of his property by larceny, or by illogical or flagitious reasoning." What Lincoln had taken from Euclid (or what, already existing in Lincoln, harmonized with what he found in Euclid) was *integrity*, the principle that one does not say a thing unless one has justified, fair and square, that one has the right to say it. Geometry is a form of honesty. They might have called him Geometrical Abe.

The one place I'll part ways with Lincoln is in his shaming the author of the fallacy. Because the hardest person to be honest with is yourself, and it's our self-authored fallacies we need to spend the most time and effort unmasking. You should always be prodding your beliefs as you would a loose tooth, or, better, a tooth whose looseness you're not quite sure about. And if something's not solid, shame is not required, just a calm retreat to the ground you're sure about, and a reassessment of where you can get to from there.

That, ideally, is what geometry has to teach us. But the "frozen formality" Sylvester complained about is far from gone. In practice, the lesson we often teach kids in geometry class is, as math writer–cartoonist–raconteur Ben Orlin puts it:

> A proof is an incomprehensible demonstration of a fact that you already knew.

Orlin's example of such a proof is the "right angle congruence theorem," the assertion that any two right angles are congruent to each other. What might be asked of a ninth grader presented with this assertion? The most typical format is the *two-column proof*, a mainstay of geometry education for more than a century, which in this case would look something like this:

angle 1 and angle 2 are both right angles	**given**
the measure of angle 1 equals 90 degrees	**definition of right angle**
the measure of angle 2 equals 90 degrees	**definition of right angle**
the measure of angle 1 equals the measure of angle 2	**transitivity of equality**
angle 1 is congruent to angle 2	**definition of congruence**

"Transitivity of equality" is one of Euclid's "common notions," arithmetic principles he states at the beginning of the *Elements* and treats as prior even to the geometric axioms. It is the principle that two things which are equal to the same thing are thereby equal to each other.*

I don't want to deny that there's a certain satisfaction in reducing everything to such tiny, precise steps. They snap together so satisfyingly, like Lego! That feeling is something a teacher truly wants to convey.

And yet . . . isn't it *obvious* that two right angles are the same thing, just placed on the page in a different place and pointing in a different direction? Indeed, Euclid makes the equality of any two right angles the fourth of his axioms, the basic rules of the game that are taken to be true without proof and from which all else is derived. So why would a modern high school require students to manufacture a proof of this fact when even Euclid said, "Come on, that's obvious?" Because there are many different sets of starting axioms from which one can derive plane geometry, and proceeding exactly as Euclid did is generally no longer considered the most rigorous or the most pedagogically beneficial choice. David Hilbert rewrote the whole foundation from scratch in 1899, and the axioms used in American schools today typically owe more to those laid down by George Birkhoff in 1932.

Whether it's an axiom or not, the fact that two right angles are equal is something the student just plain knows. You can't blame someone for

* Tony Kushner's screenplay for Steven Spielberg's *Lincoln* movie has Lincoln invoke this in a dramatic moment.

being frustrated when you tell them, "You *think* you knew that, but you didn't *really* know it until you followed the steps in the two-column proof." It's a little insulting!

Too much of geometry class is devoted to proving the obvious. I remember well a course in topology I took my first year of college. The professor, a very distinguished elder researcher, spent two weeks proving the following fact: if you draw a closed curve in the plane, no matter how squiggly and weird it may be, the curve cuts the plane into two pieces; the part outside the curve and the part inside.

Now, on the one hand it's quite difficult, it turns out, to write a formal proof of this fact, known as the Jordan Curve Theorem.* On the other hand, I spent those two weeks in a state of barely controlled irritation. Was *this* what math was truly about? Making the obvious laborious? Reader, I zoned out. So did my classmates, among them many future mathematicians and scientists. Two kids who sat right in front of me, very serious students who would go on to earn PhDs in math at top-five universities, would start vigorously making out every time Distinguished Elder Researcher turned back to the board to chalk out yet another delicate argument on a perturbation of a polygon. I mean just really going at it, as if the force of their teen hunger for each other could somehow rip them into another part of the continuum where this proof was not still taking place.

A highly trained mathematician such as my current self might say, standing up a little straighter: well, young people, you are simply not sophisticated enough to know which statements are truly obvious and which conceal subtleties. Perhaps I would bring up the feared Alexander Horned Sphere, which shows that the analogous question in three-dimensional space is not as simple as one might imagine.

But pedagogically, I think that's a pretty bad answer. If we take our time in class to prove things that seem obvious, and insist that those statements are *not* obvious, our students will stew in resentment, just like I did, or find something more interesting to do while the teacher isn't looking.

* Different Jordan.

I like the way master teacher Ben Blum-Smith describes the problem: for students to really feel the fire of math, they have to experience the *gradient of confidence*—the feeling of moving from something obvious to something not-obvious, pushed uphill by the motor of formal logic. Otherwise, we're saying, "Here is a list of axioms that seem pretty obviously correct; put these together until you have another statement that seems pretty obviously correct." It's like teaching somebody about Lego by showing them how you can make two little bricks into one big brick. You can do that, and sometimes you need to, but it is definitely not the point of Lego.

The gradient of confidence is perhaps better experienced than just talked about. If you want to *feel* it, think for a moment about a right triangle.

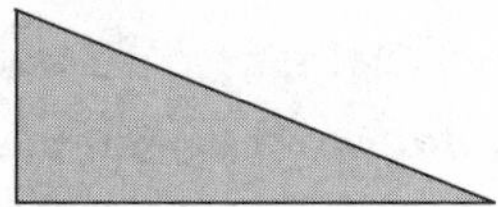

One starts with an intuition: if the vertical and horizontal sides are determined, so is the diagonal side. Walking 3 km south and then 4 km east leaves you a certain distance from your starting point; there is no ambiguity about it.

But what *is* the distance? That's what the Pythagorean Theorem, the first real theorem ever proved in geometry, is for. It tells you that if a and b are the vertical and horizontal sides of a right triangle, and c is the diagonal side, the so-called hypotenuse, then

$$a^2 + b^2 = c^2$$

In case a is 3 and b is 4, this tells us that c^2 is $3^2 + 4^2$, or 9 + 16, or 25. And we know what number, when squared, yields 25; it is 5. That's the length of the hypotenuse.

Why would such a formula be true? You could start climbing the gradient of confidence by literally drawing a triangle with sides 3 and 4 and measuring its hypotenuse—it would look really close to 5. Then draw a triangle with sides 1 and 3 and measure *its* hypotenuse; if you were careful enough with the ruler, you'd get a length really close to 3.16 . . . whose square is 1 + 9 = 10. Increased confidence derived from examples isn't a proof. But this is:

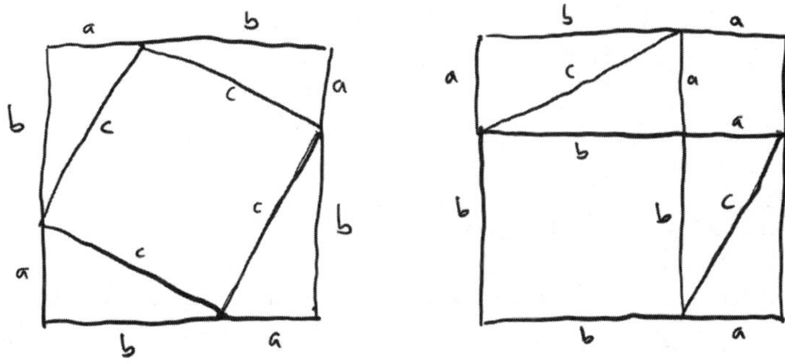

The big square is the same in both pictures. But it's cut up in two different ways. In the first picture, you have four copies of our right triangle, and a square whose side has length c. In the second picture, you also have four copies of the triangle, but they're arranged differently; what's left of the square is now two smaller squares, one whose side has length a and one whose side has length b. The area that remains when you take four copies of the triangle out of the big square has to be the same in both pictures, which means that c^2 (the area left over in the first picture) has to be the same as $a^2 + b^2$ (the area left over in the second).

If we are to be persnickety, we might complain that we have not exactly *proved* that the figure in the first picture is actually a square (that its sides are all the same length is not enough; squeeze opposite corners of a square between your thumb and forefinger and you get a diamond shape called a *rhombus* that's definitely not a square but still has all four sides the same length). But come on. Before you see the picture, you have no reason to think the Pythagorean Theorem is true; after you see it, you know *why* it's true. Proofs like this, where a geometric figure is cut up and rearranged, are called *dissection proofs,* and are prized for their clarity and ingenuity. The twelfth-century mathematician-astronomer Bhāskara* presents a proof of Pythagoras in this form, and finds the picture a demonstration so convincing as not to require verbal explanation, merely a caption that reads "Behold!"† The amateur mathematician Henry Perigal came up with his own dissection proof of Pythagoras in 1830, while trying to square the circle, like Lincoln; he esteemed his diagram so highly that he had it carved into his tombstone some sixty years later.

ACROSS THE BRIDGE OF ASSES

We need to know how to do geometry by purely formal deduction; but geometry isn't *merely* a sequence of purely formal deductions. If it were, it would be no better a way to teach the art of systematic reasoning than a thousand other things. We could teach chess problems, or Sudoku. Or we could make up a system of axioms with no relation to any known human practice at all and force our students to derive their consequences. We teach geometry instead of any of those things because geometry is a formal system that's not *just* a formal system. It's built into the way we think about space, location, and motion. We can't help being geometric. We have, in other words, intuition.

The geometer Henri Poincaré, in a 1905 essay, identifies intuition and logic as the two indispensable pillars of mathematical thought. Every mathematician leans in one direction or the other, and it is the intuition-leaners, Poincaré says, that we tend to call "geometers." We need both pillars. Without logic, we'd be helpless to say anything about a thousand-sided polygon, an object we cannot in any meaningful sense imagine. But without intuition, the subject loses all its savor. Euclid, Poincaré explains, is a dead sponge:

---

* Often known in mathematical histories as Bhāskara II, to distinguish him from an earlier mathematician with the same name.

† Some sources believe Bhāskara's proof of the Pythagorean Theorem to have been taken from an earlier Chinese source, the *Zhoubi suanjing*, but this is controversial; for that matter, so is the claim that the Pythagoreans themselves had anything we'd now call a proof.

> You have doubtless seen those delicate assemblages of silicious needles which form the skeleton of certain sponges. When the organic matter has disappeared, there remains only a frail and elegant lace-work. True, nothing is there except silica, but what is interesting is the form this silica has taken, and we could not understand it if we did not know the living sponge which has given it precisely this form. Thus it is that the old intuitive notions of our fathers, even when we have abandoned them, still imprint their form upon the logical constructions we have put in their place.

Somehow we need to train students to deduce without denying the presence of the intuitive faculty, the living spongy tissue. And yet we don't want to let our intuition completely drive the bus. The story of the parallel postulate is instructive here. Euclid, as one of his five axioms, listed this one: "Given any line L and any point P not on L, there is one and only one line through P parallel to L."*

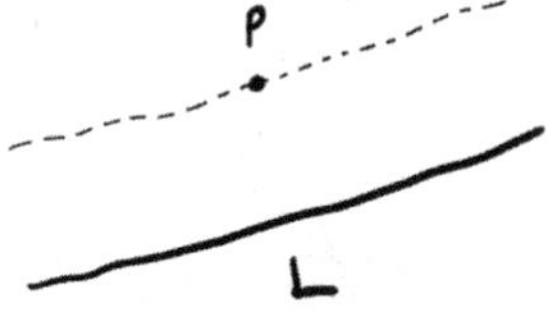

This is complicated and chunky compared to his other axioms, which are sleeker things like "any two points are connected by a line." It would be nicer, people thought, if the fifth axiom could be proven from the other four, which felt somehow more primal.

But why? Our intuition, after all, shouts out loud that the fifth axiom is true. What could possibly be more useless than trying to prove it? It's like asking whether we can really prove that 2 + 2 = 4. We *know* that!

And yet mathematicians persisted, trying and failing and trying and failing to show that the fifth axiom followed from the others. And finally they showed that they'd been doomed to fail from the start;

* This isn't exactly the way Euclid formulated it, but it's equivalent to his fifth axiom, which he phrased in an even chunkier and more complicated way.

because there were *other* geometries, in which "line" and "point" and "plane" meant something other than what Euclid (and probably you) mean by those words, but that nonetheless satisfied the first four axioms while failing the last. In some of these geometries, there were infinitely many lines through P parallel to L. In some, there were none.

Isn't that cheating? We weren't asking about *other* bizarro-world geometric entities which we perversely refer to as "lines." We were talking about *actual lines,* for which Euclid's fifth is certainly true.

Sure, that's a tack you're free to take. But by doing so, you're willfully closing off access to a whole world of geometries, just because they're not the geometry you're used to. Non-Euclidean geometry turns out to be fundamental to huge regions of math, including the math that describes the physical space we actually inhabit. (We'll come back to that in a few pages.) We *could* have refused to discover it on uptight Euclid-purist grounds. But it would be our loss.

Here's another place where a careful balance between formal logic and intuition is called for. Suppose a triangle is isosceles

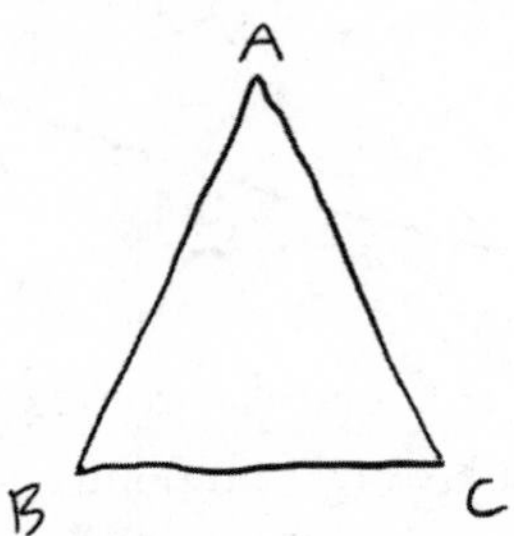

which is to say the sides AB* and AC are equal in length. Here's a theorem: the angles at B and C are equal as well.

This statement is called the *pons asinorum,* the "bridge of asses," because it's something almost all of us have to be carefully led across. Euclid's proof has somewhat more to it than the business with the right angles above. We're a little in medias res here, since in a real geometry class we'd arrive at the ass-bridge only after several weeks of prep; so

* In geometry we like to refer to the line segment joining points named A and B simply as AB, like the Baltimore-Washington Parkway but without the Parkway.

let's take for granted Euclid's Proposition 4 of book I, which says that if you know two side lengths of a triangle and you know the angle between those two sides, then you know the remaining side length and the remaining two angles, too. That is, if I draw this:

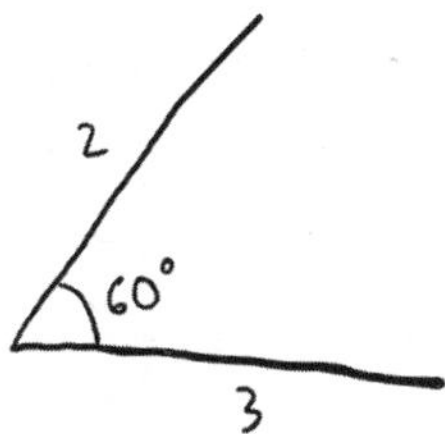

there's only one way to "fill in" the rest of the triangle. Another way to say the same thing: if I have two different triangles that have two side lengths and the angle between them in common, then the two triangles have *all* their angles and *all* their side lengths in common; they are, as the geometer's lingo has it, "congruent."

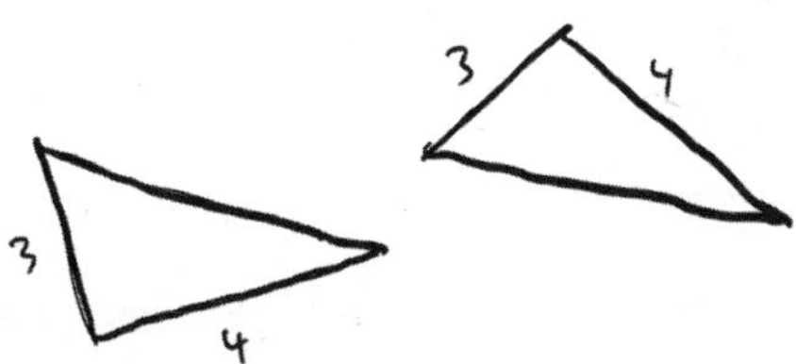

We invoked this fact already in the case where the angle between the two sides is a right angle, and I think the fact feels just as clear to the mind whatever the angle is.

(It's also true, by the way, that if the three side lengths of two triangles match up, the two triangles must be congruent; if the lengths are 3, 4, and 5, for instance, the triangle *must* be the right triangle I drew above. But this is less obvious, and Euclid proves it only a bit later, as Proposition I.8. If you think it *is* obvious, consider this: What about a four-sided figure? Remember the rhombus we just encountered; same four side lengths as a square, but definitely not a square.)

Now for the pons asinorum. Here's how a two-column proof might look.

Let L be a line through A which cuts angle BAC in half	**okay, I'll let you**
Let D be the point where L intersects BC	**still no objection**

Hey, me again, I know we're in midproof here, but we made a new point and invoked a new line segment AD, so we'd better update our picture! By the way, remember our hypothesis that our triangle is isosceles, so AB and AC have the same length; we're about to use that.

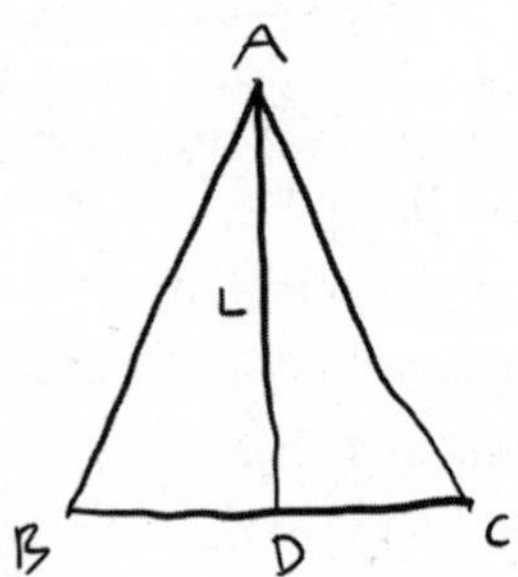

AD and AD have the same length	**a segment is equal to itself**
AB and AC have the same length	**given**
angles BAD and CAD are congruent	**We chose AD to cut angle BAC in half**
triangles ABD and ACD are congruent	**Euclid I.4, told you we'd need this**
angle B and angle C are equal	**corresponding angles in congruent triangles are equal**

QED.*

This proof has more to it than the first one we saw, because you actually have to *make* something; you made up a new line L and give the name D to the point where L hits BC. That allows you to identify B and C with edges of two newborn triangles ABD and ACD, which we then show are congruent.

* For "*Quod Erat Demonstrandum*," meaning "That which was to be proved," a little Latin bat-flip we like to execute at the end of our proofs for pizzazz. In my high school math crew we often substituted "AYD," meaning "And You're Done."

But there's a slicker way, written down about six hundred years after Euclid by Pappus of Alexandria, another North African geometer, in his compendium *Synagogue* (which in the ancient world could refer to a collection of geometrical propositions, not just a collection of Jews at prayer).

AB and AC have the same length	**given**
the angle at A equals the angle at A	**an angle is equal to itself**
AC and AB have the same length	**you already said that, what are you up to, Pappus?**
Triangles BAC and CAB are congruent	**Euclid I.4 again**
angle B and angle C are equal	**corresponding angles in congruent triangles are equal**

Wait, what happened? It seemed like we were doing nothing, and then all at once the desired conclusion appeared out of that nothing, like a rabbit jumping out of the absence of a hat. It creates a certain unease. It was not the sort of thing Euclid himself liked to do. But it is, by my lights at any rate, a true proof.

The key to Pappus's insight is that penultimate line: triangles BAC and CAB are congruent. It seems as if we're merely saying that a triangle is the same as itself, which looks like a triviality. But look more carefully.

What, really, are we saying when we say that two different triangles, PQR and DEF, are congruent?

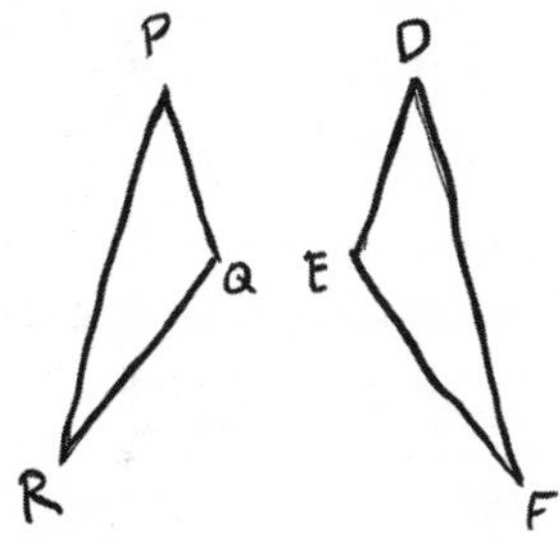

We're saying six things in one: the length of PQ is the same as the length of DE, the length of PR is the length of DF, the length of QR is the length of EF, the angle at P is the same as the angle at D, the angle at Q is the angle at E, and the angle at R is the angle at F.

Is PQR congruent to DFE? Not in this picture, no, because PQ does *not* have the same length as the corresponding side DF.

If we take the definition of congruence seriously—and we're being geometers, so taking definitions seriously is kind of our thing—then DEF and DFE are not congruent to each other, *despite being the same triangle.* Because DE and DF don't have the same length.

But in the proof of the pons asinorum, we're saying that our isosceles triangle, when you think of it as triangle BAC, is the same as the triangle when thought of as CAB. That is *not* an empty statement. If I tell you the name "ANNA" is the same backward and forward, I'm really telling you something about the name: that it's a palindrome. To object to the very concept of a palindrome by saying, "Of course they're the same, it consists of two *A*'s and two *N*'s whichever order you write it in," would be pure perverseness.

In fact, "palindromic" would be a good name for a triangle like BAC, which is congruent to the triangle CAB you get when you write the vertices in the opposite order. And it was by thinking this way that Pappus was able to give his faster path across the pons, without having to invoke any extra lines or points at all.

And yet even Pappus's proof doesn't quite capture *why* an isosceles triangle has two equal angles. It does come closer. This notion that the isosceles triangle is a palindrome, that it stays the same when written backward, records something I'll bet your intuition also tells you—that the triangle is unchanged when you pick it up, flip it over, and lay it back down again in the same spot. Like a palindromic word, it has a *symmetry.* That, one feels, is why the angles have to be the same.

In geometry class we are usually not allowed to talk about picking up shapes and turning them over.* But we ought to be. As abstract as we may try to make it, math is something we do with our body. Geometry

* The Common Core standards, once expected to provide a universal scaffolding for K–12 math education in the United States, but now decidedly in retreat, did ask for the symmetry point of view to be covered in geometry class. One hopes some symmetry arguments will be left behind as the Common Core recedes, like glacial moraine.

most of all. Sometimes literally; every working mathematician has found themself drawing invisible figures with hand gestures, and at least one study has found that children asked to act out a geometric question with their body become more likely to arrive at the correct conclusion.* Poincaré himself was said to rely on his sense of motion when reasoning geometrically. He was not a visualizer, and his recollection for faces and figures was poor; when he needed to draw a picture from memory, he said, he remembered not what it looked like but how his eyes had moved along it.

EQUAL ARMS

What does the word "isosceles" really mean? Well, it means two sides of the triangle are equal. Literally, in Greek, it refers to the two σκέλη (*skeli),* or "legs." In Chinese, 等腰 means "equal waists"; in Hebrew an isosceles triangle is one with "equal calves," in Russian "equal arms." In every case, we seem to agree that what it means to be isosceles is to have two sides equal. But why? Why not define an isosceles triangle to be one that has two angles equal? You can probably see (and indeed the whole point of the pons asinorum is to prove!) that two sides being equal means two angles are equal, and vice versa. In other words, the two definitions are equivalent; they pick out the same collection of triangles. But I wouldn't say they're the *same* definition.

Nor are they the only option. It would be more modern in flavor to define an isosceles triangle as a palindromic one: a triangle you can pick up, flip over, and place back down, only to find it unchanged. That such a triangle has two equal sides and two equal angles is just about automatic. In this geometric world, Pappus's proof would be the means of showing that a triangle with two equal sides was isosceles; that the triangles BAC and CAB are the same.

A good definition is one that sheds light on situations beyond the ones it was devised for. The idea that "isosceles" means "unchanged

* Though no more likely to be able to construct a formal proof of that conclusion!

when flipped over" gives us a good idea of what we should mean by an isosceles trapezoid, or an isosceles pentagon. You *could* say that an isosceles pentagon is one that has two sides equal; then you'll be admitting saggy, lopsided pentagons like this one into the fold:

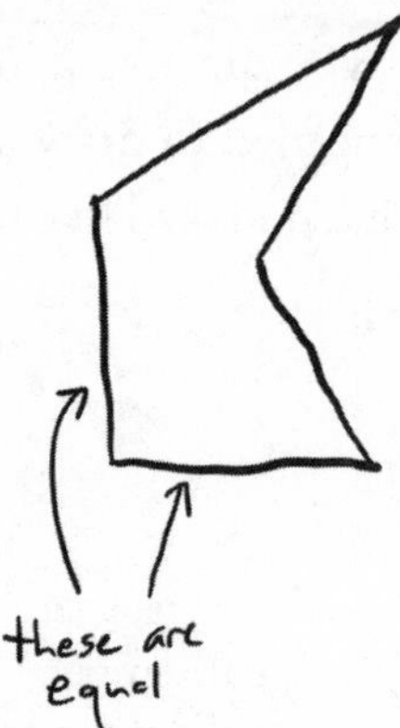

But do you want to? Surely a pentagon like this handsome figure

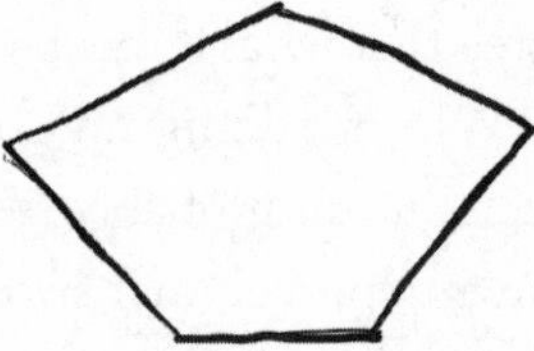

is more what one means by isosceles. Indeed, in your schoolbook, an "isosceles trapezoid" isn't one with two equal sides, or with two equal angles; it is one that can be flipped without changing it. The post-Euclidean notion of symmetry has crept in, and it's there because our minds are built to find it. More and more geometry classes are placing the idea of symmetry at the center, and building structures of proof starting from there. It's not Euclid, but it's where geometry is now.

Chapter 2

How Many Holes Does a Straw Have?

It is always a pleasure, for those of us in the mathematical professions, when the internet spends a day or two tying itself in a knot over a math problem. We get to watch other people discovering and enjoying the mode of thought we spend our whole lives taking pleasure in. When you have a really nice house, you like it when people unexpectedly come over.

The problems that catch on in this way are usually good ones, though they might appear frivolous at first. What hooks and holds the attention is the sensation of encountering an *actual mathematical issue.*

For example: How many holes does a straw have?

Most people I've asked this question see the answer as obvious. And they are extremely surprised, sometimes even a little aggrieved, to learn that there are people whose obvious answer differs from their own. It's the math version of "You've got another think coming" vs. "You've got another thing coming."*

As far as I can tell, the hole-in-the-straw question first appears in a 1970 paper in the *Australasian Journal of Philosophy* by the husband-and-wife team of Stephanie and David Lewis, where the tubular object

* It's "think," of course. If you think otherwise, you've got . . . well, you know.

under discussion is a paper towel roll. The question then reappears in 2014 as a poll on a bodybuilding forum. The arguments presented on the bodybuilding forum are different in tone from those that appear in the *Australasian Journal of Philosophy*, but the outline of the controversy is quite consistent; the answers "zero holes," "one hole," and "two holes" all command substantial support.

Then a Snapchat clip of two college friends getting angrier and angrier over two holes vs. one hole appeared, and started spreading, eventually drawing more than a million and a half views. The straw question showed up all over Reddit and Twitter and in *The New York Times*. A group of young, attractive, extremely-confused-about-holes BuzzFeed staffers shot a video, and that too racked up hundreds of thousands of hits.

You've probably already started formulating the main arguments in your mind. Let's recount them here:

Zero holes: A straw is what you get when you take a rectangle of plastic and roll it up and glue it closed. A rectangle doesn't have any holes in it. You didn't *put* any holes in it when you wrapped it up—so it still has no holes.

One hole: The hole is the empty space in the center of the straw. It extends all the way from the top to the bottom.

Two holes: Just look at it! There's a hole at the top and a hole at the bottom!

My first goal is to convince you you're confused about the holes, even if you think you're not. Each of these views has serious problems.

I'll dispense with the zero-holers first. Something can have a hole in it without any substance being removed. You don't make a bagel by first baking a bialy and then punching out the center. No—you roll out a snake of dough and join the ends together to form the bagel. If you denied that a bagel has a hole, you'd be laughed out of New York City, Montreal, and any self-respecting deli worldwide. I consider this final.

What about the two-hole theory? Here's a question to think about:

If there are two holes in the straw, where does one hole stop and the other begin? If that doesn't bug you, consider a slice of swiss cheese. Someone asks you to count the holes. Do you count the holes in the top of the slice and the bottom of the slice separately?

Or this: fill in the bottom of the straw, thus eliminating what you, two-holer, refer to as the bottom hole. Now the straw is basically a tall, thin cup. Does a cup have a hole in it? Yes, you say—the opening at the top is a hole. Okay, what if the cup gets stubbier and stubbier, until it's an ashtray? Surely we wouldn't call the upper rim of the ashtray a "hole." But if the hole is lost in the passage from cup to ashtray, *when* exactly is it lost?

You might say, here, that the ashtray still has a hole, because it has a depression in it, a negative space where material that might be there actually isn't. A hole doesn't have to "go all the way through," you insist—think about what we mean by a hole in the ground! This is a fair objection, but I think if we're going to be so laid back about what counts as a hole that *any* concavity or divot counts, we are going to expand the concept beyond usefulness. When you say a bucket has a hole in it, you don't mean it has a dent in the bottom, you mean it won't hold water. When you take a bite out of a bialy it doesn't turn into a bagel.

This leaves us with "one hole." That's the most popular of the three choices. Now let me ruin it for you. When I asked my friend Kellie about the straw, she rejected the one-hole theory very simply: "Does that mean the mouth and the anus are the same hole?" (Kellie is a yoga teacher, so she tends to see things anatomically.) It's a fair question.

But let's say you're one of those bold enough to accept the "mouth = anus" equation. There are still challenges. Here's a scene from the college dudes' Snapchat (but seriously, go watch this yourself, I can't fully capture the beautifully mounting frustration in words and stage directions). Bro 1 is an advocate of the one-hole theory, while Bro 2 is a two-holer.

> ***Bro 2*** [holds up a vase]: "How many holes does this have? So this has one hole, right?"
>
> [Bro 1 assents nonverbally.]

Bro 2 [holds up a paper towel roll]: "So how many holes does this have?"

Bro 1: "One."

Bro 2: "How?" [Holds up the vase again] "Do these look the same?

Bro 1: "Because if I put a hole right here" [gestures at bottom of vase] "it's still gonna be one hole!"

Bro 2 [exasperated]: "You just said, *if I put a hole right there*."

[emits a kind of frustrated keening noise]

Bro 1: "If I put another hole in here it's gonna be—"

Bro 2: "Right—*another* hole including this hole! Two! Period!"

The two-hole bro in this scene is expressing a very plausible principle: making a new hole in something should increase the number of holes in it.

Let's make this harder still: How many holes are there in a pair of pants? Most people would say three: the waist and the two leg holes. But if you sew the waist shut, what you've got left is a very big denim straw with a bend in it. If you started with three holes, and you closed one up, you should have two holes left, not one—right?

If you're committed to one hole in the straw, maybe you then say the pants have just two holes, so that when you close up the waist you're down to one. This is an answer I hear a lot. But this answer suffers from the same problem as the two-hole straw theory: If there are two holes in the pants, where are they, and where does one stop and the other begin?

Or perhaps your take is that the pair of pants has only *one* hole, because what you mean by a hole is the region of negative space inside the pants. So what if I rip my jeans at the knee and make a new hole? Does that not count? No, you insist, there's still just one hole; all you've done with your artful rip is create a new *opening* to the hole. And when you sew up the bottom of your pants, or plug the bottom of a straw, you're not removing the hole, just closing off an entrance to or exit from the hole.

But this brings us back to the problem of having to say there's a hole in an ashtray. Or even worse: suppose I have a blown-up balloon. According to you, that balloon has a hole in it—the hole is the zone of pressurized air inside the balloon. Now I take a pin and make a hole in the balloon, so it pops. What's left is a round disc of rubber, maybe with a knot tied into it. A circular piece of rubber obviously doesn't have a hole. So you took something that had a hole, you poked a hole in it, and now it *doesn't* have a hole.

Are you confused now? I hope so!

Math doesn't answer this question, not exactly. It can't tell you what you should mean by the word "hole"—that's between you and your idiolect. But it can tell you something you *could* mean, which at least keeps you from tripping over your own assumptions.

Let me start with an annoyingly philosophical slogan. The straw has two holes; but they are the *same* hole.

REASONING WELL FROM BADLY DRAWN FIGURES

The style of geometry we're adopting here is called topology, and it's characterized by the fact that we don't really care about how big things are or how far apart they are or how they might be bent and deformed, which first of all may seem like a wrenching departure from the themes of this book and second of all may leave you wondering whether I am proposing some kind of geometric nihilism where we don't care about *anything.*

No! A lot of math is figuring out what we can, temporarily or for all time, get away with not caring about. This kind of selective attention is a basic part of our reason. You're crossing the street and a car blows through a red light and comes straight at you—there are all kinds of things you *might* consider as you plan your next move. Can you get a good enough look through the windshield to see if the driver seems incapacitated? What model of car is it? Did you put on clean underwear today in case you end up splayed in the street? These are all questions you *do not* ask—you license yourself not to care about them and you devote your whole consciousness to the task of gauging the car's path and jumping out of its way as fast and as far as you can.

The problems of mathematics are usually less dramatic, but they invoke in us the same process of abstraction, willful ignorance of every feature that doesn't touch the question immediately before us. Newton was able to do celestial mechanics when he understood that heavenly bodies weren't driven by their idiosyncratic whims but by universal laws that applied to every chunk of matter in the universe. To do this he had to steel himself to fail to care about what a thing was made of and what its shape was; all that mattered was its mass and its location relative to other bodies. Or go back even further, to the very beginning of math. The very idea of *number* is that for purposes of reckoning you can treat seven cows or seven rocks or seven people using exactly the same rules of enumeration and combination—and from there it's a short step to seven nations, or seven ideas. It doesn't matter (for those purposes) *what* things are—only how many there are.

Topology is like that but for shapes. In its modern form it comes to us from the French mathematician Henri Poincaré. Him again! It's a name we'll be hearing a lot, because Poincaré had a hand in an astonishingly broad range of geometric developments, from special relativity to chaos to the theory of card shuffling. (Yes, there's a theory, and that's geometry, too; we'll get to it.) Poincaré was born in 1854 into a well-off academic family in Nancy, the son of a professor of medicine. At the age of five he fell seriously ill with diphtheria and was, for several months, completely unable to speak; he recovered fully, but remained physically weak throughout his childhood. Even as an adult, one student described him: "I recall above all his unusual eyes: myopic, yet luminous and penetrating. Otherwise, my memory is that of a man small in stature, stooped and ill at ease, as it were, in limb and joint." When Poincaré was a teenager, the Germans conquered Alsace and Lorraine, though Nancy stayed under French rule. France's unexpected and thorough defeat in the Franco-Prussian War was a national trauma; not only did France resolve to win back the territory it had lost, it undertook to imitate the bureaucratic efficiency and technological adeptness it believed had given Germany the advantage. Just as the surprise launch of Sputnik created a huge wave of funding for scientific education in the United States in the late 1950s, the loss of Alsace and Lorraine (or Elsass-

Lothringen, as it now had to be called) spurred France to catch up with Germany's more fully realized scientific institutions. Poincaré, who had learned to read German during the occupation, became one of the new vanguard of French mathematicians trained in the modern way who would make Paris one of the mathematical centers of the world, with Poincaré at the center of the center. Poincaré was an excellent student, but not a prodigy; his first work of importance began to appear in his mid-twenties, and it was only in the late 1880s that he became an internationally famous figure. In 1889 he won the prize offered by King Oscar of Sweden for the best essay on the "three-body problem," concerning the motion of three celestial bodies subject only to the forces of each other's gravity. That problem remains only incompletely understood in the twenty-first century, but the theory of dynamical systems, the method by which modern mathematicians study the three-body problem and a thousand other problems like it, was launched by Poincaré in his prize paper.

Poincaré was a man of precise habits, who worked on research mathematics exactly four hours per day, from ten in the morning until noon and from five to seven in the evening. He was a believer in the critical importance of intuition and unconscious work, but his career was in some sense quite methodical, characterized not so much by blazing moments of insight as by a systematic and steady expansion of the realm of the understood against the territory of darkness, four hours each weekday and never during holidays. On the other hand, Poincaré had famously terrible handwriting. He was ambidextrous, and the joke in Paris circles was to say he could write equally well—that is, poorly—with either hand.

He was not only the most distinguished mathematician of his time and place, but a popular writer about science and philosophy for the general public; his books popularizing au courant topics like non-Euclidean geometry, the phenomena of radium, and novel theories of infinity sold tens of thousands of copies, and were translated into English, German, Spanish, Hungarian, and Japanese. He was a skilled writer, especially able at capturing a mathematical idea in a finely tuned epigram. Here's one that's very relevant to the question before us:

Geometry is the art of reasoning well from badly drawn figures.

That is: if you and I are going to talk about a circle, I need us to have something to look at, so I'm going to take out a piece of paper and draw one:

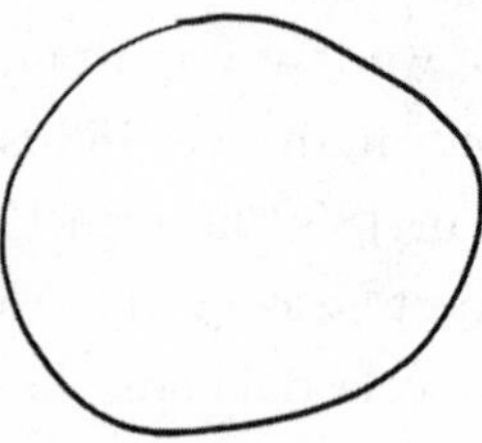

And you might, if you are in a pedantic mood, complain that this is *not* a circle; maybe you have your ruler on you and you check that the distance from my purported center isn't exactly the same for every point on the purported circle. Fine, I say, but if what we're talking about is how many holes there are in the circle, that doesn't matter. In this respect I'm following the example of Poincaré himself, who, true to his epigram and his crappy handwriting, was terrible at drawing figures. His student Tobias Dantzig remembers: "The circles he drew on the board were purely formal, resembling the normal variety only in that they were closed and convex."*

For Poincaré, and for us, these are all circles:

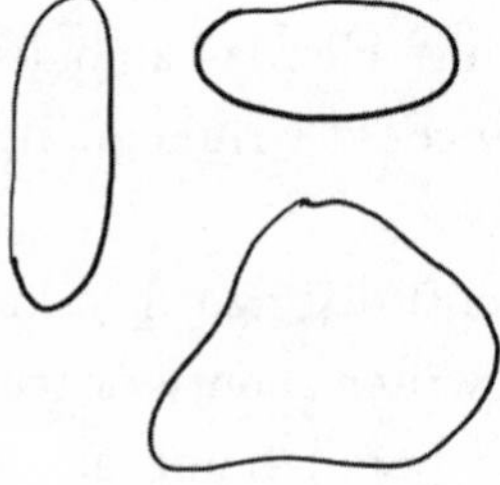

* "Convex" here is a term of art meaning, roughly, "only bends outward, never inward"—more on this in chapter 14, when we meet the even skronkier shapes of legislative districts.

Even a square is a circle!*

This goofy squiggle, too:

But this isn't a circle—

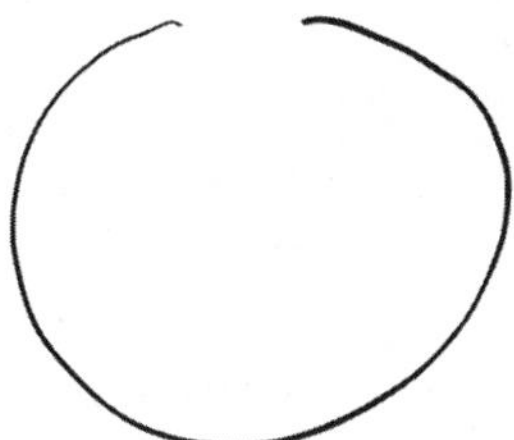

—because it's broken. And by breaking it, I've done something more irrevocably violent than mashing it or bending it or even kinking some corners in it; I have *truly* changed its shape, making it a badly drawn line segment instead of a badly drawn circle. And I have changed it from a thing with a hole in it to a thing without one.

The question of the hole in the straw *feels* like a topological question. Do the two mathematical bros, presented with it, demand to know the

* Or rather: A square is a circle if we care about topological questions about curves like how many holes they are, or how many pieces they fall into. If you care about questions like "how many tangent lines can the curve have at a single point?" a square and a circle are spectacularly different.

precise dimensions of the straw, or whether it's exactly straight, or whether its cross-section is a perfect circle of the kind Euclid would endorse? They do not. On some level they understand these are questions that, for the purpose at hand, can be safely laid aside.

And once you've laid those things aside, what's left? Poincaré counsels us to take the straw and shorten it, shorten it, shorten it. It is the same straw, as far as Henri P. is concerned. Pretty soon it's just a narrow band of plastic:

You could go further still, and bend the walls of the band outward in order to flatten the shape onto the page of a book.

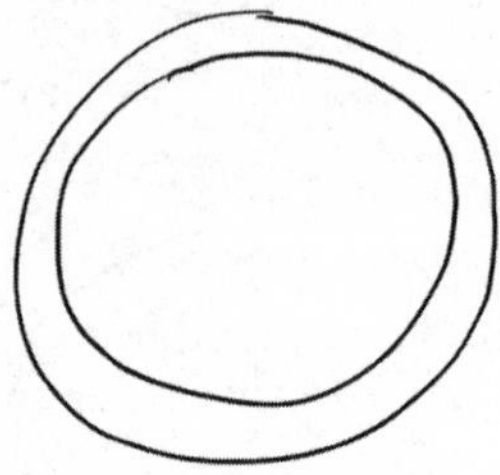

Now it's a shape bounded between two circles, whose official geometry name is an *annulus* but which you might also know as a seven-inch single or an Aerobie or, if you can imagine it with a razor-sharp outer edge and being flung at you in combat in sixteenth-century India, a chakram.

Whatever you call it, it's still a badly drawn figure of a straw, and it has just one hole in it.

If topology insists we say a straw has just one hole, what can we say about pants? We can make those pants shorter, just as we did the straw. First they become shorts, then short-shorts, finally a thong. And when I press that thong flat against the pages of the book you're reading, you see a double annulus:

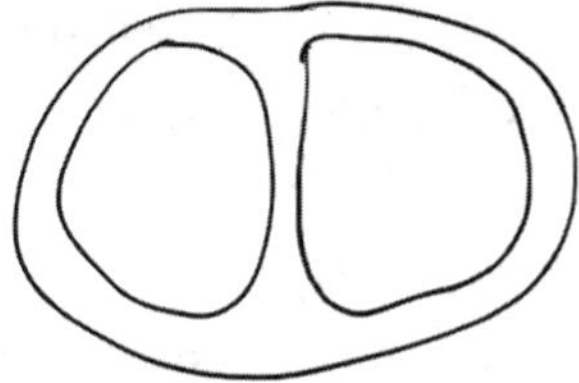

which visibly has two holes. So that's where we end up, for now: a straw has one hole, and pants have two.

NOETHER'S PANTS

But our problems aren't quite over. If the pants have two holes, *which holes are they?* The shortening process we described seems to identify the two holes in the pants as the legs, while the waist has become the outer rim. But as you might have noticed folding laundry, you could just as well have flattened the thong in a different way, with a leg hole on the outside and the other leg and waist making up the two "holes."

My daughter, without benefit of formal education in Poincaré's work, says pants have two holes, her argument being that a pair of pants is really just two straws. The hole in the waist, she says, is the combination of the two leg holes. She's right! And the best way to grasp this is to take the analogy between pants and straw seriously. Imagine, if you will, a straw shaped like a pair of pants, through which one endeavors to drink a malted milkshake. You might dip one leg in the shake and sip; then the same amount of milkshake is passing into the leg as passes out of the waist and into your mouth. Or you could do the same with the other leg; or you could let both legs penetrate the shake. But whatever you do, by the law of conservation of milkshake, the amount of milkshake coming out of the waist hole is the *sum* of the amount coming in through each leg. If three milliliters of milkshake per second is coming into the left leg and five milliliters into the right, then eight milliliters of milkshake is flowing out the top.*

* No, I don't know how you drink a milkshake in a way that draws 1 2/3 as much shake through one side of the straw as through the other, but you've already granted me a pants-shaped straw, so you might as well just keep rolling with this thought experiment.

This is why my daughter is right to say that the waist hole isn't really a new hole at all, but the combination of the two leg holes.

So does that mean the two leg holes are the "real" holes? Not so fast. Just a second ago, when we were folding the just-laundered thong, it seemed like there was no true difference between the waist and the leg. But now the waist seems to be playing a special role again; 3 + 5 = 8 but not 5 + 8 = 3 or 8 + 3 = 5.

This is a matter of being careful about positives and negatives. Outflow is the opposite of inflow, so we should keep track of it with a negative sign; rather than saying 8 milliliters of milkshake is flowing out the waist of the straw, we say that—8 milliliters is flowing in! And now we have a beautifully symmetric description; the sum of the milkshake flow through all three openings is zero. In order to give a complete picture of the flow of milkshake through the pants, I just have to tell you two of those three numbers; but it doesn't matter *which* two numbers. Any pair would do.

Now we're ready to correct the lie we told earlier. It is not quite right to say the hole in the top of a straw (a straw-shaped straw, that is) is the same hole as the one at the bottom. But it's not really a brand-new hole, either. The hole at the top is the *negative* of the hole in the bottom. What flows into one must flow out the other.

Mathematicians before Poincaré, especially the Tuscan geometer and politician Enrico Betti, had wrestled with the question of assigning a shape a number of holes, but Poincaré was the first to grasp the issue that some holes could be combinations of others. And even Poincaré didn't really think about holes the way mathematicians do today; that would have to wait for the work of the German mathematician Emmy Noether in the mid-1920s. Noether introduced the notion of the *homology group* into topology, and it is her notion of "holes" we've been using ever since.

Noether expressed her ideas in the language of "chain complexes" and "homomorphisms," not pants and milkshakes, but I'll stick with our current notation to avoid a wrenching stylistic shift. Noether's innovation was to see that it wasn't right to think of holes as discrete objects, but rather as something more like directions in space.

How many directions can you move on a map? In some sense, you can move in infinitely many different directions: you can go north, south, east, or west, you can go southwest or northeast by east, you can travel at an angle precisely 43.28 degrees eastward from due south, whatever. The point is, for all this infinitude of choice, there are only two *dimensions* in which you can travel; you can get anywhere you want to go by combining just two directions, north and east (as long as you're willing to express a ten-mile journey west as a negative-ten-mile journey east, that is).

But it doesn't make sense to ask which two directions are *the* fundamental ones from which all others derive. Any pair would be as good as any other; you can choose north and east, you can choose south and west, you can choose NW and NNE. The only thing you can't do is choose two directions that are either the same or directly opposite each other; try that and you're confined to a single line on the map.

The top and bottom of a straw are like that: exact opposites, one north, the other south. There's only one dimension to be found here. The waist and two legs of a pair of pants, by contrast, fill out two dimensions, like so:

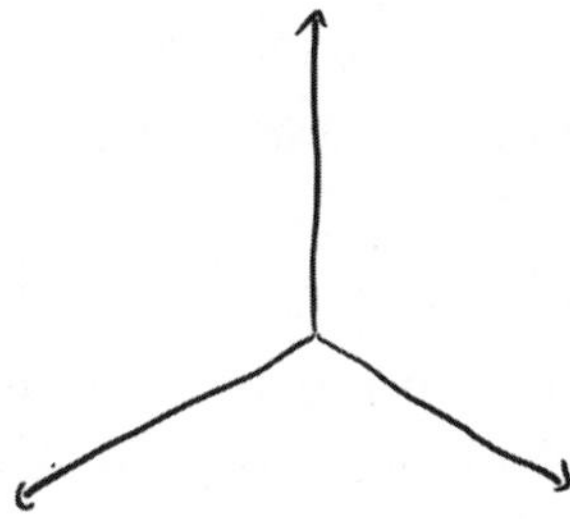

Traveling a mile in one of these directions, then in the second, then in the third, brings you back to your starting point:

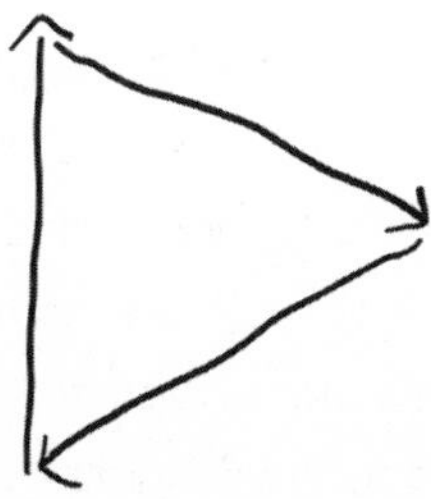

The three directions cancel each other out, combining into a zero.

"Nowadays this tendency is taken as self-evident," Paul Alexandroff and Heinz Hopf wrote in their foundational topology textbook of 1935, "but it wasn't so eight years ago. It required the energy and personality of Emmy Noether to make it common knowledge among topologists. Because of her it came to play the role in the problems and methods of topology that it does today."

"NOBODY DOUBTS NOWADAYS THAT THE GEOMETRY OF N DIMENSIONS IS A REAL OBJECT"

Poincaré created modern topology, but he didn't call it "topology"—he used the more cumbersome "analysis situs" (analysis of position). Good thing that didn't catch on! The word "topology" is actually sixty years older, a word made up by Johann Benedict Listing, a scientific jack-of-all-trades who also invented the word "micron" to mean a millionth of a meter, made major developments in the physiology of sight, dabbled in geology, and studied the sugar content of diabetic patients' urine. He traveled the world measuring the Earth's magnetic field with the magnetometer his PhD advisor, Carl Friedrich Gauss, had invented. He was a convivial and well-liked companion, maybe a little too convivial, because he was constantly running just ahead of his debts. The physicist Ernst Breitenberger called him "one of the many minor universalists who lend so much colour to the history of 19th century science."

Listing accompanied his well-heeled friend Wolfgang Sartorius von Waltershausen on a surveying trip to the volcanic Mount Etna, in Sicily, in the summer of 1834, and during his downtime, while the volcano slumbered, he thought about shapes and their properties, and gave topology its name. His approach wasn't systematic, like Poincaré's or Noether's. In topology, as in science and as in life, he was a sort of magpie, alighting where his interest carried him. He drew lots of pictures of knots, and he drew the Möbius strip before August Ferdinand Möbius did (though there's no evidence that Listing understood, as Möbius did, its curious property of being a surface with only one side).

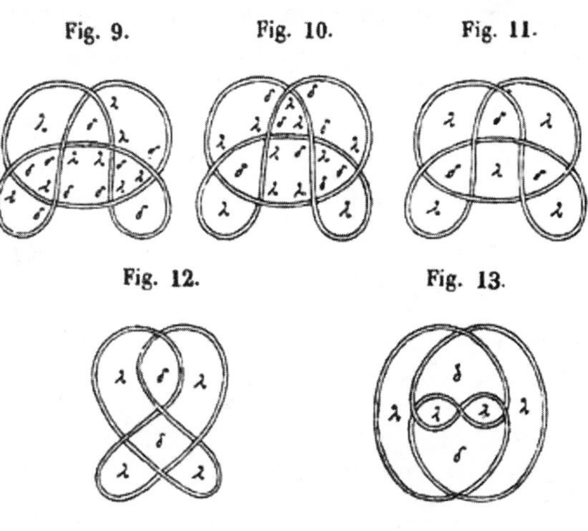

Late in his life he constructed an elaborate "Census of Spatial Aggregates," a menagerie in book form of all the shapes he could think of. He was a kind of geometric Audubon, cataloging the richness of natural variety.

Is there a reason to go beyond Listing's listing? Debating the number of holes in a straw is diverting, but what makes it more important than debating the number of angels who can scrum on a pinhead?

You can find the answer in the very first sentence of Poincaré's *Analysis Situs*, which sternly begins:

> Nobody doubts nowadays that the geometry of n dimensions is a real object.

It's easy to visualize the straw and the pants, and we don't need a mathematical formalism to distinguish between them. Shapes in higher dimensions are a different story. Our inner eye is helpless to glimpse them. And we want more than a glimpse; we want to gaze. As we'll see, in the geometry of machine learning, we'll be searching a space with hundreds or thousands of dimensions, trying to find the highest peak in that unvisualizable landscape. Even in the nineteenth century, Poincaré, studying the three-body problem, needed to keep track of both the location and movement of chunks of matter in the sky; that meant recording, for each heavenly body, three coordinates for position and three for velocity,* or six dimensions in all. And if he wanted to keep track of all

* Because velocity, to a physicist, doesn't just mean speed (which would be a single number) but the *direction* of motion too; you need to keep track of the rate at which you're moving up, north, and east, three numbers in all.

three moving bodies at once, that required six dimensions for each one, a total of eighteen. A picture on the page won't help you understand how many holes there are in an eighteen-dimensional straw, let alone tell apart an eighteen-dimensional straw from eighteen-dimensional pants. A new, more formal language is needed, which will inevitably have to decouple itself from our innate notions of what counts as a hole. This is how geometry always works: we start from our intuitions about shapes in the physical world (where else could we start?), we analyze closely our sense of the way those shapes look and move, so precisely that we can talk about them without relying on our intuition if we need to. Because when we rise up from the shallow waters of the three-dimensional space we're accustomed to inhabit, we *will* need to.

And we can already see the beginning of this process. There's one troubling example from the beginning of this discussion that we're only now ready to return to. Remember the balloon? It has no hole. You poked a hole in it. A loud sound ensues and now it's a rubber disk. Plainly it has no hole now. But didn't we just give it one?

Here's one way to unwind the apparent paradox. If you made a hole in the balloon, and as a result it now has no holes, *it must have had −1 holes to start with*.

We are at a decision point. We can either junk the very appealing idea that putting a hole in a thing increases the number of holes by one, or we can junk the very appealing idea that it's nuts to talk about a negative number of holes. The history of mathematics is a long story of painful decisions like this. Two ideas both feel comfortable to the intuition and we discover by careful consideration that they're logically incompatible. One must go.*

There is no abstract eternal truth about how many holes there are in a balloon, or a straw, or my pants. We have to *choose* a definition, when we come to one of the forking paths math presents us with. You shouldn't think of one path as true and one path as false; you should think of one

---

* Compare Jefferson, one of the most forceful exponents of the ideas of liberty and equality that animated the Revolution, who at the same time doubted a Black person could be "capable of tracing and comprehending the investigations of Euclid" and who, despite verbal opposition to the practice, participated in chattel slavery all his life. Fond of Euclid though he was, he was never able to look directly at the contradiction here.

path as better and one path as worse. What is better is what proves to be more explanatory and enlightening in a wider range of cases. And what mathematicians have found, over the many centuries we've been at this, is that it's generally better to accept something that feels "weird," like negative numbers of holes, than something that breaks a general principle, like the principle that punching a hole in something should bump up by one the number of holes in it. So I'll plant my flag: the best thing is to say the unpopped balloon has –1 holes in it. In fact, there's a way of measuring spaces called the *Euler characteristic,* which is an invariant for topology, unchanged by any kind of continuous smooshing. You can think of it as one minus the number of holes:

pants: Euler characteristic –1, 2 holes
straw: Euler characteristic 0, 1 hole
popped balloon: Euler characteristic 1, 0 holes
unpopped balloon: Euler characteristic 2, –1 holes

A way to describe the Euler characteristic, if you want it to seem less weird, is as a difference between two numbers, the number of even-dimensional holes and the number of odd-dimensional holes. An unpopped balloon, which is to say a sphere, *does* have a hole, in the same sense a brick of swiss cheese has holes—the interior of the balloon is itself a hole. But one feels it to be a hole of a different sort than the hole in a straw. True! It's what we'd call a two-dimensional hole. A balloon has one two-dimensional hole and no one-dimensional holes. That might make it seem like the Euler characteristic should be 1 minus 1, or 0, which doesn't match our table. What's missing is that the balloon has a *zero*-dimensional hole, too.

What is that supposed to mean?

This is where the theory of Poincaré and Noether kicks in. As its name suggests, the Euler characteristic was first systematically investigated by the Swiss omnimath Leonhard Euler, but only for two-dimensional surfaces. Many people, including Johann Listing, worked on extending Euler's idea to the three-dimensional case. But it was not until Poincaré that people began to understand how to bring Euler into dimensions beyond those of our three-dimensional space. Rather than deliver a first course in algebraic topology pressed into a single page, I am just going to say: Poincaré and Noether provide a general theory of holes of any dimension, and in their setup, the number of zero-dimensional holes in a space is just the number of pieces it breaks into. A balloon, like a straw, is a single connected piece, so it has just one zero-dimensional hole. But *two* balloons have two zero-dimensional holes.

This might seem like a weird definition, but it makes everything work. The balloon has

(1 zero-dimensional hole + 1 two-dimensional hole) –
(0 one-dimensional holes)

for an* Euler characteristic of 2.

A capital *B* has one zero-dimensional hole and two one-dimensional holes, so it has Euler characteristic –1.† Snip the bottom loop of the *B* and it becomes an *R*, which has Euler characteristic 0; there's one fewer (one-dimensional) hole, so the Euler characteristic goes up. Snip the loop of the *R* and you get a *K*, which has Euler characteristic 1. Or you could have used your snip to sever the *R*'s lower leg, leaving you with a *P* and an *I*; now there are two separate pieces, so two zero-dimensional holes, and the lone one-dimensional hole in the *P*, which gives you an Euler characteristic of 2 – 1 = 1 again. Each time you snip, you increase the Euler characteristic by 1, and this persists even when what you're doing isn't in any way cutting open a one-dimensional hole anymore. An *I* has Euler characteristic 1; snip it and you get two *I*'s, which has Euler characteristic 2; another snip makes the Euler characteristic 3, and so on.

What if you sew the two leg holes of your pants together, ankle to ankle? It's a bit too grainy to explain in this space, but in Poincaré's system the resulting shape has one zero-dimensional hole and two one-

* If you're wondering why this says "an" and not "a," it's because Euler is pronounced "oiler," not "yooler."

† This is the lowest Euler characteristic of anything on your keyboard, unless you have the double-barred dollar sign, which has Euler characteristic –3, or the Apple "command" symbol, which has Euler characteristic –4. The highest Euler characteristic is 2, for symbols like *!* that have two zero-dimensional holes and no other holes of any kind.

dimensional holes, for an Euler characteristic of –1; in other words, the vandalized pants have the same number of holes as the original ones. You got rid of one when you sewed the two ankle holes together, but created a new one encircled by the two conjoined legs. Is that convincing? It's a Snapchat argument I'd love to see.

Chapter 3

Giving the Same Name to Different Things

Symmetry is the basis of geometry as geometers now see it. More than that: what we decide to count as a symmetry is what determines what kind of geometry we're doing.

In Euclidean geometry, the symmetries are the *rigid motions*: any combination of sliding things around (translations), picking them up and flipping them over (reflections), and rotating them. The language of symmetry provides us a more modern way of talking about congruence. Rather than saying two triangles are congruent if all their sides and all their angles agree, we say they are congruent if you can apply a rigid motion to one that makes it coincide with the other. Isn't that more natural? Indeed, reading Euclid, you can feel the man himself straining (not always successfully) to avoid expressing himself this way.

Why take rigid motions as the fundamental symmetries? One good reason for this choice is that (though this is not so easy to prove!) the rigid motions are exactly those things you can do to the plane that keep every line segment the same length—thus, *symmetry*, from the Greek for "with measure." A better Greekism would be to use the phrase for "equal measure," or *isometry*, and that is indeed what we call a rigid motion in modern math.

The two triangles below are congruent:

and so we're inclined to do as Euclid did and declare them to be equal, even though they're not *really* the same; they're two different triangles about three inches apart.

This brings us to another slogan of the always-quotable Poincaré: "Mathematics is the art of giving the same name to different things." Definitional collapses like this are part of our everyday thinking and speaking. Imagine if, when someone asked you if you were from Chicago, you said, "No, I'm from the Chicago of twenty-five years ago"—that would be absurdly pedantic, because when we talk about cities we implicitly invoke a symmetry under translation in time. In Poincaréan fashion, we call Chicago-then and Chicago-now by the same name.

Of course, we could be stricter than Euclid about what counts as a symmetry; we could, for instance, forbid reflections and rotations, allowing only sliding around the plane without spinning. Then those two triangles above wouldn't be the same anymore, because they're pointing in different directions.

What if we allow rotations but not reflections? You might think of that as the class of transformations we're allowed to carry out if we're stuck in the plane with the triangle, able to slide things and spin them but never to *pick them up and flip them over,* because that would involve making use of the three-dimensional space we're now barred from exploring. Under these rules, we still can't call those two triangles by the same name. In the triangle on the left, ordering the edges from shortest to longest takes us on a counterclockwise path. No matter how we may slide and spin the figure, that fact never changes; which means it can never be made to coincide with the triangle on the right, in which shortest-middle-longest is a clockwise path. Reflection switches clockwise and counterclockwise; rotations and translations don't. Without reflections, the clock-direction of the shortest-middle-longest path is a feature of a triangle that cannot be changed by any symmetry. It is what we call an *invariant*.

Every class of symmetries has its particular invariants. Rigid motions can never change the area of a triangle, or of any figure; in physical terms, we might say there's a "law of conservation of area" for rigid motions. There's a "law of conservation of length," too, for a rigid motion can't change the length of a line segment.*

Rotations of the plane are easy to understand, but going up to three-dimensional space elevates the challenge considerably. It was understood as far back as the eighteenth century (Leonhard Euler again!) that any rotation of three-dimensional space can be thought of as rotation about some fixed line, or axis. So far, so good: but that leaves a lot of questions unanswered. Suppose I rotate 20 degrees about a vertical line and then 30 degrees about a line pointing horizontally to the north. The resulting rotation must be a rotation by some number of degrees around some axis, but what? It's approximately 36 degrees around an axis that points up and off somewhere to the north-northwest. But that's not easy to see! The person who developed a much handier way of thinking about these rotations—thinking of a rotation as a kind of *number* called a *quaternion*—was Wordsworth's young friend William Rowan Hamilton. As the famous story goes, on October 16, 1843, Hamilton and his wife were on a walk along the Royal Canal in Dublin when—well, I'll let Hamilton tell it:

> [A]lthough she talked with me now and then, yet an *undercurrent* of thought was going on in my mind, which gave at last a *result*, whereof it is not too much to say that I felt at once the importance. An electric circuit seemed to close; and a spark flashed forth. . . . Nor could I resist the impulse—unphilosophical as it may

* I can't resist adding: in fact, conservation of length *implies* conservation of triangle areas, because two triangles with the same side lengths are congruent, and thus have the same area; or you can use a beautiful old formula of Hero of Alexandria which just *tells* you what the area is in terms of the side lengths.

have been—to cut with a knife on a stone of Brougham Bridge, as we passed it, the fundamental formula . . .

Hamilton spent much of the rest of his life working through the consequences of his discovery. Needless to say, he also wrote a poem about it. ("Of high Mathesis, with her charm severe / Of line and number, was our theme; and we / Sought to behold her unborn progeny . . ." You get the idea.)

SCRONCHOMETRY

We can also turn the knob toward loosey-goosey and consider a wider range of transformations. We could allow magnification and shrinking, so that these two figures are the same:

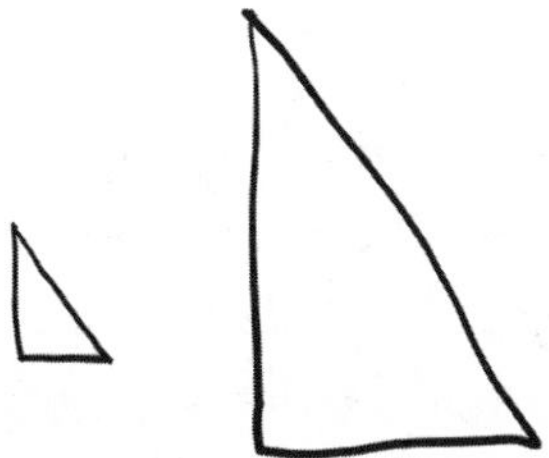

Things about triangles that were invariants before, like area, are no longer invariant under this more forgiving notion of sameness. Other things, like the three angles, do remain invariant. In your high school geometry class, shapes that are the same in this looser sense were called *similar.*

Or we can invent entirely new notions, never seen in the classroom. We might, for instance, allow a kind of transformation we'll call a *scronch,* which stretches a figure vertically by some factor, and compensates by shrinking it horizontally the same amount:*

* Transformations of this kind are well-known to animators, who call them "squash and stretch" and have been using them to make objects look satisfyingly "cartoony" on-screen for almost a century.

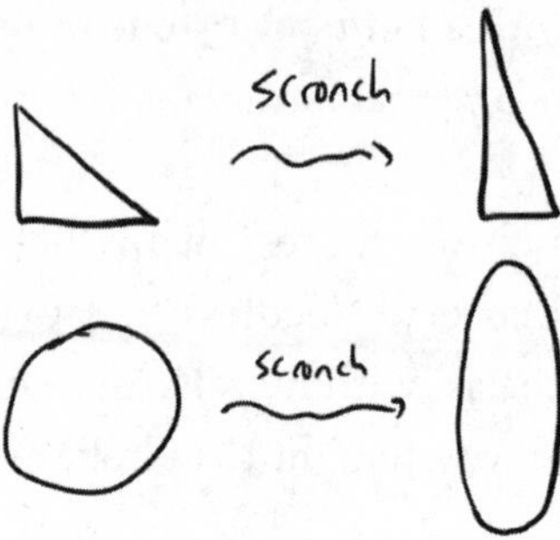

When I scronch a figure, its area doesn't change. This is straightforward for rectangles oriented to have vertical and horizontal sides, since their area is given by width times height; the scronch multiplies the height by something and divides the width by the same thing, so their product, the area, remains the same. See if you can prove the same fact for a triangle, which is a little harder!

In scronch geometry, we call two figures the same if you can get from one to the other by translating and scronching. Two scronch-same triangles have the same area, but two triangles with the same area need not be scronch-same; for instance, any horizontal line segment is still horizontal after scronching, so a triangle with a horizontal side cannot be the scronch-same as a triangle that does not.

The possible types of symmetry, even in the plane, are too numerous to cover here anything like exhaustively. To give a modest idea of this menagerie, the next page shows a diagram from H. S. M. Coxeter and Samuel Greitzer's magisterial textbook *Geometry Revisited*.

This is a tree, much like a family tree, where each "child" is a special case of its "parent"—so an isometry, what we've called a "rigid motion," is a special kind of a similarity, and reflections and rotations are special kinds of isometries. A "Procrustean stretch" is Coxeter and Greitzer's vivid term for a scronch. The "affinities" are what you get if you allow scronches and similarities. The language of symmetry gives us a natural way of organizing the many definitions in plane geometry. Exercise: satisfy yourself that an ellipse is any figure with an affinity to a circle. Harder exercise: show that a parallelogram is any figure with an affinity to a square.

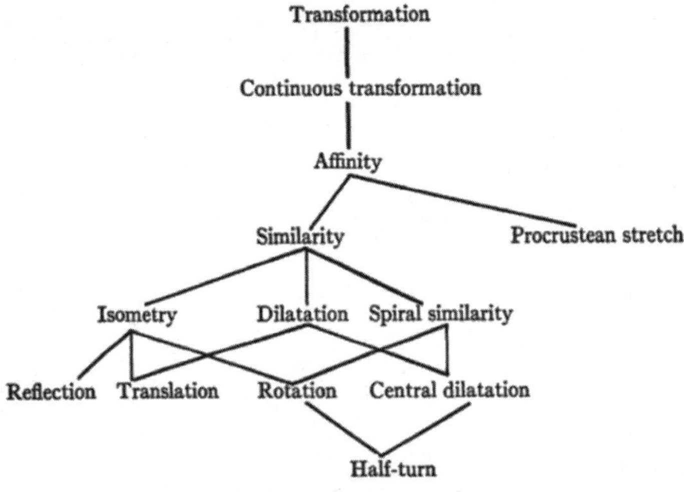

There's no right answer to the question of which pairs of figures are "really" the same. It depends what we're interested in. If we're interested in area, similarity is not good enough, because area isn't invariant under similarity. But if we're only interested in angles, there's no reason to insist on congruence; maybe that's just too demanding. Similarity would be good enough. Each notion of symmetry induces its own geometry, its own way of deciding which things are different enough that we'd better not give them the same name.

Euclid didn't write much about symmetry directly, but his disciples couldn't help thinking about it, even in contexts very far from plane figures. The idea that quantities of importance should be preserved by symmetries sits naturally in the mind. Here, for instance, is Lincoln, writing in his own private notes in 1854, in a very geometrical style:

> If A. can prove, however conclusively, that he may, of right, enslave B.—why may not B. snatch the same argument, and prove equally, that he may enslave A.?

Moral permissibility, Lincoln suggests, should be invariant, like the area of a Euclidean triangle; it ought not change merely because you've reflected the figure to point in the opposite direction.

We can go still further, if we like, leaving the high school classroom entirely behind. No more pencils, no more books, no more Euclid's dirty looks! We could allow completely arbitrary stretching and smooshing of a figure, as long as we never break it, so that the triangle could bulge out into a circle or even fold itself into a square;

but it could *not* become a line segment, because that would require ripping the triangle open somewhere.* Sound familiar? This extravagantly forgiving kind of geometry, where a triangle, a square, and a circle are all the same thing, is exactly the field of topology that Poincaré founded in order to count the holes in a straw. (Okay, he may have had other reasons.) These symmetries, which include every other type of symmetry we've talked about, are the "continuous transformations" one notch below the very top of Coxeter and Greitzer's diagram. In this floppy geometry, notions like angles and area are not conserved. All those inessentialities Euclid cared about fall away; only a pure notion of shape remains.

---

* If you want to see a true definition of the idea I've somewhat imprecisely sketched here, the word to look up is "homeomorphism"—but I'll warn you, there's a bit of a notational on-ramp to the formal definition.

HENRI, I SCRONCHED THE SPACETIME

In 1904, the city of St. Louis held a Louisiana Purchase Exposition, to commemorate the one-hundredth anniversary of the massive land purchase that had brought its territory under American dominion, 101 years earlier. (You try launching an event that big on time!) More than 20 million people visited the fair, which shared the city that summer with the Olympics and the Democratic National Convention. The goal was to show off the United States, and especially its midsection, as ready for the world stage. The song "Meet Me in St. Louis" commemorated the event ("Meet me in St. Louis, Louis / Meet me at the fair / Don't tell me the lights are shining / Any place but there"). The Liberty Bell rolled in from Philadelphia. There were paintings by James McNeill Whistler and John Singer Sargent. A baby born in a construction tent was named "Louisiana Purchase O'Leary." The city of Birmingham, Alabama, commissioned a fifty-six-foot cast-iron statue of Vulcan to promote its steel industry. Geronimo was there signing pictures of himself, and Helen Keller appeared before an overflow crowd. Some say the ice-cream cone was invented on-site. And in September, there was the International Congress of Arts and Sciences, which brought distinguished foreign professors from all quarters to mingle with their American counterparts in what would become the campus of Washington University. Sir Ronald Ross, the British doctor who had just won the Nobel Prize in Medicine for discovering the means of transmission of malaria, was there. Also present were the rival German physicists Ludwig Boltzmann and Wilhelm Ostwald, who were in the midst of their great battle over the fundamental nature of matter: Did it consist of discrete atoms, as Boltzmann thought, or was Ostwald right that continuous fields of energy were the fundamental stuff of the universe? And there was Poincaré, by that time fifty years old and the most famous geometer in the world. The subject of his lecture, delivered on the last day of the congress, was "The Principles of Mathematical Physics." His tone was one of great caution; for those principles, at that moment, were under extraordinary pressure.

"[T]here are symptoms of a serious crisis," Poincaré said, "which would seem to indicate that we may expect presently a transformation. However, there is no cause for great anxiety. We are assured that the patient will not die, and indeed we may hope that this crisis will be salutary."

The crisis physics faced was a problem of symmetry. The laws of physics, one would hope, don't change if you take a step to the side, or turn your eyes in a different direction; that is to say, they are invariant under rigid motions of three-dimensional space. Even more, the laws as Poincaré saw them shouldn't change if he were to step onto a moving bus; this is a slightly more complicated kind of symmetry, involving the coordinates of both space and time.

It might not be obvious at first that nothing about physics should change from the standpoint of a moving observer; it *feels* different to be moving and to be standing still, right? Wrong. Even if Henri doesn't board the bus, he's standing on the Earth, which is careening at great speed around the sun, which itself is on some kind of mad trajectory with respect to the galactic core, etc. If there's no such thing as an observer who can be said to be unmoving, we had better not adopt physical laws that are true only from such an observer's viewpoint. They should hold independent of the observer's motion.

Now here's the crisis: physics didn't actually seem to work that way. Maxwell's equations, which splendidly unified the theories of electricity, magnetism, and light, were not invariant under the symmetries, as they ought to have been. The most popular way to resolve this queasy-making situation was to postulate that there *was* an absolutely still standpoint, that of an unmoving and invisible background called the *ether*, the felt on which all the universe's billiard balls are rolling and clacking. The true laws of physics would be the way physics worked when seen from the ether's point of view, not that of planet-riding humans. But clever experiments designed to detect the ether, or to measure the speed of Earth's passage through it, had all failed. Attempts to account for these failures had taken the form of unpleasantly ad hoc extra postulates, like Hendrik Lorentz's "contraction"—the idea that all moving objects grow foreshortened in their direction of velocity. Fundamental physics was in an unsound condition. Poincaré closed his lecture with an attempt to envision a way past the danger:

> Perhaps too we shall have to construct an entirely new mechanics, which we can only just get a glimpse of, where, the inertia increasing with the velocity, the velocity of light would be a limit beyond which it would be impossible to go. The ordinary, simpler mechanics would remain a first approximation since it would be valid for velocities that are not too great, so that the old dynamics would be found in the new. We should have no reason to regret that we believed in the older principles, and indeed since the velocities that are too great for the old formulas will always be exceptional, the safest thing to do in practice would be to act as though we continued to believe in them. They are so useful that a place should be saved for them. To wish to banish them altogether would be to deprive oneself of a valuable weapon. I hasten to say, in closing, that we are not yet at that pass, and that nothing proves as yet that they will not come out of the fray victorious and intact.

Just as Poincaré had predicted, the patient did not die. On the contrary: it was about to bolt up from the table in bizarrely altered form. In 1905, less than a year after the St. Louis conference, Poincaré would show that Maxwell's equations were symmetric after all. But the symmetries involved, the so-called Lorentz transformations, were of a novel kind, which intermingled space and time in a much more subtle way than "I've been on this bus two hours so I'm forty kilometers north of where I was." (The difference is especially noticeable when the bus is moving at 90% of the speed of light.) From this new point of view, the Lorentz contraction was not a weird, inelegant kluge, but a natural symmetry; that the same object can change its length when hit with a Lorentz symmetry is no stranger than the fact that the same triangle can change its shape when you scronch it. Once you know the symmetries, you know the whole story of how different two things called "the same" are allowed to be. Poincaré was well prepared to make this leap, because he was already one of the innovators in pure mathematics developing forms of plane geometry distinct from Euclid's, having in particular a different group of symmetries. And Poincaré's "fourth geometry," which he had formulated back in 1887, was none other than the scronch plane.

Scronch geometry has laws of "conservation of horizontal and vertical"; if two points are joined by a horizontal or vertical line segment, so are their respective scronches. Lorentzian spacetime is much the same. A point in spacetime is a location *and* a moment; the special line segments conserved by Lorentz symmetries are those joining two location-moments whose two locations are separated by the exact distance light would cover in the amount of time between the two moments. The speed of light, in other words, is built into the geometry. The question of whether light can reach location-moment A to location-moment B has a definite answer, which is the same whether you get on the bus or not.

The scronch plane is like a baby version of Lorentz spacetime. You might think of it as what relativistic physics would look like if, instead of three dimensions of space, there were only one, joining with the one dimension of time to make a two-dimensional spacetime.

But Poincaré did not invent the theory of relativity. The very last sentence of his St. Louis lecture shows why. Poincaré hoped *not* to fundamentally change physics. He had discovered, by mathematical inspection, the strange geometry that Maxwell's equations pointed to, but he was not quite bold enough to follow the finger all the way to the strange point at the horizon it indicated. He was willing to accept that physics might not be what he and Newton had thought, but not that *the geometry of the universe itself* might not be what he and Euclid had thought.

What Poincaré saw in Maxwell's equations, Albert Einstein saw, too, in that same year of 1905. The younger scientist was bolder. And it was Einstein, out-geometrizing the world's foremost geometer, who remade physics as symmetry instructed.

Mathematicians were quick to understand the importance of the new developments. Hermann Minkowski was the first to work out Einstein's theory of spacetime all the way to its geometrical bottom (thus, what we call here the "scronch plane" is actually called the Minkowski plane, if you need to look it up). And in 1915, Emmy Noether established the fundamental relation between symmetries and conservation laws. Noether lived for abstraction; as a senior mathematician, she would describe her 1907 PhD thesis, a computational tour de force involving the determination of 331 invariant features of polynomials of degree four in three variables, as "crap" and "Formelngestrupp" ("formula-thicket"). Too messy and specific! Modernizing Poincaré's theory of "holes" to make it refer to the space of holes instead of just counting how many was very much in her line, and so was cleaning up the welter of conservation laws in mathematical physics. Finding quantities that are conserved by the symmetries of interest is almost always an important physical question; Noether proved that *every* flavor of symmetry comes with an associated conservation law, tying up what had been a messy bundle of computations into a neatly finished mathematical theory, and solving a mystery that had baffled Einstein himself.

Noether was expelled from the mathematics department at Göttingen in 1933, along with all the other Jewish researchers; she made it to the United States and joined the faculty at Bryn Mawr, but not long afterward died, at only fifty-three, of an infection following an apparently successful surgery to remove a tumor. Einstein wrote a letter to *The New York Times*, honoring her work in words the great abstractionist would surely have appreciated:

> [S]he discovered methods which have proved of enormous importance in the development of the present-day younger generation of mathematicians. Pure mathematics is, in its way, the poetry of logical ideas. One seeks the most general ideas of operation which will bring together in simple, logical, and unified form the largest possible circle of formal relationships. In this effort toward logical beauty spiritual formulas are discovered necessary for the deeper penetration into the laws of nature.

Chapter 4

A Fragment of the Sphinx

Back to the St. Louis exposition. Among the scientific bigshots present, remember, was Sir Ronald Ross, who in 1897 had discovered that malaria was carried by the bite of the anopheles mosquito. By 1904 he was a global celebrity, and getting him to Missouri for a public lecture was a coup. "Mosquito Man Coming," read a headline in the *St. Louis Post-Dispatch*.

Ross's lecture was titled "The Logical Basis of the Sanitary Policy of Mosquito Reduction," which does not, I'll concede, sound like a barn burner. But in fact the talk was the first glimmer of a new geometric theory that was about to explode into physics, finance, and even the study of poetic style: the theory of the random walk.

Ross spoke on the afternoon of September 21, while elsewhere at the exposition Governor Richard Yates of Illinois reviewed a parade of award-winning livestock. Suppose, Ross began, you eliminate propagation of mosquitoes in a circular region by draining the pools where they breed. That doesn't eliminate all potentially malarial mosquitoes from the region, because mosquitoes can be born outside the circle and fly in. But a mosquito's life is brief and it lacks focused ambition; it won't set a course straight for the center and stick to it, and the odds seem against its meandering far into the interior in the short time it has to fly. So some region around the center would hopefully be malaria-free, as long as the circle is large enough.

How large is large enough? That depends how far a mosquito is likely to wander. Ross said:

> Suppose that a mosquito is born at a given point, and that during its life it wanders about, to or fro, to left or to right, where it wills. . . . After a time it will die. What are the probabilities that its dead body will be found at a given distance from its birthplace?

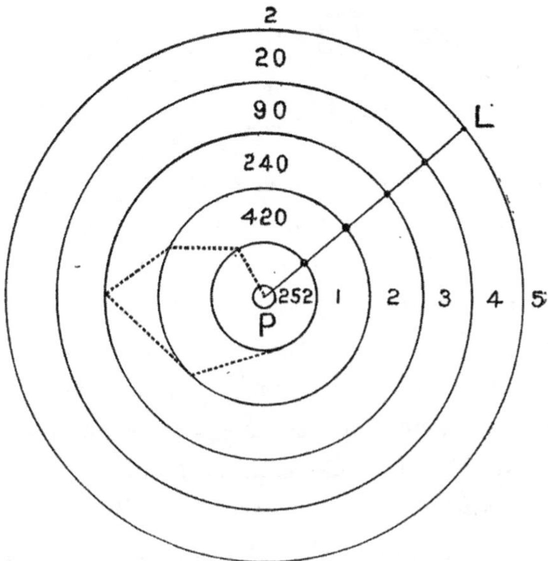

Here's the diagram Ross provided. The dotted line is the wandering mosquito; the straight line is the path a more goal-directed mosquito would take, covering a much greater distance before its demise. "The full mathematical analysis determining the question is of some complexity," Ross said, "and I cannot here deal with it in its entirety."

In the twenty-first century, you can easily simulate a mosquito moving on a Rossian path, so you can improve Ross's diagram to see what happens when the mosquito flits ten thousand times instead of five:

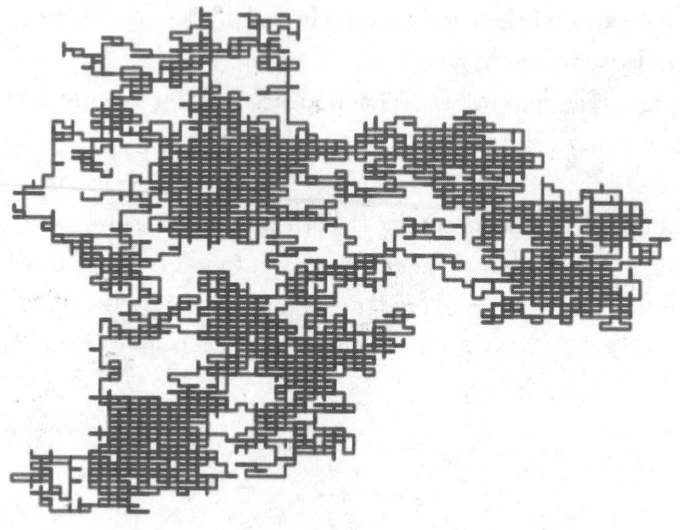

This is typical—sometimes the mosquito sticks around one area for a while, its path crisscrossing itself so much it almost fills up space; sometimes the mosquito appears to acquire a brief sense of purpose and manages to cover some distance. Watching animation of this process, I have to tell you, is unreasonably captivating.

Ross was only able to handle the much simpler case where the mosquito is fixed to a straight line, choosing merely whether to flit northeast or southwest. We can handle this, too! Suppose the mosquito lives for ten days, choosing each day whether to fly a kilometer to the northeast or a kilometer to the southwest. Each day it makes one of two choices, so the total number of career paths a mosquito can have is 2 × 2 × 2 × 2 × 2 × 2 × 2 × 2 × 2 × 2 = 1,024, and—assuming an unbiased mosquito—each of these paths is equally likely. In order for the mosquito to expire 10 km to the northeast of its hatching ground, it would have to make the choice to fly northeast ten times in a row, which only 1 in 1,024 mosquitoes will manage to do. The same tiny proportion end up 10 km southwest; so, in all, 2 out of 1,024 make it 10 km from home. How many travel 8 km? That requires the mosquito to make a sequence of choices like

NE, NE, NE, SW, NE, NE, NE, NE, NE, NE

with nine of one choice and one of the other. The lone "SW" can be in any one of the ten spaces, so there are 10 out of 1024 paths which end 8 km northeast, and 10 that end 8 km southwest, for a total of 20. If you squint, you can see that Ross has written a little "20" and a "2" on the outer two rings of his circle. If you want to, you can write down the 45 paths that end up 6 km northeast of home, or the 210 that end up 2 km northeast, or the 252 that lead the mosquito right back to the fetid pond it spawned from. The mosquito's starting point is its most likely grave. Which makes sense, because this random mosquito problem is really the same thing as flipping ten coins, counting heads as northeast and tails as southwest. Ending up 8 km away amounts to getting nine heads and one tail; ending up at home means getting five of each, which indeed is the most likely thing to happen when you flip ten coins. If you made a bar chart of the different locations, you'd get the good old bell curve, showing the propensity of the mosquito to stay close to its roots.

But we can say more. With a little work, one can compute that in ten days, the average mosquito will travel 2.46 km. That's a typical life span for the fellas; female mosquitoes live more like fifty days, and in that time they'd travel an average of 5.61 km. A Methuselasquito with a two-hundred-day life span, which in principle could cover 200 km of ground, would end up on average just 11.27 km away from home. Four times the life span yields twice the distance covered. We're encountering here a principle first observed in the eighteenth century by Abraham de Moivre, in the context of coin flips, not mosquitoes: the deviation from 50% heads when n coins are tossed is typically around the *square root* of n. A mosquito with a life span a hundred times the norm would likely only wander about ten times as far as its short-lived cousins. A mosquito *can* fly farther than you expect, but it probably won't. The chance that a mosquito on its two-hundredth day of life is at least 40 km from home is just under 3 in 1000.*

* The exact computation here, if you want to look up the words and do it yourself, is: "What is the probability that a binomial random variable with p = 0.5 and n = 200 takes value at least 120?"

CUT!

But 2.46 isn't the square root of 10 and 11.27 isn't the square root of 200! Good, I'm glad you're reading with a pencil in your hand. A better approximation would be that the mosquito in its first n days on earth travels, on average, about $\sqrt{2N/\pi}$ km. Check it: for ten days of mosquito flight, we get

$$\sqrt{2 \times 10/\pi} = 2.52$$

Pretty close! And for two hundred days,

$$\sqrt{2 \times 200/\pi} = 11.28\ldots$$

which also pretty much matches what we saw above.

The presence of π here might get your geometry antennae pinging—is π present because the mosquito is traversing a circular region? Sadly, no. After all, in Ross's simple model, the mosquito is really moving back and forth on a single line. Yes, we first meet π as a ratio arising from the geometry of a circle, but, like most of your better class of mathematical constants, it arises all over the place—you are always rounding a corner and encountering it. One of my favorite examples: Choose two random whole numbers and ask how likely it is they have no factor other than 1 in common. The chance is $6/\pi^2$, though there's no circle in sight.

The mosquito's π comes from calculus, in particular from the value of a certain integral, which has a π in it for its own idiosyncratic reasons. Its calculation was a difficult problem for the French analysts of the eighteenth and nineteenth centuries—now it's something we can teach in third-semester calculus, though it would be an unusual student who could work the integral out without being shown the trick. You can see it computed in full in the 2017 movie *Gifted*, where the integral is presented as a puzzle for Mary Adler, a seven-year-old mathematical prodigy played by nine-year-old Mckenna Grace.

I know this not because I saw the movie on a plane (though I did—for a while you could almost always see this movie on the plane) but because I was there on the set while this scene was being filmed, as a consultant to make sure everything on screen was mathematically on the level. If you've ever seen a movie with mathematical content you may wonder how much effort is taken to get the details right. A lot, it turns out. Enough to pay a mathematician to spend most of a day sitting in the back of what's purported to be an MIT lecture hall (actually Emory University) while a professor, played by a veteran character actor more frequently cast as a Slavic heavy in cop shows, puts the child prodigy through her paces. There *was* something for me to do, it turned out. In a line of dialogue Mary delivers to her grandmother (who is British for some reason, and also is estranged from Mary's single-dude uncle, her guardian, over a proof of the Navier-Stokes conjecture the prodigy's dead mother may or may not have secretly written—you know what, it's kind of a lot to explain, let's move on), she says "negative," where "positive" is what matched the writing on the chalkboard. Off-set, I approached Grace's mom, the only person I felt confident I was allowed to talk to. Should I tell someone, I asked her? Is this important? It was important. She marched me straight up to the director, Marc Webb, at a no-nonsense pace, and ordered me to tell him what I'd just told her. And everything instantly stopped. They changed the word, Grace went off to learn the new line, everybody else stood around and ate snacks off the snack table. How much money is being lost per second when the several dozen extremely specialized professionals it takes to make a major-studio feature film are all idly eating macadamia nuts at once? That amount is a lower bound for how much the studio cares about the mathematical fine print. I asked the director: Does anybody really care? Would anybody even *notice*? In a weary yet somehow admiring voice he told me, "The people on the internet will notice."

Making a movie, I learned, has something in common with writing a math paper: the underlying idea is not that hard to get down, but a huge proportion of the time is spent locking down extremely fine-grained details most people would look right past.

Since I was already on set, Webb offered me the chance to get in front of the camera, where I played the role of "Professor," talking about number theory for about six seconds while Grace looks on studiously. I spent an hour in wardrobe getting ready for those six seconds on-screen.

There was one exception to the obsessive devotion of the crew of *Gifted* to getting details exactly right, it turns out; they put me in shoes much nicer and more expensive than any math professor has ever worn to teach class. And this was something else I learned about the movie industry, something sad: they don't let you keep the shoes.

THE SIP TASTES LIKE THE SOUP

A question people ask me a lot: How can a poll of two hundred people possibly tell me anything reliable about the preferences of millions of voters? It does sound untrustworthy when you put it that way. It's like trying to figure out what kind of soup is in your bowl by tasting a single spoonful.

But in fact you can totally do that! Because you have every reason to think what's in your spoon is a random sample of the soup. You'll never dip into a bowl of clam chowder and get one sip of minestrone.

The soup principle is what makes polls so effective. But it doesn't tell you how closely you can expect the poll to reflect the city, state, or country being surveyed. The answer to that lies in the slow, disorderly progress of the mosquito from its pond. Take a state like the one I live in, Wisconsin, whose population is just about exactly evenly distributed between Democrats and Republicans. And now imagine a mosquito whose motion is determined as follows: I call a random Wisconsinite on the phone, ask them their political leanings, and instruct the mosquito to fly northeast if my respondent is a Democrat or southwest if they vote GOP. That's exactly Ross's model; the mosquito moves randomly in one direction or its opposite, two hundred times. How do we know we won't just happen to call two hundred Democrats and get a totally skewed view of how Wisconsin votes? Sure, that could happen—and the mosquito *could* have just gone for it and flown single-mindedly northeast from its birth to its demise. But it probably won't. We've already seen that the mosquito's distance from home after two hundred days, which in kilometers is exactly the difference between the number of Democrats and the number of Republicans in our poll, is about 11 km on average. So to find 106

Republicans and 94 Democrats in our poll wouldn't be at all strange. Something as far from political reality as a 120–80 split is another story. That's like dipping into a bowl of Wisconsin and getting a spoonful of Missouri. Finding 40 more Republicans than Democrats is the equivalent of the mosquito wandering 40 km from home, and we've already computed the chance of that to be just 3 in 1000.

In other words, it's quite unlikely the two hundred poll respondents will differ substantially from Wisconsinites as a whole. The sip tastes like the soup. There's about a 95% chance the proportion of Republicans in our sample will wind up between 43% and 57%, which is why a poll like this would be reported as having a margin of error of ± 7%.

But: that's assuming there's no bias lurking in our choice of whom to poll. Ross understood very well that bias could confound his mosquito model; before he gets down to calculating and circle drawing, he stipulates a landscape so homogeneous that "every point of it is equally attractive to them [the mosquitoes] as regards food supply, and that there is nothing—such, for instance, as steady winds or local enemies—which tends to drive them into certain parts of the country."

Ross insists on this assumption for a really good reason: without it, everything goes to hell. Suppose it *is* windy. Mosquitoes are little and even a light breeze can sway them in their course. Maybe a northward wind gives the mosquito a 53% chance instead of 50% of flying northeast. That's like an unnoticed bias in our poll that causes each random voter I call to have a 53% chance of being a Republican; maybe because Republicans are more likely than Democrats to agree to answer our survey questions, or to answer their phone in the first place, or to have a phone at all. That makes it a lot more likely that our poll deviates from the truth about the electorate. With an unbiased poll, the chance of finding 120 Republicans and 80 Democrats was only 3 in 1000. With this Republican wind, that chance jumps to 2.7%, almost ten times larger.

In real life, we never know a poll is perfectly unbiased. So we probably ought to be modestly skeptical about what polls say their margin of error is. If polls are routinely nudged in one direction or the other by soft winds of bias, in one direction or another, we'd expect to see real-life election results falling outside reported margins of error a lot more often than advertised. And guess what? They do. A 2018 paper found that actual election results typically deviated from polls about twice as much as the margin of error would suggest. Elections are windy.

Here's another way to think of the presence of the unknown wind. It means that the mosquito's movements, instead of being completely independent from one day to the next, are *correlated* with each other. If the mosquito moves northeast on the first day, that makes it a little more likely that the wind is blowing northeast, which makes it more likely the mosquito moves northeast the next day, too. It's a small effect, but as we've seen, it adds up.

There's a famous fallacy, the so-called law of averages, which holds that after a coin lands heads a few times in a row, the next flip becomes more likely to be a tail, in order to make things "average out." That's not true, the wise person says, because coin flips are independent from one another: the next flip has exactly a 50–50 chance of being a head, no matter what happened before.

But it's worse than that! Unless you are absolutely certain the coin is fair, there's a law of *anti-averages*. If you get a hundred heads in a row, you might just marvel at your unusual streak of luck—or you might quite reasonably start to entertain the possibility you were flipping a two-headed coin. The more heads you get in a row, the *more* you should start to expect heads in the future.*

Which brings us to Donald Trump. As the 2016 U.S. presidential election neared, one thing everyone agreed on was that Hillary Clinton was running ahead. But just how much of a chance Donald Trump had was in great dispute. The newsmagazine *Vox* wrote on November 3:

> Just last week, Nate Silver's polls-only forecast gave Hillary Clinton an overwhelming 85 percent chance of winning. But as of Thursday morning, her odds have fallen down to 66.9 percent—suggesting that while Donald Trump is still the underdog, there's a one-in-three shot he'll end up the next president.

* Take care, though. Superficially similar reasoning like "I drive drunk plenty and haven't hit anyone yet, it must not be that dangerous" can lead to some bad places.

> Liberals have tried to comfort themselves with the knowledge that FiveThirtyEight is an outlier among the six major forecasts, and that the other five give Trump between a 16 percent and a sub-1 percent chance of winning.

Sam Wang at Princeton had Trump's chances at 7% and was so confident in a Clinton victory that he promised to eat a bug if she lost. A week after the election, he choked down a cricket live on CNN. Mathematicians* make mistakes sometimes, but we are people of our word.

How did Wang get it so wrong? He assumed, like Ross, that wind didn't exist. All forecasters agreed the results of the election would hinge on a small collection of swing states, including Florida, Pennsylvania, Michigan, North Carolina, and, of course, Wisconsin. Trump would likely have to take a majority of those states to win; but in each one, it looked like Clinton was clinging to a modest lead. Silver's election-morning estimates for Trump's chances were:

Florida 45%
North Carolina 45%
Pennsylvania 23%
Michigan 21%
Wisconsin 17%

Trump *could* win all these states, but the chance seemed pretty small, just as the chance is pretty small of the mosquito hopping the same direction five times running. You might estimate this chance—or Sam Wang, the cricket-eater-to-be, might estimate this chance—as being

$$0.45 \times 0.45 \times 0.23 \times 0.21 \times 0.17$$

which is about 1 in 600. Trump's chance of winning even three or four of these states, under the same kind of computation, is pretty small.

Nate Silver saw things differently. His model built in a healthy

* Wang is professionally a neuroscientist, not a mathematician, but by me a mathematician is anyone who's doing mathematics at the moment in question.

amount of correlation between the different states, based on the undeniable fact that pollsters could unknowingly be making design choices that biased the sample toward one candidate or the other. Yes, our best estimate was that Trump was behind in Florida, and in North Carolina, and in each of the other swing states. But if he won one of those states, it was evidence that the bias in our polls was making Clinton's standing look better than it was, which made a Trump win in one of the other states more likely. It's the law of anti-averages in action, and it means a Trump sweep of the swing states was more likely than you'd expect from the individual numbers. That's why Silver gave Trump a healthy chance of winning the election. And it's the same reason he rated Clinton as having a better than 1 in 4 chance of winning in a double-digit landslide, an outcome Wang also considered highly unlikely.*

Election-following types freaked out after the 2016 surprise, issuing headlines of weepy betrayal: "After 2016, Can We Ever Trust the Polls Again?"

Yes. We can. Polls are still a much better way of gauging public opinion than a pundit's rating of abstract presidentiality or debate zingers' zinginess. Silver's assessment was that the race was very close and that either candidate could win. He was right! If you consider that a cop-out, ask yourself this: Is it better, sounder mathematical analysis to pretend you know with near certainty who's going to win when neither you nor anybody else actually does?

A LETTER TO *NATURE*

Ronald Ross had fully worked out the behavior of a mosquito fixed to a northeast-southwest track. But the more realistic situation where a mosquito can fly in any direction was beyond the mathematics he knew. So, in that same summer of 1904, he wrote to Karl Pearson.

---

* I've oversimplified a little: Wang didn't really assume *no* correlation, just too little. After the election, he wrote: "The failure was in the general election—and even there, polls told us clearly about just how close the race was. The mistake was mine, in July: when I set up the model, my estimate of the home-stretch correlated error (also known as the systematic uncertainty) was too low. To be honest, it seemed like a minor parameter at the time. But in the final weeks, this parameter became important."

Pearson was a natural person to consult if you had a really new idea that didn't fit neatly in an academic box. He was a well-established professor of applied mathematics at University College, London, a position he had attained in his late twenties after reading law, abandoning that, studying medieval German folklore in Heidelberg, being offered a professorship in Cambridge in the subject, and then abandoning that, too. He was in love with Germany, which compared to England seemed a paradise of fiery intellectual life unencumbered by social convention in general and religion in particular. A fan of Goethe, Pearson wrote a romantic novel called *The New Werther* under the pen name "Loki." The University of Heidelberg misspelled "Carl" as "Karl" in his paperwork and he found he preferred that spelling to the one he'd been born with. Impressed that German had a gender-neutral word *Geschwister*, meaning "brother or sister," he invented the word "sibling."

Back in England, he advocated for irreligious rationalism and women's liberation and gave scandalous lectures on such topics as "Socialism and Sex." *The Glasgow Herald* wrote of one of his talks: "Mr. Pearson would nationalise land and nationalise capital: he at present stands alone in proposing to nationalise women also." His charisma enabled him to get away with moderate outrages of this kind; he was remembered by one former student as a "typical Greek athlete, with finely cut features, crisp curly hair and a magnificent physique." Photographs from the early 1880s show a man with a towering forehead, an intense gaze, and a jaw set in a way that suggests he's about to set you straight about something.

In his adulthood, he returned to mathematics, the subject he'd excelled at in college. He wrote that he "longed to be working with symbols rather than words." He applied to two professorships of mathematics and was rejected; when he finally got the appointment in London, his friend Robert Parker wrote to Pearson's mother:

> Knowing Karl as I do, I always felt sure that he would some day make his value felt and get something which would *really* suit him, however disheartened his friends might be at momentary failure. And now we can realise too what a great thing it has been for him to have three or four years quite free and occupied in other studies

> than mathematics; I do not mean that it has at all conduced to his present success, but no doubt it will make him a happier and more useful man and enable him to avoid any taint of that narrowness which one sees so often in, and fears so much for, men who have devoted themselves exclusively to one absorbing pursuit. And besides, great ideas are often suggested outside the range of the special subjects to which they relate, and Karl is returning to science with a fund of such ideas to be worked out and to make him some day as famous as Clifford* or any of his predecessors.

Pearson himself wasn't so sure: he wrote to Parker, in November of his first semester, "[I]f I only had a spark of originality or was a genius, I would *never* have settled down to the life of a teacher, but instead would have wandered through life† in the hope of producing something that might survive me." But Parker had the better side of the argument. Pearson became one of the founders of the new discipline of mathematical statistics, not because he proved theorems as magnificent as his physique, but because he understood how to bring the wider world in contact with the language of mathematics.

It was with that end in mind that Pearson, in 1891, took up the Gresham Professorship in Geometry, a position whose sole duty since its founding in 1597 has been to deliver a series of evening lectures on mathematics aimed at the general public. The lectures were to be on geometry, but Pearson, in typical fashion, had something more convention-busting in mind than stolid math-appreciation talks on Euclid's circles and lines. He had become a popular teacher by bringing vivid real-life demonstrations into the classroom. Once he flung ten thousand pennies to the floor and made the students count the heads and tails, so that they might witness and not just learn from a book the law of large numbers that pulled the proportion of heads inexorably toward 50%. In Pearson's application for the professorship, he wrote: "I believe that, by legitimate interpretation of the wide meaning of the

* The geometer W. K. Clifford, who might not be famous to everyone but was and is a bigshot in math and physics circles. He has an algebra named after him, a sure sign you've made it.

† Like Ross's mosquito!

word *Geometry* as used in Sir Thomas Gresham's time for one of the seven branches of knowledge, courses of lectures on the elements of the exact sciences, on the geometry of motion, on graphical statistics, on the theory of probability and insurance might be given, in addition to purely geometrical courses, which would supply a want felt by clerks and others engaged during the day in the City." He gave lectures entitled *Geometry of Statistics*, on what would now be called data visualization. He introduced his ideas of standard deviation and the histogram for the first time. Before long, he would develop the general theory of correlation; this is perhaps the most geometric of all Pearson's work, since it reveals that a robust way to understand the way two observed variables are yoked together is by means of the cosine of an angle in a very high-dimensional space!*

By the time Ross was thinking about mosquitoes, Pearson had become a world leader in the application of mathematics to biological problems. In 1901, he cofounded the journal *Biometrika*, back issues of which filled whole bookshelves in my childhood home.† (I didn't grow up in an academic library, I just have two biostatisticians for parents.)

Pearson found that biologists already working on those problems weren't totally convinced: "I felt sadly out of place in such a gathering of biologists, and little capable of expressing opinions, which would only have hurt their feelings and not have been productive of any real good. I always succeed in creating hostility without getting others to see my views; infelicity of expression is I expect to blame."

I have some sympathy for the biologists here. Mathematicians are prone to an imperial tendency; we often see other people's problems as consisting of a true mathematical core surrounded by an irritating amount of distracting domain-specific knowledge, which we impatiently tear away to get as quickly as possible to "the good stuff." The biologist Raphael Weldon wrote to Francis Galton: "Here, as always

* And it would be a perfect idea to explain in a book about geometry, except that I already wrote about it in another book; if you have a copy of *How Not to Be Wrong*, go read pp. 336–43, then come right back.

† Around this time, Pearson's interest in grand social schemes turned to the advocacy of eugenic "improvements" in the British population, and heredity of mental characteristics. An early issue of *Biometrika* features an exhaustive study by Pearson of sibling comparisons among thousands of schoolchildren, each of whom he rated on measures of vivacity, assertiveness, introspection, popularity, conscientiousness, temper, and handwriting.

when he emerges from his clouds of mathematical symbols, Pearson seems to me to reason loosely, and not to take any care to understand his data . . ." and, in another letter, "But I am horribly afraid of pure mathematicians with no experimental training. Consider Pearson." Weldon wasn't just any biologist; he was one of Pearson's closest colleagues, and Galton was their revered older mentor. These were the three men who would later found *Biometrika*. The letters here have the flavor of two friends out of a triad talking behind the third friend's back—we like him, of course we like him, but he's *so annoying* sometimes . . .

Pearson must have been pleased to get a geometrical query from one of the most distinguished medical scientists of his time. He wrote back to Ross:

> The mathematical statement of the simplest case of your mosquito problem is not difficult, but the solution is another thing! I have spent more than a whole day over it & only succeeded in getting the distribution after two flights. . . . It is, I fear, beyond my powers of analysis & wants a strong mathematical analyst. If you set such men the thing as a mosquito problem, however, they will not look at it. I must restate it as a chessboard problem or something of that sort in order to get mathematicians to work at it!

A contemporary mathematician trying to arouse interest in an unfamiliar problem might post a question to social media, or to a public Q&A website like MathOverflow. The 1905 analogue was the letters column of *Nature,* which is where Pearson posed the question, removing all mention of mosquitoes, as promised, but also, to Ross's irritation, all mention of Ross. On the same page of the July 27 edition we find a letter from the physicist James Jeans vainly attempting to beat back Max Planck's newfangled theory of quanta. Between Jeans and Pearson comes a notice from one John Butler Burke, who believed he had observed spontaneous generation of microscopic life in a vat of beef bouillon by exposure to the recently discovered element radium. This is not, perhaps, where you'd expect to find the very beginnings of a mathematical field that flourishes to the present day.

Ross's question was answered very quickly. In fact, it took about negative twenty-five years. The very next issue of *Nature* included a letter from Lord Rayleigh, the previous year's Nobel Prize winner in physics, who informed Pearson that he had solved the random walk problem in 1880, in the course of some investigations of the mathematical theory of sound waves. Pearson responded, rather defensively I think, "Lord Rayleigh's solution . . . is most valuable, and may very probably suffice for the purposes I have immediately in view. I ought to have known it, but my reading of late years has drifted into other channels, and one does not expect to find the first stage in a biometric problem provided in a memoir on sound." (You'll note that, despite Pearson's concession that the problem's origin is in biology, Ronald Ross has still been entirely erased.)

What Rayleigh had shown was that a mosquito that could fly in any direction wasn't so different from Ross's simpler one-dimensional model. It's still true that the mosquito tends to wander only very slowly from its starting point, its typical distance from home being proportional to the square root of the number of days it's been in flight. And it's still true that the very most likely location for the mosquito to be is at the place where it started. This led Pearson to remark, "The lesson of Lord Rayleigh's solution is that in open country the most probable place to find a drunken man who is at all capable of keeping on his feet is somewhere near his starting point!"*

It's from this offhand comment of Pearson that we get the customary metaphor of the random walk as the path of an intoxicated human instead of a disease-carrying insect. It was once often called a "drunkard's walk," though in the kinder present era most people no longer think of a life-ruining addiction as an amusing peg to hang a mathematical concept on.

* But didn't we just say the typical distance from home is proportional to the square root of the number of days traveled, which isn't zero? Yes, that's a subtlety. If the mosquito has flown for a while, the most likely distance from home might be ten miles, but the locations ten miles from home form a great big circle, while the locations zero miles from home form a circle so tiny it's just a point; the odds of being roughly on the big circle are better than the odds of being more or less at home, but the odds of being near *any particular point* on that big circle are worse than the odds of being back at the starting point.

A RANDOM WALK TO THE BOURSE

Ross and Pearson weren't the only people thinking about random walks as the new century rolled in. In Paris, Louis Bachelier, a young man from Normandy, was working at the Bourse, the great stock exchange at the financial center of France. He began studying mathematics at the Sorbonne in the 1890s, taking great interest in the probability courses, which were taught by Henri Poincaré. Bachelier was not a typical student; an orphan, he had to work for his living, and hadn't received the lycée training that had molded most of his peers in the styles and mores of French mathematics. He struggled to get through his exams, scraping by with close to the lowest passing score. And his interests were just plain weird. High-status math at the time was celestial mechanics and physics, like the three-body problem Poincaré had wrestled with to win King Oscar's prize. What Bachelier wanted to study was the fluctuations of bond prices he'd observed at the Bourse; he proposed to treat those motions mathematically, just as his professors were treating the motions of the heavenly bodies.

Poincaré was deeply skeptical about applying mathematical analysis to human actions, dating back at least as far as his reluctant participation in the Dreyfus affair, the fiery controversy over a Jewish soldier accused of spying for the Germans. Poincaré had little taste for political battles and had somehow managed to stay largely neutral as the conflict engulfed French society. But his colleague Paul Painlevé, a fervent Dreyfusard (also the second Frenchman to fly in an airplane and, much later, briefly prime minister of France under the presidency of Poincaré's cousin Raymond), was able to convince him to wade in. Police Chief Alphonse Bertillon, the founder of "scientific policing," had presented a case against Dreyfus arguing that Dreyfus's innocence was ruled out by the laws of probability. The most distinguished mathematician in France, Painlevé argued, could not be silent now that the matter had become a question of numbers. Poincaré, won over, wrote a letter assessing Bertillon's calculations, to be read to the jury at Dreyfus's 1899 retrial at Rennes. Just as Painlevé had hoped, when Poincaré read the police chief's analysis he found crimes against mathematics. Bertillon had found many "coincidences" that he believed pointed irrefutably to Dreyfus's guilt. Poincaré observed that Bertillon's methods allowed him so many opportunities to locate coincidences that it would have been unusual for him *not* to find some. Bertillon's case, Poincaré concluded, was "absolutely devoid of scientific value." But Poincaré went further, declaring that "the application of the calculus of probability to the moral sciences"—what we would now call the *social* sciences—"is the scandal of mathematics. To wish to eliminate moral elements and replace them with numbers is as dangerous as it is pointless. In short, the calculus of probabilities is not, as people seem to believe, a marvelous science which excuses those who have mastered it from having common sense."

Dreyfus was convicted anyway.

Poincaré's student Bachelier set out in his thesis, a year later, to establish the appropriate price for an option, a financial instrument that allows you to purchase a bond at a specified price at some fixed time in the future. Of course, the option has value only if the market price of the bond exceeds the price you've locked in. So to understand the worth of the option you need to have some sense of *how likely* it is the bond's price will end up above or below that crucial line. Bachelier's idea for analyzing this question was to treat the price of the bond as a random process, which each day ticked up or ticked down, without any reference to what it had done before. Sound familiar? It's Ross's mosquito, but now it's money. And Bachelier came to the same kind of conclusions that Ross would five years later (and that Rayleigh had twenty years before). The distance a price wanders in a certain amount of time is typically proportional to the square root of the amount of time that passes.

Poincaré swallowed his skepticism and wrote a warm report on Bachelier's thesis, emphasizing the modesty of his student's goals: "[O]ne might fear that the author has exaggerated the applicability of Probability Theory as has often been done. Fortunately, this is not the case . . . he strives to set limits within which one can legitimately apply this type of calculation." But the thesis was graded "honorable," good enough to pass, not the "very honorable" he would have needed to launch himself in French academia. His work was just too far away from the mainstream—or so it seemed, before the random-walk revolution began. Bachelier did end up getting a job as a professor at Besançon, and lived until 1946, long enough to see the originality of his work appreciated by other mathematicians, but not to see the random walk become a standard tool in mathematical finance. Word has even diffused to the general public: Burton Malkiel's book on investing, *A Random Walk Down Wall Street*, has sold over a million copies. Malkiel's message is a sobering one. The constant up-and-down wandering of a stock price *looks* like events are driving it, but it might well be as random as the mosquito's endless flitting. Don't waste your time trying to time the market's ups and downs; instead, Malkiel says, park your money in an index fund and forget about it. No amount of thought can predict the mosquito's next move and provide you an advantage. Or, as Bachelier wrote in 1900, asserting what he calls a "fundamental principle":

L'espérance mathématique du spéculateur est nulle.

("Mathematically, the expected gain of a speculator is zero.")

A VERY UNEXPECTED FACT OF SEEMING VITALITY

In July 1905, the very same month Pearson was posing Ross's question in *Nature*, Albert Einstein published his paper "On the Motion of Small Particles Suspended in a Stationary Liquid, as Required by the Molecular Kinetic Theory of Heat," in the *Annalen der Physik*. The paper concerned "Brownian motion," the mysterious jittering of small particles floating in a liquid. Robert Brown had first noticed the motion while studying pollen particles under a microscope, and wondered whether this "very unexpected fact of seeming vitality" represented some principle of life that remained in the pollen even after its separation from the plant. But in further experiments he witnessed exactly the same effect in particles with no living origin: shavings of glass from his window, powders of manganese, bismuth, and arsenic, asbestos fibers, and—

Brown just drops this in casually as though it's a normal thing for a botanist to have around the house—"a fragment of the Sphinx."

The explanation of Brownian motion was hotly fought over. One popular theory was that the pieces of pollen or Sphinx were being kicked around by innumerable even smaller particles, the molecules of the fluid, too small to be seen in a nineteenth-century microscope. The molecules were constantly buffeting the pollen at random, forcing it into its lifelike Brownian dance. But remember, not everyone believed that matter was made of tiny invisible particles! This was the substance of a great dispute, with Ludwig Boltzmann on the "tiny particles" side and Wilhelm Ostwald on the other. To the Ostwaldians, "explaining" a physical phenomenon by postulating tiny undetectable molecules doing the work was little better than invoking invisible demons to push the pollen around. Karl Pearson himself had written, in his 1892 book *The Grammar of Science,* "No physicist ever saw or felt an individual atom." But Pearson was an atomist, in his way; whether atoms could ever be detected by instruments or not, he wrote, the hypothesis of their existence could bring clarity and unity to physics and generate experiments that could be tested. In 1902, Einstein hosted an occasional scholarly discussion society and dinner club, "The Olympia Academy," in his apartment in Bern. The frugal dinner typically consisted of "one slice of bologna, a piece of Gruyere cheese, a fruit, a small container of honey and one or two cups of tea." (Einstein, who had not yet gotten his position at the Swiss patent office, was scraping out a living tutoring physics at three francs an hour, and was contemplating a side hustle as a street violinist to keep himself fed.) The Academy read Spinoza, they read Hume, they read Dedekind's *What Are Numbers and What Should They Be?,* and they read Poincaré's *Science and Hypothesis.* But the very first book they studied was Pearson's *The Grammar of Science.* And Einstein's breakthrough, three years later, was very much in the spirit Pearson had imagined.

Invisible demons are unpredictable; there's no mathematical model for what those rascals will do next. Molecules, on the other hand, are subject to the laws of probability. If a particle is struck by a tiny water molecule moving in a random direction, the particle is moved by the impact to travel a tiny distance in that direction. If there are a trillion such impacts every second, then the pollen moves a small fixed distance in a randomly chosen direction every one-trillionth of a second. What does the pollen do in the long term? That might be predictable, even if the individual impacts can't be seen.

This is exactly the question Ross had asked. Instead of a pollen particle, Ross had a mosquito, and instead of a trillion motions a second, he had one per day, but the mathematical idea is the same. Just as Rayleigh had done, Einstein worked out mathematically how particles would tend to behave under a sequence of motions in random directions. This made the molecular theory something you could test experimentally, as Jean Perrin subsequently did, with complete success; this was the decisive blow for Boltzmann's side of the battle. Molecules were invisible, but the accumulated effect of a trillion randomly jostling molecules was not.

To analyze Brownian motion and the stock market and mosquito all at once, with the mathematics of the random walk, is to follow Poincaré's slogan and give the same name to different things. Poincaré formulated his famous advice in his 1908 address to the International Congress of Mathematicians in Rome. He spoke movingly of the way doing complex computations can feel like "blind groping," until that moment when you encounter something more: a common mathematical understructure shared by two separate problems, illuminating each in the light of the other. "[In] a word," Poincaré says, "it has enabled me to perceive the possibility of a generalization. Then it will not be merely a new result that I have acquired, but a new force."

FREE WILL VS. ANDREI THE FURIOUS

Meanwhile, in Russia, two mathematical cliques were feuding viciously over the relationship between probability, free will, and God. The Moscow school was led by Pavel Alekseevich Nekrasov, who had originally trained as an Orthodox theologian before turning to mathematics. Nekrasov was an archconservative, a devoted Christian to the point of mysticism, and, according to some, a member of the ultranationalist Black Hundred movement. He was in every respect a man of the czarist establishment. "Nekrasov sharply opposes political changes in which the masses participate," one source records. "He considers private property a prime principle, which it is the czarist regime's province to protect." His conservative credentials made him popular with anti-revolutionary politicians who wanted to keep a lid on student radicalism, and he steadily rose in administrative ranks, becoming rector of Moscow University and then superintendent of the Moscow Educational District.

His opposite number in the St. Petersburg school was his contemporary Andrei Andreyevich Markov, an atheist and a bitter enemy of the Orthodox Church.* He wrote a lot of angry letters to the newspapers on social matters and was widely known as Neistovyj Andrei,† "Andrei the Furious." In protest of Leo Tolstoy's excommunication, Markov demanded in 1912 that the Most Holy Synod of the Russian Orthodox Church excommunicate him as well (and got his wish, though the church stopped short of imposing an anathema, its harshest punishment).

Nekrasov, as one might imagine, fell out of favor after the Revolution—he wasn't purged, but his role as a mathematical power broker was over, and he was said to seem a "queer shadow of the past." When he died in 1924, *Izvestia* ran a mildly complimentary obituary praising Nekrasov for "determinedly striving to understand the Marxist system," a final insult to the deceased.

Perhaps surprisingly, Markov didn't do much better. Nekrasov had accused Markov of Marxist sympathies in the czar's days, but Markov had little more use for Communist ideology than he did for the Most Holy Synod; his furious spirit just found a new target. In 1921, a year before his death, Markov informed the St. Petersburg Academy of Sciences that he would no longer be able to attend meetings, as he had no shoes. The Communist Party sent him a pair of shoes, which Markov thought so poorly made as to require one final angry public statement:

---

* Markov's father, Andrei Grigorievich Markov, was, like Nekrasov, a seminary graduate and a government functionary. Make of that what you will, psychoanalysis fans.

† Неистовый Андрей, in the original.

> Finally, I received footwear; not only, however, is it stupidly stitched together, it does not in essence accord with my measurements. Thus, as before, I cannot attend meetings of the Academy. I propose placing the footwear received by me in the Ethnographic Museum as an example of the material culture of the current time, to which end I am ready to sacrifice it.

The drastic differences between Markov and Nekrasov might have stayed amicable had they not leaked out from religious and political topics into the more serious subject of mathematics. Both Markov and Nekrasov were interested in probability, and in particular the so-called Law of Large Numbers, the theorem Karl Pearson had demonstrated in class by throwing ten thousand pennies on the floor. The original version of this theorem, proved some two hundred years before Markov's time by Jakob Bernoulli, says roughly this: if you flip a coin enough times, the proportion of heads will get closer and closer to 50%. Of course, no physical law forces this to be the case; a coin *could* land heads as many times in a row as you like. But it's not very likely, and any fixed percentage of imbalance, whether it be 60% heads, 51% heads, or 50.00001% heads, becomes more and more outlandishly improbable as the number of tosses grows. As with coin flips, so with human existence. Statistics of human behavior and action, like frequencies of various crimes and age of first marriages, tend to settle down to fixed averages, too, as if people in the aggregate were just a bunch of mindless coins.

In the two centuries since Bernoulli, many mathematicians, including Markov's mentor Pafnuty Chebyshev, had refined the Law of Large Numbers to cover more and more general cases. But their results all required a hypothesis of *independence*. The flip of one coin had to be independent from the flip of another.

The earlier example of the 2016 election shows us why this hypothesis matters. In each state, the difference between our best polling estimate and the final vote can be thought of as a random variable, called an error. If those errors were independent from each other, the chance is very low that all the errors favored one candidate; much more likely is that some would go in one direction, some in the other, their average would be close to zero, and our overall assessment of the election would be close to right. But if the errors are correlated, as they often are in real life, that assumption could be wrong; it becomes much more likely our whole polling apparatus is systematically biased toward underrating one candidate, in Wisconsin and Arizona and North Carolina alike.

Nekrasov was troubled by the observed statistical regularity of human behavior. The idea that humans were fundamentally *predictable,* no more able to choose their own course through the universe than a comet or an asteroid, was incompatible with church doctrine, and therefore unacceptable to him. In Bernoulli's theorem he saw a way out. The Law of Large Numbers said that averages behaved predictably when the individual variables were independent from each other. Well, there you go, Nekrasov said! The regularities we see in nature don't mean we're all just deterministic particles running along nature's prelaid track, but only that we're *independent* from each other, able to make our own choices. The theorem, in other words, amounted to a mathematical proof of free will. He spelled out his theory in a series of rambling papers, hundreds of pages long, published in a journal edited by his advisor and fellow nationalist Nikolai Vasilievich Bugaev, culminating in a hefty book in 1902.

To Markov, this was mystical nonsense. Worse, it was mystical nonsense wearing mathematical clothes. Nekrasov's work, Markov complained bitterly to a colleague, was "an abuse of mathematics." He had no means of fixing what he saw as Nekrasov's metaphysical errors. But the math—that he could take an axe to. And so he got swinging.

There's almost nothing I think of as more inherently intellectually sterile than verbal warfare between true religious believers and movement atheists. And yet, this one time, it led to a major mathematical advance, whose echoes have been bouncing around ever since. Nekrasov's mistake, Markov saw at once, was in reading the theorem backward. What Bernoulli and Chebyshev knew was that averages settled down whenever the variables in question were independent. And from this Nekrasov had concluded that the variables were independent whenever the averages settled down. But that doesn't follow! Whenever I eat goulash I get heartburn, but that doesn't mean that whenever I have heartburn I've been eating goulash.

For Markov to really knock down his rival, he needed to come up with a *counterexample*: a family of variables whose average was completely predictable, but which weren't independent from each other. It was with this in mind that he invented what we now call the Markov chain. And guess what—it's the same idea that Ross came up with to model the mosquito, that Bachelier had applied to the stock market, and that Einstein had used to explain Brownian motion. Markov's first paper on the Markov chain appeared in 1906; he was fifty years old and had retired the year before from his academic position. It was the perfect time to really lean into an intellectual beef.

Markov considered a mosquito leading a very constrained life; it only has two places it can fly. Call them Bog 0 and Bog 1. Wherever the mosquito is, it prefers to stay, if it can get enough blood to drink. Let's say that, on any given day, if the mosquito is in Bog 0, there's a 90% chance it stays put, and a 10% chance it flits over to Bog 1 to see if the blood is redder on the other side of the fence. At Bog 1, which is perhaps a slightly less promising hunting ground, the mosquito is only 80% likely to stay, and has a 20% chance of heading over to Bog 0. We can capture the situation in a diagram.

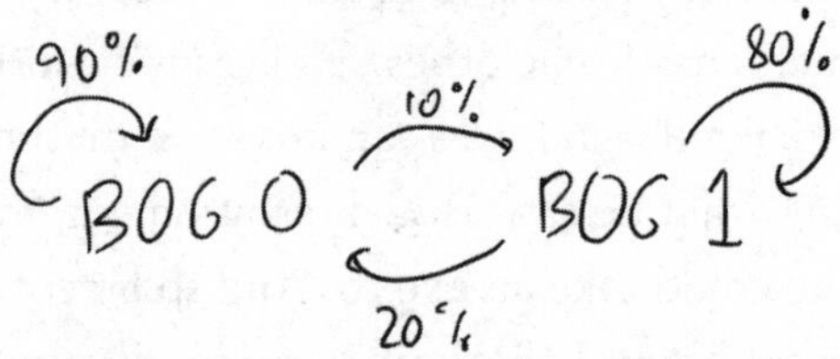

We carefully track the mosquito's progress, keeping track of where it spends each day. You'll very likely see long strings of consecutive Bog 0's and Bog 1's, since bog-hopping is a low-probability event. The sequence might look something like

0, 0, 0, 0, 1, 1, 1, 1, 1, 1, 1, 1, 1, 0, 0, 0, 0, 0, 0, 0, 0, 0, 0, 0, 0, 0, 1, 1, 0, 0, 0, 0, 0, 0, 0 . . .

What Markov showed was this. If you watch the mosquito for a long time, and average all these numbers—this amounts to computing the proportion of its life the mosquito has spent in Bog 1—that average settles down to a fixed probability, just as the fraction of heads in a sequence of coin flips does. You might think the mosquito, flitting around at random, is going to end up equally likely to be in either bog. But no! The asymmetry we built into the problem persists. In this case, the average of all those numbers is going to settle down to 2/3. The mosquito spends two-thirds of its life in Bog 1, and only a third in Bog 0.

That is *not supposed to be obvious*. But let me try to at least convince you it's reasonable. On any given day in Bog 0, the mosquito's chance of leaving is 1 in 10; so you might expect a typical residency in Bog 0 to last ten days. By the same reasoning, a typical stay at Bog 1 should be five days long. That suggests that the mosquito should overall spend twice as much time in Bog 0 as it does at Bog 1, and that turns out to be correct.

But—and here is the killing blow to Pavel Alekseevich—the numbers in this sequence are *not independent* from one another. Anything but! Where the mosquito is today and where the mosquito is tomorrow are very highly correlated; indeed, they're overwhelmingly likely to be the same. And yet the Law of Large Numbers still applies. Independence was not required. So much for the mathematical proof of free will.

We call a list of variables like this a Markov *chain*, because the order in which the variables appear matters a lot. Each one depends on the one before it, but in some sense *only* on that one; if you want to know where the mosquito is likely to be tomorrow, it doesn't matter where it was yesterday or the day before that, only where it is today.* Each variable connects to the next, like the links in a chain. Even if the network of bogs and the paths between them is more complicated than that (as long as it remains a *finite* network), the proportion of the time the mosquito spends at each bog tends to settle down to a fixed ratio, just as successive coin flips or dice rolls do. Where we once only had the Law of Large Numbers, there was now a Law of Long Walks.

The global scientific community we now enjoy didn't exist in the first decade of the twentieth century. It was neither easy nor common

* The way we express this in technical terms is to say each variable is independent of all earlier ones, *conditionally* on the most recent value.

for mathematical work to cross national and linguistic borders. Einstein didn't know about Bachelier's work on the random walk. Markov didn't know about Einstein's. And none of them knew about Ronald Ross. Yet they all arrived at the same insight. One can't help feeling that, in those opening years of the 1900s, something was in the air—a painful recognition of some unavoidable bubbling randomness at the very bottom of things. (This is not even to speak about the development of quantum mechanics, which would eventually twine probability into physics in an entirely different way.) To talk about the geometry of a space, whether that space is a vial of fluid, the space of market conditions, or a mosquito-ridden marsh, is to talk about how one moves through it—and there seems to be no space in the whole world of geometry where the random walk has not proved an illustrative tool. We'll see later in this book that Markov chains are critical in exploring ways to carve a state into legislative districts; and we'll see, right now, how they apply to the purely abstract space of the English language itself.

PONDENOME OF DEMONSTURES OF THE REPTAGIN

Markov's original work was a purely abstract exercise in probability theory. Were there applications? "I am concerned only with questions of pure analysis," Markov wrote in a letter. "I refer to the question of the applicability of probability theory with indifference." According to Markov, Karl Pearson, the eminent statistician and biometrician, had "not done anything worthy of note." Apprised some years later of the prior work of Bachelier on random walks and the stock market, he responded, "I, of course, have seen Bachelier's article but strongly dislike it. I do not attempt to judge its significance for statistics but with respect to mathematics, it has no importance in my opinion."

But finally Markov gave in and applied his theory, moved by the one passion that united Russians atheist and Orthodox, the poetry of Alexander Pushkin. The meaning and art of Pushkin's poetry certainly wasn't capturable by the mechanics of probability. So Markov contented himself to think of the first twenty thousand letters of Pushkin's verse novel

Eugene Onegin as a sequence of consonants and vowels: 43.2% vowels, 56.8% consonants, to be precise. One might naively hope that the letters were independent of one another, which would mean a letter following a consonant was just as likely to be a consonant as any other letter in the text—that is, 56.8% likely.

Not so, Markov found. He laboriously classified every pair of consecutive letters as consonant-consonant, consonant-vowel, vowel-consonant, or vowel-vowel, and ended up with the following diagram:

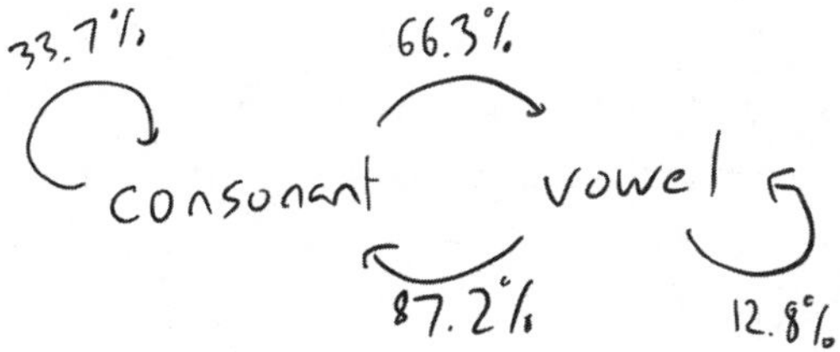

This is a Markov chain just like the one that governed the two-bog mosquito; only the probabilities have changed. If the present letter is a consonant, switching is more likely than staying; the next letter has a 66.3% chance of being a vowel, and only a 33.7% chance of being a consonant. Double vowels are even rarer; there's only a 12.8% chance that one vowel will be followed by another. These numbers are statistically stable throughout the text. You might think of them as a statistical signature of Pushkin's writing. And indeed, Markov later went back to the problem, analyzing one hundred thousand letters out of Sergey Aksakov's novel *The Childhood Years of Bagrov, Grandson*. Aksakov's vowel percentage wasn't much different from Pushkin's: his text consisted of 44.9% vowels. But the Markov chain looks totally different:

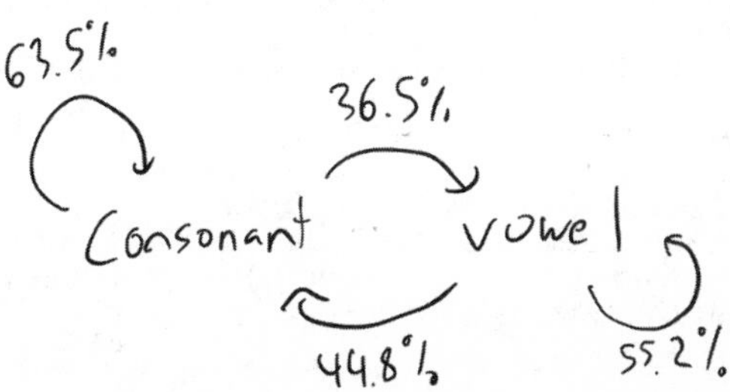

If for some reason you needed to determine whether an unknown Russian text was by Aksakov or Pushkin, one good way—especially if you can't read Russian—would be to count the pairs of consecutive vowels, which Aksakov seemed to savor but Pushkin avoided.

You can't blame Markov for reducing literary texts to a binary sequence of consonants and vowels; he had to do everything on paper. Once electronic computers existed, much more was possible. Instead of just having two bogs, you could have twenty-six, one for each letter of the English alphabet. And given a suitably large body of text to work with, one could estimate all the probabilities necessary to define the Markov chain on letters. Peter Norvig, a director of research at Google, used a text corpus about 3.5 trillion letters long to work out those probabilities. Some 445 billion of the letters, 12.5% of the total, were *E*, the most commonly used letter in the English language. But the letter following one of those 445 billion *E*'s was another *E* in only about 10.6 billion cases, a chance of a little over 2%. Much more common was to follow an *E* with an *R*, which happened 57.8 billion times; so the proportion of *R*'s among *E*-following letters was almost 13%, more than twice as high as the frequency of *R*'s among all letters. In fact, the two-letter sequence, or "bigram," *ER* is the fourth most common among *all* bigrams in English. (The top three appear in this footnote, if you want to guess before looking.)*

I like to think of the letters as places on the map, and the probabilities as walkways that are more or less inviting and easy to traverse. From *E* to *R* there's a wide, well-paved road. From *E* to *B* the way is much narrower and more brambly. Oh, and the roads are one-way; it's more than twenty times easier to get from *T* to *H* than it is to get back. (English speakers say "the" and "there" and "this" and "that" a lot, "light" and "ashtray" not so much.) The Markov chain tells us what kind of winding path through the map an English text is likely to traverse.

Once you're here, why not go deeper? Instead of a sequence of

* *TH* is number 1, followed by *HE* and *IN*. But note that these aren't laws of nature; in a different corpus Norvig gathered in 2008, *IN* edges out *TH* for the top spot, with *ER*, *RE*, and *HE* rounding out the top five. Each corpus has a slightly different bigram frequency therein.

letters, we might think of a text as a sequence of bigrams; for instance, the first sentence of this paragraph would start

ON, NC, CE, EY, YO, OU . . .

Now, there are some restrictions on the paths. *ON* can't go to just *any* bigram; what follows has to be a bigram starting with *N*. (The most common follow-up, Norvig's tables tell us, is *NS*, which happens 14.7% of the time, followed by *NT* at 11.3%.) This gives a yet more refined picture of the structure of English text.

It was the engineer and mathematician Claude Shannon who first realized that the Markov chain could be used not only to analyze text, but to *generate* it. Suppose you want to produce a passage of text with the same statistical properties as written English, and it starts with *ON*. Then you can use a random number generator to select the next letter; there should be a 14.7% chance it is *S*, an 11.3% chance it is *T*, and so on. Having chosen your next letter (*T*, say), you have your next bigram (*NT*) and you can proceed as before, as long as you like. Shannon's paper "A Mathematical Theory of Communication" (the one that launched the entire field of information theory) was written in 1948 and thus did not have access to 3.5 trillion letters of English text in a modern magnetic storage system. So he estimated the Markov chain in a different way. If the bigram before him were *ON*, he would take a book off the shelf and look through it until he found the letters O and *N* in succession. If the next letter after that were a *D*, the next bigram is *ND*; so you open a new book, look for an *N* followed by a *D*, and so on. (If what follows the O and the *N* is a space, you can keep track of that, too, which gives you word breaks.) You write down the sequence of letters thus produced and you get Shannon's famous phrase

IN NO IST LAT WHEY CRATICT FROURE BIRS GROCID PONDENOME OF DEMONSTURES OF THE REPTAGIN IS REGOACTIONA OF CRE.

This simple Markov process produces something that isn't English but recognizably kind of *looks* like English. That's the spooky power of the chain.

Of course, the Markov chain depends on the body of text you use to learn the probabilities: the "training data," as we say in the machine learning business. Norvig used the massive body of text Google harvested from websites and your emails; Shannon used the books on his shelf; Markov used Pushkin. Here's some text I generated using a Markov chain trained on a list of names given to babies born in the United States in 1971:

> Teandola, Amberylon, Madrihadria, Kaseniane, Quille, Abenellett . . .

That's using the Markov process on bigrams. We might go further still and ask, for a sequence of *three* letters (a trigram), with what frequency does each letter appear immediately following it? That requires you to keep track of more data, because there are a lot more trigrams than there are bigrams, but it gives you more recognizably namelike results:

> Kendi, Jeane, Abby, Fleureemaira, Jean, Starlo, Caming, Bettilia . . .

If we go up to five-letter strings, the fealty becomes so good that we often just reproduce full names from the database, but some novelty remains:

> Adam, Dalila, Melicia, Kelsey, Bevan, Chrisann, Contrina, Susan . . .

If we use the trigram chain on names of babies born in 2017, we get

> Anaki, Emalee, Chan, Jalee, Elif, Branshi, Naaviel, Corby, Luxton, Naftalene, Rayerson, Alahna . . . ,

a decidedly more modern feel. (In fact, about half of these are actual names little kids are walking around with right now.) For babies born in 1917:

> Vensie, Adelle, Allwood, Walter, Wandeliottlie, Kathryn, Fran, Earnet, Carlus, Hazellia, Oberta . . .

The Markov chain, simple as it is, is somehow capturing something of the *style* of naming practices of different eras. And there's a way in which one almost experiences it as creative. Some of these names aren't bad! You can imagine a kid in elementary school named "Jalee," or, for a retro feel, "Vensie." Maybe not "Naftalene."

The ability of a Markov chain to produce something like language gives one pause. *Is* language just a Markov chain? When we talk, are we just producing new words based on the last few words we said, based on some probability distribution we've come to learn based on all the other utterances we've ever heard?

It's not *just* that. We do, after all, choose our words to make some reference to the world around us. We're not just riffing on things we've already said.

And yet, modern-day Markov chains can produce something remarkably like human language. An algorithm like Open AI's GPT-3 is the spiritual descendant of Shannon's text machine, only much bigger. The input, instead of being three letters, is a chunk of text hundreds of words long, but the principle is the same: given the passage of text most recently produced, what is the probability that the next word is "the," or "geometry," or "graupel"?

You might think that this would be easy. You could take the first five sentences from your book and run them through GPT-3, and you'd get back a list of probabilities for every possible combination of words in those sentences.

Wait, why would you think it would be easy? You wouldn't, actually. That paragraph above is GPT-3's attempt to continue on from the three paragraphs before it. I picked the most sensible output out of about ten tries. But all the outputs somehow do *sound* like they come from the book you're reading, which, let me tell you, is somewhat unsettling for the human being writing the book, even when the sentences make no literal sense at all, as in this GPT-3 output:

> If you're familiar with the concept of Bayes' theorem, then this should be easy for you. If there's a 50% chance that the next word will be "the" and a 50% chance that it'll be "geometry," then the probability that the next word is either "the geometry" or "graupel" is $(50/50)^2 = 0$.

There's a really big difference between this problem and Shannon's text machine. Imagine a Claude Shannon with a much bigger library, trying to produce English sentences using this method, starting with five hundred words of what you've just read. He looks through his books until he finds one where those exact words appear in that exact order, so that he can record what word comes next. But of course he *doesn't* find one! No one (I hope!) has ever written the five hundred words I just wrote. So Shannon's method fails at the first step. It's as if he were trying to guess the next letter when the two letters before him are *XZ*. There really might not *be* a book on his shelf where those two letters appear in succession. So does he just shrug and give up? Let's impute to imaginary Claude a little more stick-to-itiveness than that! One might instead say: Given that we've never encountered *XZ* before, what bigrams that are in some way *like XZ* have we seen, and what letters followed those bigrams? Once we start thinking this way, we're making judgments about which strings of letters are "close to" which other strings of letters, which means we're thinking about a geometry of letter strings. It's not obvious what notion of "closeness" we should have in mind, and the problem is all the more difficult if we're talking about five-hundred-word passages. What does it mean for one passage to be close to another? Is there a geometry of language? Of style? And how is a computer supposed to figure it out? We'll come back to this. But first, the greatest checkers player in the world.

Chapter 5

"His Style Was Invincibility"

The greatest champion of any competitive enterprise in the history of the human species—better at his game than Serena Williams is at tennis, better than Babe Ruth was at hitting home runs, better than Agatha Christie was at cranking out bestsellers, better than Beyoncé is at staging spectacular concerts—was a mild-tempered math professor and occasional preacher who lived with his elderly mother in Tallahassee, Florida. His name was Marion Franklin Tinsley, and he played checkers. He played checkers as no one did before him and no one ever will again.

Tinsley grew up in Columbus, Ohio, where he learned competitive checkers from a boarder in his family's house, one Mrs. Kershaw, who delighted in her dominance over the boy. "Oh, how she'd cackle as she jumped my men," Tinsley remembered. It was Tinsley's good luck that the world champion at the time, Asa Long, lived nearby in Toledo. Starting in 1944, the teenager studied checkers with Long on weekends, and two years later, at nineteen, he was good enough to finish second in the United States championship, though he never did beat Mrs. Kershaw, who had moved away years before. He won the U.S. title in 1954, by which time he was a math PhD student at Ohio State. The next year he took the world championship, which he would hold, on and off, for the next forty years. When he wasn't champion, it was because he was on hiatus from the game. Tinsley defended his title in 1958 against the UK's Derek Oldbury, winning nine games, drawing twenty-four, and losing just one. He won another world title match against his old mentor, Asa Long, in 1985, winning six, losing one, and drawing twenty-eight. In 1975, he lost one game to Everett Fuller on his way to winning the Florida Open.

Out of more than a thousand tournament games Tinsley played from 1951 to 1990, against the greatest checkers players the world could produce, those were the three he lost.

He didn't have an intimidating manner; he didn't bully or taunt or lord it over his opponents. He just won, and won, and won. Burke Grandjean, the secretary of the American Checker Federation, said, "His style was invincibility." Interviewed before a championship match in London in 1992, Tinsley said, "I am just free of all stress and strain because I *feel* I can't lose."

But he did lose. You guessed where this was going, didn't you? Tinsley won that championship in 1992, but was eventually dethroned by his London opponent, the one player greater than the greatest checkers player who ever lived. It was a computer program called Chinook, developed at the University of Alberta by computer scientist Jonathan Schaeffer, and it is, as you read this, the checkers champion of the world. I don't know when you're reading this, of course. But I'm safe, because Chinook is going to be checkers champion of the world for the rest of time. Marion Tinsley felt he couldn't lose. For Chinook, it's not just a feeling. It cannot lose. There's a mathematical proof. Game over.

Tinsley and Chinook had faced off before. In 1990, he played a fourteen-game exhibition against Chinook in Edmonton. Thirteen times the game ended in a draw—but once, in the tenth move, Chinook made a critical error. "You're going to regret that," Tinsley said, when he saw what Chinook had done. But it took twenty-three more moves for Chinook to understand that it had lost the game.

By 1992, the balance had started to shift. That's when Tinsley lost his first game to Chinook, in the first Man versus Machine World

Checkers/Draughts Championship in London. "No one was happy," Schaeffer recalls. "I'd expected to be jumping up and partying." Instead, there was melancholy. If Tinsley could lose, it meant the era of human supremacy in checkers was soon to be over for good.

But not quite yet. Chinook took one more game from Tinsley. When Tinsley rose to shake hands with Schaeffer on resigning, spectators thought the two had agreed to a draw. No one in the room besides Tinsley and Chinook were capable of seeing that Chinook had won the game. Then Tinsley rallied, winning three more games and the match. Tinsley remained world champion, but Chinook became the first opponent to take two games from Tinsley since the Truman administration.

If it makes you feel better, puny human, Tinsley never did lose to Chinook, not exactly. In August 1994, Tinsley, now sixty-seven, agreed to face Chinook once again. At this point, Chinook had gone ninety-four games without a loss against the rest of the top checkers field. It was running on hardware upgraded to a gigabyte of RAM: an impressive armament at the time, now about a quarter of what a cheap Android phone has. Tinsley and Chinook met at the Computer Museum in Boston, on a wharf overlooking the harbor. Tinsley wore a green suit and a tie pin that spelled out JESUS. They played their games before a sprinkling of spectators, mostly other checkers masters. The match started with six straight draws over three days, most of the games involving little tension or danger for either player. On the fourth day, Tinsley asked for a postponement; he'd had an upset stomach in the night that had kept him from sleeping. Schaeffer took him to the hospital for a checkup. Tinsley, clearly troubled, gave Schaeffer his sister's contact information, in case a next of kin was needed. He talked about his time on earth and what came after. He told Schaeffer, "I'm ready to go." Tinsley saw the doctor, got an X-ray, and spent the afternoon relaxing, but the next morning he reported that once again he hadn't been able to sleep. "I resign the match and the title to Chinook," he told the assembled officials. That's how human domination of checkers ended. And that afternoon, the X-ray results came back. There was a lump on Tinsley's pancreas. Eight months later, he was dead.

AKBAR, JEFF, AND THE TREE OF NIM

How can you prove, absolutely prove, that you can't lose a game? No matter how good you are, surely there could be some tiny channel of strategy you've overlooked, some way, as in a 1980s skiing movie, for the underdog to one-up the sneering Members Only–wearing king of the hill.

But no. We can prove things about games, just as we can prove things about geometry, because games *are* geometry. I could draw the geometry of checkers for you, except actually I couldn't, because it would cover millions of pages and our weak human sensory apparatus wouldn't be capable of making sense of it. So we'll start with a simpler game: the game of Nim.

Here's how you play. Two players sit before some piles of stones. (How many piles and how many stones in each pile can vary, but whatever choices you make about this, it's still Nim.) The players take turns taking stones away. You can take as many stones as you want, but—here it is, the one and only rule of Nim—you can only take stones from one pile at a time. No passing allowed; you've got to take at least one stone. Whoever takes the last stone wins.

So let's say Akbar and Jeff play Nim. And to make things simple, let's start with just two piles, each with two stones. Akbar goes first. What should he do?

Akbar could take two stones, totally emptying out one pile. But that's a bad idea, because then Jeff clears the other pile and wins. So Akbar should just take one stone from one pile. That's no better, because Jeff has a killer move—he takes one stone from the other pile, leaving each pile with one stone. Akbar, seeing the inevitable approaching, sullenly takes a stone. From which pile? Doesn't matter and Akbar knows it. Jeff takes the last remaining stone and wins.

No matter how Akbar chooses his first move, he can't escape. Unless Jeff blunders, Jeff wins.

Now what if there are three piles with two stones each? Or with ten stones each, or a hundred? Suddenly it's a lot harder to play out the game in your head.

So let's take out pencil and paper and diagram out the course of the game that starts with two piles of two stones. At the start, Akbar has two choices; he can take one stone or he can take two. Here's a little sketch of his options, showing the outcome of each. The bottommost picture is the way the game looks at the start, and when you play the game you're moving upward in the diagram, choosing one of the branches ascending from your current position.

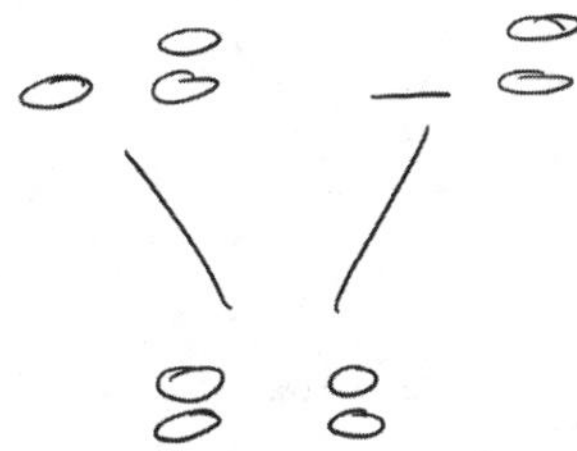

Okay, I hear you—technically Akbar has *four* choices, since he can take one stone from the first pile, one stone from the second pile, both stones from the first pile, or both stones from the second pile. We're doing a little Poincaré-style "calling different things by the same name" here. Nim has a perfect symmetry, or at least it does when you start; whichever pile Akbar reaches for first, you *call* that pile the pile on the left. All the argument that follows would proceed in exactly the same way if we called that pile the pile on the right, only with the words "left" and "right" swapped everywhere they appear. This is the point in math where we like to say "without loss of generality," which is just a fancy way of saying "I'm going to make an assumption now, but if you don't like my assumption, make the opposite assumption, and everything is going to go exactly the same except with the words 'left' and 'right' switched." If it really bugs you, turn the book upside down.

Now it's Jeff's turn. And the choices he has depends on what Akbar did. If Akbar took one stone, then there's one stone on the left and two stones on the right. So there are three things Jeff can do: clear the left pile, clear the right pile, or take one stone from the right pile. But if

Akbar took two stones, then there's only one pile left, and Jeff has only two choices; he can take one stone away, or take both.

Did you find that paragraph a little hard to read? I found it a little dull to write. A picture is better!

And we can just go on expanding our picture until we've explored every possible course this game can take. It doesn't take long. After all, each player has to take at least one stone at each turn, and there are only four stones to start with, so the game has to end in four moves or fewer. Here it is, the whole megillah, the game of two-pile-two-stone Nim in geometric form:

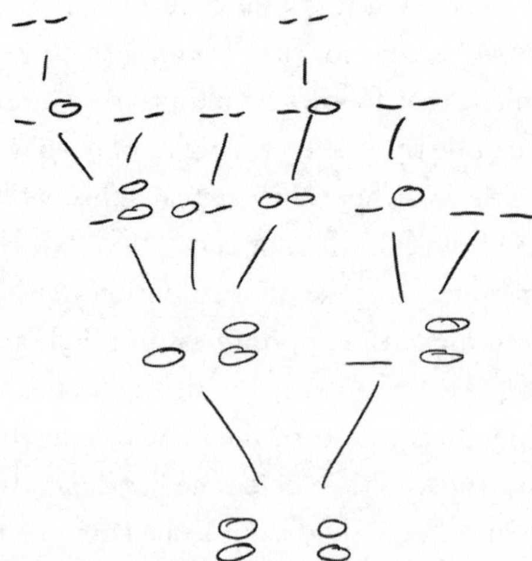

This diagram is what mathematicians call a *tree*. You might have to squint a little to make the botanical metaphor work. The bottommost spot, the starting point of the game, is the *root*—the base from which everything else grows. The upward paths are called *branches*. Some people like to call the point where a branch ends and branches no further a *leaf*.*

The tree is a picture of the game—a complete picture, which depicts all possible states of the game and the paths between them. The picture tells a story. You make a choice and that choice sends you upward along one of the branches. Once you choose, you're on that branch and its offshoots forevermore. No going back. All you get to do is make further choices, traverse narrower branches, get closer and closer to the inevitable end when your choices finally run out.

Your *life* is a tree, is what I'm basically saying.

THE ARDOR OF ARBOREALITY

Geometric objects are interesting to a broad spread of humanity just insofar as they resonate with real things we encounter in our lives with some frequency. If the only triangular things in the universe were the little metal percussion instruments, we wouldn't care as much as we do about triangles.

A tree is a picture of a game, but not *only* that. The same geometry shows up everywhere. In literal bark-having carbon-absorbing trees, of course. But also in family trees, where in place of the branching of choices in a game we have the branching of children. The root of the family tree is the founding couple. The leaves are family members who didn't produce children, or haven't yet. Family trees are usually written with the root at the top—we call ourselves "descendants" of our ancestors, not twigs that sprout up from them.

The arteries in your body form a tree, too. The root is the aorta, the big pipe that carries oxygenated blood out of your heart; from there blood branches out to the right and left coronary arteries, the brachiocephalic

* This is just how mathematicians are; once we get a metaphor in our teeth we wring every last drop of blood out of it. But that's about as far as the forestry goes: a mathematical tree doesn't have bark or knots or xylem and phloem. A group of them is indeed called a forest, though.

trunk, the left carotid artery, the left subclavian artery, the bronchial arteries, the esophageal arteries . . . and each of these in turn branches into smaller arteries; the brachiocephalic trunk splits into the right carotid and the right subclavian arteries, the right carotid branches into the external and internal carotids right about where your chin hits your neck, and so on, all the way down to the tiny network of arterioles, a hairsbreadth or two in diameter, the blood's last stop before it drops off its oxygen and starts its journey back to the lungs to get more.

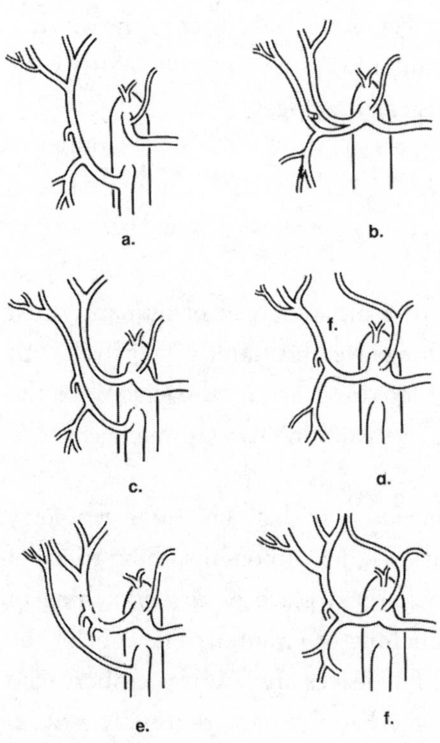

We don't all have the same blood-tree inside us! This looks like a multiple-choice test for aliens but is actually a picture of different ways the branching of the arteries feeding our liver might look.

A river is a tree. As long as you remember to move *against* the flow. The root is whatever gulf or sea the river dumps its water into, and from there you proceed upstream, branching into tributaries and then into subtributaries until you come to the beginning of the flow, the source.

The same goes for any kind of hierarchical classification, like the Linnaean classification of living beings. Kingdoms break into phyla, phyla into classes, classes into orders, orders into families, families into genera, genera into species. So there's a tree of trees:

white oaks
red oaks
Canyon live oaks
Quercus
Siebold's beech
Taiwan beech
chestnuts
chinquapins
Fagoideae
Quercoideae
FAGACEAE

Good, evil: those are trees too! The *Speculum Virginum* ("Mirror for virgins") was a kind of moral self-help book for medieval nuns, traditionally held to have been compiled by the Benedictine monk Conrad of Hirsau, deep in the Black Forest in the early twelfth century, though this far back in literary history the provenance questions get pretty hard. We have the book, though, and in it, the Tree of Virtue and the Tree of Vices. Evil's more interesting, so here are the vices.

The root of the tree, source of all sins, is *superbia*, pride, sprouting from the head of a richly garbed gentleman. Pride's descendants include *ira*, anger, and *avaritia*, greed, and, at the top of the page, *luxuria*, lust, the word helpfully inscribed on a smirking guy's pelvis. And each of these sins has its own children: *ira*'s seven offspring include blasphemy and contumely, while *luxuria* spawns *libido*, *fornicario,* and *turpitudo.* (I can't say I grasp the fine distinctions here, which is one of the reasons why I would make a bad medieval nun.)

As you move forward in time, and people's concerns get less moralistic and more corporate, the tree returns in the form of the org chart, a diagram showing the chains of command inside a business. The tree tells you who reports to whom and who does whose bidding. On the next page is what might be the first such chart ever created, by Daniel McCallum, a Scottish-American engineer who made this diagram for the New York and Erie Railroad in 1855, and who would later serve as superintendent of military railroads for the Union forces in the Civil War.*

Information flows from the leaves back to the root, the president of the railroad, while authority flows in the other direction, from the president through chains of subordinates out to the tiny leaves and buds, labeled in print too small to read on this page as "LABORERS" and "ENGINEMEN" and "CARPENTERS" and "WIPERS."† This chart isn't exactly a pure tree; it combines the organizational structure with visual depictions of the railroad lines the organization oversees. At the center it looks much like the Tree of Vices, while at the outskirts it resembles a late-twentieth-century American cul-de-sac suburb viewed from above. The tree represents the geometry of hierarchy for the same reason it represents the geometry of Nim, or the geometry of the garden of forking paths that makes up our lives; there are no cycles, no infinite regress. If I am in charge of you, you can't be in charge of me; that's the principle of command and control in business. If a position in Nim follows from an earlier one, no further move can bring you back to the previous state; that's what keeps games from going on forever.‡

But the trees I like best of all, better than arteries, rivers, and sins, are trees of numbers. Here's how to make them. You start with a number, say 1,001. And then you start hacking it up with an axe. By which

* When you hear "nineteenth-century Scotsman" and "Civil War officer" you probably think, "I'll bet that guy had a truly magnificent beard," and you are not wrong.

† I had to look this up, too. "Wipers" are entry-level railroad workers who clean and lubricate engine parts.

‡ In fact, there's a more general notion than a tree, called a *directed acyclic graph,* that captures this notion a little more precisely: a DAG is like a tree where some branches are allowed to fuse together, but there are still no cycles because branches can be traversed in only one direction. Think of a particularly aristocratic family tree where your parents may share a great-grandparent or two. The analysis of DAGs can be a little nastier than that of trees, but most of what we say in this chapter still applies.

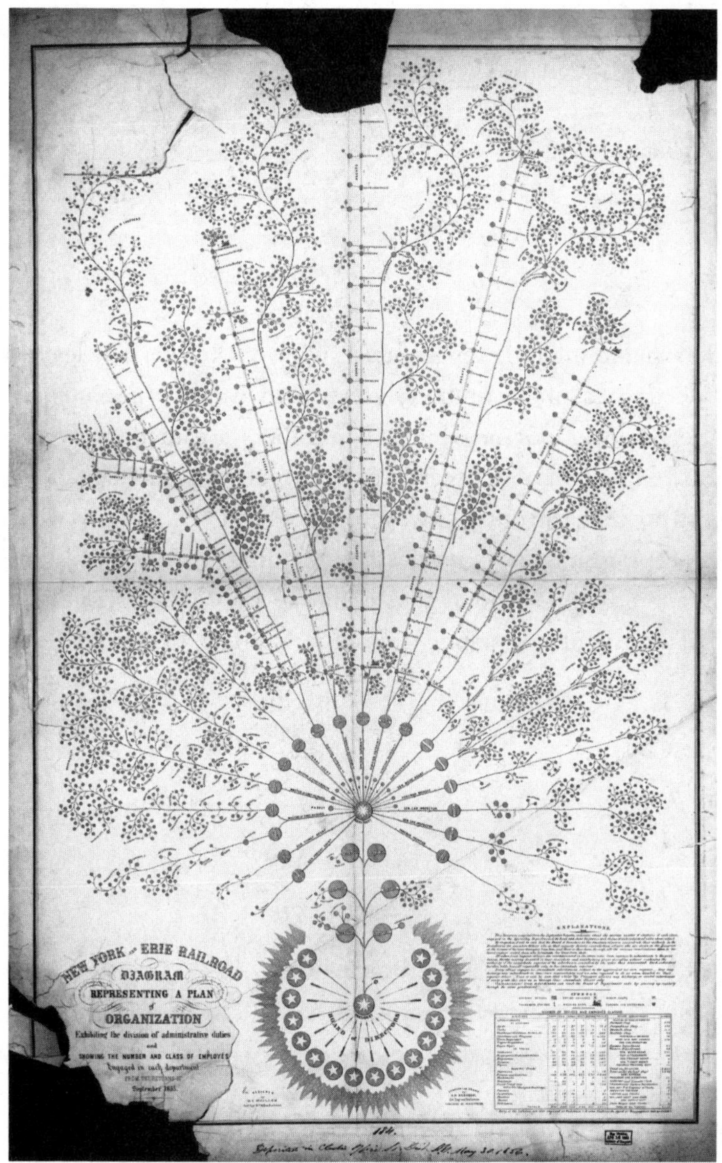

The first org chart ever created

I mean: You find two smaller numbers whose product is 1,001. Say, 1,001 = 13 × 77. And now we take the axe to each factor separately. We can break 77 up as 7 × 11. And 13? Well, with 13 we're stuck. There's no way to express it as a product of two smaller numbers. Whack it with the axe as hard as you like; it won't break. The same is true for 7 and 11. And we can record what we just did in a tree

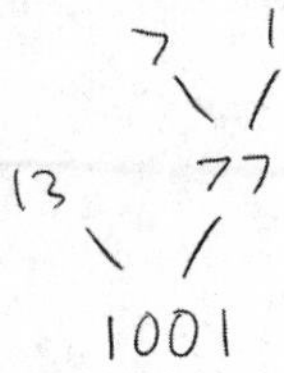

7 11
\ /
13 77
\ /
1001

in which each branching represents a swing of the axe. The leaves of the tree, the unbreakable numbers, are what we call *prime numbers,* the basic Lego blocks out of which all numbers are made. *All* numbers? How do I know? I know because of the tree. At each stage of our axe-swinging process, the number we attack either branches into two smaller factors or it doesn't, and if it doesn't, it's prime. We keep on axing until we can axe no more. And at that point, *all* the remaining numbers are prime. That might take a long time, if we started, say, with 1024:

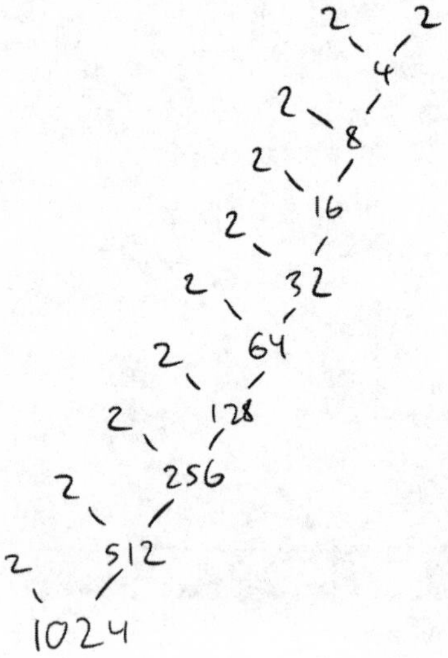

2 2
4
2 8
2 16
2 32
2 64
2 128
2 256
2 512
1024

or right away, if we started at a prime number like 1009:

1009

but sooner or later, it has to happen.

The process can't go on forever, because with each swing the numbers in the tree get smaller, and a sequence of positive whole numbers that decreases at each step must, eventually, hit bottom and stop.*

At the end of the axefest we're left with a tree each of whose leaves is a number that can't be factored—that is, a prime—and those prime numbers, multiplied together, give the number we started with.

This fact, that every whole number, however big and complicated, can be expressed as a product of primes, was probably first proved around the end of the thirteenth century by the Persian mathematician (and pioneer in optics—things were less specialized back then) Kamāl al-Dīn al-Fārisī, in his treatise "Tadhkirat al-Ahbab fi bayan al-Tahabb" ("Memo for Friends Explaining the Proof of Amicability").†

Which might seem weird, given that we just proved it in a paragraph. Why did it take nearly two thousand years to get from the first recorded definition of a prime number by the Pythagoreans to the theorem of al-Fārisī? This comes down to geometry again. Euclid certainly understood facts that, to a modern number theorist, would immediately imply that any number can be factored into primes: a whole bunch of primes, like 1024, or only one, like 1009, or somewhere in between, like 1001. But Euclid didn't talk about products of long lists of primes, and our best guess is that it's because he *couldn't*. For Euclid, everything is geometry, and a number is a way of referring to the length of a line segment. To say a number is divisible by 5 is to say the segment is "measured by five"—that is, we can lay down some number of length-5 segments to cover the segment in question exactly. When Euclid multiplies two numbers, he thinks of the result as the area of a rectangle whose length and width are the two numbers we multiplied together (the "multiplicands," to use one of my favorite math words). When Euclid multiplies *three* whole numbers, he calls the result a "solid," because he

* That last sentence sounds obvious, and it sort of is, but it's worth taking a second to reflect that it wouldn't be true if I didn't say "positive"—what about 2, 1, 0, −1, −2, −3 . . . ? Or if I didn't say "whole" —what about 1, 0.1, 0.01, 0.001 . . . ?

† "Amicability" here doesn't mean "friendliness," but the property enjoyed by a pair of numbers each of whose proper divisors add up to the other. There's an interesting story here, but I don't see any geometry in it, so it'll wait for another time.

thinks of it as the volume of a rectilinear brick whose length, width, and height are given by the multiplicands.

Mathematics is a fundamentally imaginative enterprise, which draws on every cognitive and creative ability we have. When we do geometry we use what our minds and bodies know about the size and shape of things in space. Euclid made strides in number theory not as a palate-cleansing break from his work in geometry, but *because* of his work in geometry. By thinking of numbers as lengths of line segments he was able to understand numbers better than his forebears. But yoking his number theory to his geometric intuition limited him, too. A product of two numbers was a rectangle, and a product of three numbers a brick. What was the product of four numbers? That's not a quantity that can be realized in the three-dimensional space people live in. So it was a quantity Euclid had to pass over in silence. The more algebraic approach favored by the mathematicians of medieval Persia was less tied to our physical experience, and thus better able to leap out into purely mental abstract realms. But that doesn't mean it's not still geometric. Geometry, we've already seen, isn't really limited to three dimensions. There can be as many dimensions as you like. We just have to imagine a little more strenuously. We'll get there.

THE TREE OF NIM

We have seen that the game of Nim, like the organization of a railroad or our inevitable human descent into sinful abandon, is described by a tree of some finite extent. No matter what path the players take through the branches, they eventually wind up at an endpoint, a leaf; someone has won and someone has lost.

But *who*?

This, it turns out, is something the tree can tell us, too.

The trick is to start at the end of the game. That's the easiest time to tell who's winning! If there are no stones left, whoever just played has won the game. So if it's my turn, and there are no stones, I have lost. To keep track of that, I'm going to decorate the tree of Nim I drew before, writing an "L" on top of all the positions in the tree with no stones, to remind us that I lose if I encounter one of these positions when it's my turn to play.

What if there's just one stone? Then I have only one choice. I take the stone, and I win. So I write a "W" on top of that position.

What about two stones in one pile? Now things get more complicated, because I have options. I can take both stones; if I do that, I've won. But if I'm foolish or inattentive or perverse or generous enough to take just one, I've put my opponent in the winning position we just marked "W," and I lose. How to label a position like this, where who wins depends on what I do? We go by the principle that players of competitive games are *not* foolish or generous or inattentive or perverse; they want to win, and they make whatever choices will get them there, if they can. So this position gets marked a "W." To be clear, that doesn't mean I win *whatever* I do next. For most games, that'll never be the case; no matter how good your position, you can always find a real dog of a move that gives the game away. The mark of "W" just means that one of the moves available to me right now puts my opponent in a losing position. You can read it as "path to victory."

Two stones, one in each pile, is a different story. Whatever I do, I put my opponent in a "W" position, a position from which they can win. So this position gets an "L" drawn on.

Here's how our tree looks so far:

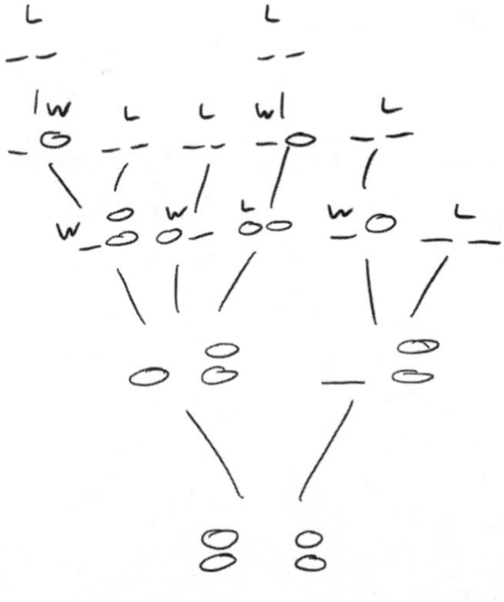

And now we can proceed, backward in time, step by step. Two piles, one with two stones, one with just one? We have three choices of move: take away the small pile, take away the big pile, or take one from the big pile. The resulting positions are already labeled W, W, L. But since *one* of the options leads to a losing position for my opponent, that's the move I should choose, and the current position gets a "W." The player presented with a 2-pile and a 1-pile is going to win, as long as they make the right move.

You win when your opponent has no choice but to lose. That sounds like a motivational poster in a CrossFit gym, but it's actually math. In the language of the tree, it says "Mark a position with a W if there's a branch starting at that position that ends in an L." And, by the same token, mark a position with an L if you *can't* do that. Because that means, whatever choice you make, you're presenting your opponent with a W. *You lose when your opponent can win no matter what you do.*

It comes down to this:

THE TWO RULES:

First Rule: If every move I make leads to a W, my current position is an L.

Second Rule: If some move I can make leads to an L, my current position is a W.

The Two Rules allow us to mark every single position on the tree with either a W or an L, systematically, all the way back to the root where we started. You never get caught in a cycle, because trees don't have cycles.

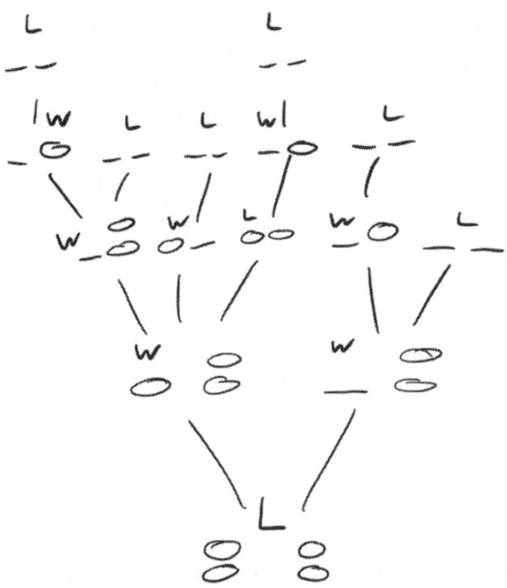

And the root is an L. Which is why Akbar, who goes first, loses, unless Jeff makes a move he shouldn't have.

I can describe this process in words on the page, but honestly? The only way to really get this down in your bones is to do it yourself. This is a partner activity, so get a friend. Offer to play Nim with two piles of two stones. Let your friend go first, because maybe you are not really that good of a friend. Now use the tree above to tell you how to move. Win and win and win again. Now you can *feel* how it works.

The tree method works for Nim with more stones, it works for Nim with more piles, it works for all kinds of Nim. You want to know who wins Nim with two piles with twenty stones each? You can draw a big tree and you can work your way down and you can figure it out. (Jeff wins.) Two piles of a hundred stones each? (Jeff still wins.) One pile of a hundred stones and one of a thousand? (This one's an Akbar win.)* What's more, the labeled tree doesn't just tell you who wins—it tells you *how* to win. If you're at a W position, you know there's at least one move that leads to an L—take it. If you're at an L, shrug philosophically, move however you like, and hope your opponent screws the pooch.

* Exercise for the reader: Can you see why this claim follows from the previous one?

For Nim with just two piles of stones, I should say, you can dodge the tedium of labeling the entire tree. There's an easier (and honestly kind of lovely) way to figure out who wins, which exploits the symmetry of left and right. Remember how much simpler Pappus's symmetry-exploiting proof of the pons asinorum was than Euclid's original argument? Nim is much the same. Suppose Akbar and Jeff start with a hundred stones each. Do you want to draw that tree? Neither do I. So here's a better way. You know that incredibly annoying thing siblings inflict on each other, where the younger sibling repeats everything the older sibling says? "Stop repeating me." "Stop repeating me." "You're *so irritating*." "You're *so irritating*." And so on. Well, imagine Jeff plays the game like that. Whatever Akbar does, Jeff responds by doing the exact same thing in the other pile. Akbar takes fifteen stones off the left pile, leaving eighty-five? Jeff takes fifteen off the right and now both piles have eighty-five. Akbar switches to the right pile and takes seventeen off, leaving sixty-eight? Jeff does the same to the left pile. Jeff always mirrors Akbar, always leaving the piles equally large. In particular, Jeff can never be the first to empty a pile, because he never does anything that's not the reflection of something Akbar just did. Akbar will be the first to empty one pile of stones, and when he does, Jeff mirrors him, empties the other pile, and wins. So Nim with two equal piles is a win for Jeff. His strategy is as unbeatable as it is exasperating.

And what if the piles aren't equal in size? Akbar, going first, will take just enough stones from the bigger pile to make the two piles equal. Now *Jeff* is the irritated older sibling, because from here on in Akbar is going to mirror his every move, and inevitably win. In the language of the Two Rules, Akbar is using his move to get to a position with two equal piles, which the paragraph above shows is an L; and if you can get to an L your current position is a W by the Second Rule.

When you have more than two piles, a simple symmetry argument like this doesn't work. But there's actually still a way to find out who wins without drawing the whole tree. It's a bit too complex to describe here, involving the base-2 expansions of the sizes of all the piles, but you can learn all about it in Elwyn Berlekamp, John Conway, and Richard Guy's astonishingly colorful, profound, and idea-rich book *Winning Ways for Your Mathematical Plays*, along with other games like Hackenbush, Snort, and Sprouts, and why every game is, in the end, a kind of number.

In a variant of Nim called the "subtraction game," you start with just one pile of stones, but at each move you're only allowed to take away 1, 2, or 3. The player who takes the last stone wins. This game is a tree too, and you can analyze it in the same way, starting from the end. This version of Nim got a big publicity boost when it appeared as a challenge for the competitors in the fifth season of the reality show *Survivor*, held in Thailand. (The game was billed neither as Nim nor as the subtraction game but as "Thai 21," though the game has no Thai roots; the usage is presumably aimed at an American audience primed to see activities of Asian origin as intricate and inscrutable. It's the same reason there's a seemingly ineradicable tradition of describing Nim as an "ancient Chinese game," though this claim appears to be wholly invented, and Nim is first attested in a sixteenth-century book of mathematical puzzles and magic tricks by Fra Luca Bartolomeo de Pacioli, a buddy of Leonardo da Vinci, a Franciscan friar, and the generally recognized "Father of Double-Entry Bookkeeping." Isn't that at least as interesting as being ancient and from China?)

The thing about *Survivor* is that the conventional wisdom holds it to be one of the dumbest programs on television when in fact it's one of the smartest. How many shows are there where you get to watch people think, actually think, in real time? Let alone *do mathematics* in real time? That's what episode six of season five of *Survivor* offers. Ted Rogers Jr., a big strong man who played, very briefly, for the Dallas Cowboys, takes the lead, telling his teammates, "At the end, we wanna make sure there's four flags." (*Survivor*'s version of the game used flags in place of stones.) "Faaaahve or four?" says Jan Gentry, the Texan lady in the huddle with Rogers. "Four," the big man insists.

Rogers is doing the same computation, in his head, that we did for the Nim tree. He's approaching the problem just the way a mathematician would—starting from the end of the game. That's no surprise; we are all mathematicians in the deep strategic parts of our brain, whether it says that on our business cards or not.

If there's one flag left, that's a W; you take that flag and you've won.

Two or three flags, same thing, since you still have the ability to take all the flags at once. What about four?

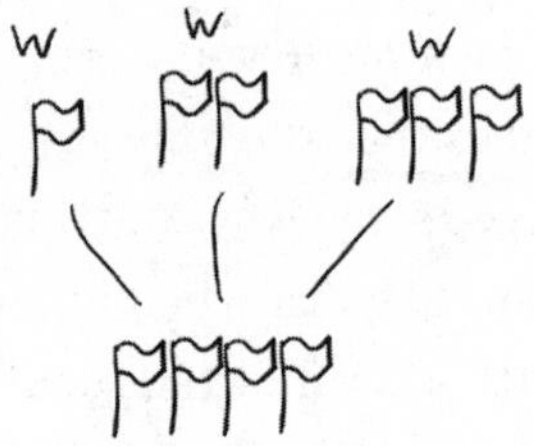

No matter what move the survivors make, they're leaving the other team a W. So by the Second Rule, four flags is an L. Big Ted was right; leave the other team with four flags and you can ensure yourself victory. The opposing team has the same realization, but much too late; after they take away three flags from the nine flags before them, leaving six, they stare at each other, stricken, and one says, "If they take two, we lose." They do, and they do.*

The insight arrived too late to save them, but it's still of use to us. Why is it so bad to be faced with four flags? Because every move available has a natural counter from your opponent. You take three, they take one. You take two, they take two. You take one, they take three. Either way, all four flags are gone, game over, winner: not you.

So it's good to leave your opponent with four flags. And if you're presented with five, six, or seven, that's exactly what you do; take away just the number of flags that leaves the fatal four. If you've got eight, though, there's a black fly in your chardonnay. You take three, they take one. You take two, they take two. You take one, they take three. And it's *you* who's faced with the four.

Sounds familiar, right? It's because starting with eight flags is, as far as strategy goes, the *same thing* as starting with four. Whatever you do, your opponent counters in a way that reduces the total number of flags by four, which means you lose. And starting with twelve is the same as starting with eight, and starting with sixteen is the same as starting with twelve, and, so on. . . .

* Exercise 2: How many of the nine flags should they have taken? We'll answer this in a second.

If you start with a number of flags that's a multiple of four, you lose, and if not, you win—as long as you take the right number of flags to leave your opponent one of the fatal numbers.

We just proved a theorem!

The reasoning we're carrying out here is exactly the sort we are meant to be teaching in math class, and especially in geometry class—the reasoning of proof. We observe (perhaps by pure thought, perhaps by repeated play) that four and eight flags are losing positions. We analyze our understanding of why those positions are such losers. We begin to comprehend why it is not just four, not just eight, but *any* multiple of four that spells defeat. Having done so brings us to a mental position where we can, if we feel like it, construct a more formal chain of reasoning, demonstrating that you lose whenever the number of flags is four times whatever.

A proof is crystallized thought. It takes that brilliant buoyant moment of "getting it" and fixes it to the page so we can contemplate it at leisure. More importantly, we can share it with other people, in whose mind it springs to life again. A proof is like one of those hardy microbial spores so robust they can survive a trip through outer space on a meteorite and colonize a new planet after impact. Proof makes insight portable. We mathematicians have been known to describe ourselves as standing on the shoulders of giants, but I prefer to say we're walking up a staircase made of the frozen thoughts of regular-sized people. We get to the top, we sprinkle our own thoughts on the ice, they freeze to the mass and make the staircase that much higher. Not as pithy but truer.

AND SO ON . . .

I said we proved a theorem. Should we write it down? Let's write it down.

> **Survivor Theorem:** If the number of flags is four, or eight, or twelve, or any multiple of four, the first player loses; otherwise, the first

player can win by choosing whatever number of flags leaves the second player with a multiple of four.

And now the proof. Maybe you found my reasoning convincing already. I hope so! But there's a soft spot and it's these three words: "And so on . . ." Those three dots are a punctuation called an *ellipsis*, Greek for "falling short." They indicate there's something we've decided to leave unsaid. In a proof, that seems like kind of a bad idea.

What happens when we try to speak the unspoken? We mentioned four flags, eight flags, twelve flags, and sixteen flags, but not twenty. So we could add a discussion of why you lose when you start with twenty flags. But then we'd still need to cover twenty-four. And having done so, we'd have twenty-eight flags left to do. And so on. . . . It's a real problem! A proof that's infinitely long is not of any use. Who would read it? And yet it seems somehow like a dereliction of duty to wave one's hands and say, "I could just keep going with this, but I won't."

Let's try this another way. We can break our Survivor Theorem into two parts:

ST1: If the number of flags is four, or eight, or twelve, or any multiple of four, the first player loses.

ST2: If the number of flags is not a multiple of four, the first player wins.

Why do we think **ST1** is true? Because however many flags we take away, whether one, two, or three, we leave our opponent a number of flags which isn't a multiple of four. And according to **ST2**, that situation is to be marked a W. The Second Rule now tells me my current position is an L. So the reason **ST1** is true is because **ST2** is true. In logic lingo, we say **ST2** *implies* **ST1.**

That seems like progress! We had to prove two things, and now we only need to prove one. So why is **ST2** true? Suppose the number of flags isn't a multiple of four. Then you can reduce the number of flags to a multiple of four by removing either one, two, or three flags on your next move.* Now by **ST1** you've put your opponent in an L position, and since you can move to an L, the First Rule tells you that your current position is a W.

To sum up: **ST1** is true because **ST2** is true, and **ST2** is true because **ST1** is true.

Uh-oh!

This feels like circular reasoning, that tragic argumentative move where you take a claim to be its own justification. Most of us are too clever to talk ourselves into doing it directly, so we build a little cycle of statements, each of which implies the next:

"I don't believe anything I read in *Angry Pundit Weekly,* they're untrustworthy. How do I know not to trust them? Because they run false stories all the time. How do I know those stories are false? Because I read them in the untrustworthy *Angry Pundit Weekly.*"

This is the kind of trap math is supposed to help you evade. Yet here is our ankle, caught in its jaws.

Thankfully, there's a way out. Think again of our original argument, which apart from the pesky ellipsis was pretty convincing, and rightfully so. Because there was a *downwardness* to it—you proved a fact about sixteen using a fact about twelve, which in turn you proved from a fact about eight, which in turn you proved from a fact about four. That process can't go on forever; it has to stop, because positive whole numbers can't just go on getting smaller and smaller without end. This too is geometry! On a continuous path, we can just keep getting and closer to the path's endpoint, taking an unlimited number of smaller and daintier steps. But the whole numbers have a geometry that's *discrete,* not continuous; they're like a sequence of individual rocks you hop between. There are only so many rocks on your route and you eventually run out. If that point sounds familiar, it's because we mentioned it a few pages back, talking about why the factorization of numbers has to eventually wind up in a pile of unbreakable primes. And the method we're using here, which is called *mathematical induction,* dates back in some sense to

* A nice number-theoretic way to put this is that the number of flags you should take is the *remainder* when you divide the current number of flags by four.

that fact about prime factorization, which al-Fārisī wrote down seven centuries ago.

The argument is a proof by contradiction, which by now is an almost reflexive habit for most mathematicians. Whatever you want to prove, you presume the opposite. That sounds perverse and wrong, but it's immensely helpful. You take as your presumption that you're wrong about the state of the world, and you hold that thought in your mind, turning it over and over and following its chain of implications, until (one hopes!) you arrive at the conclusion that your unappealing presumption couldn't have been correct. It's like holding a hard candy in your mouth and letting it dissolve, dissolve until you come to the sour contradiction at its center.

So let's suppose we were wrong about the Survivor Theorem. Then there's a *counterexample*: some bad number of flags where the theorem tells us we lose, but where actually we win, or where it tells us we win, but actually we lose. Maybe there are a *lot* of bad numbers like that. But whether there's one bad number or many, there's one that's smallest.

At this moment, algebra enters. People quail, sometimes, when an "x" or a "y" heaves into the scene. It helps to think of a symbol like that as a pronoun. Sometimes you want to refer to a person, but you don't know that person's name. Maybe you don't even know who exactly the person is. Say you're talking about the next president of the United States. You refer to them with a pronoun, saying "he" or "she" or, as I just did, "them," not because that person doesn't have a name but because you don't know what it is. So let's use the pronoun "N" for that smallest bad number. Remember: "bad" here means that either N isn't a multiple of four and it's a winning position, or it *is* a multiple of four and it's a losing one. What if it's a multiple of four? Then whatever I do next, whether I take away one flag, two, or three, the result isn't a multiple of four. What's more, the number of flags is now smaller than N, so it can't be a bad number—

—this is the big moment of the proof, so stop and admire. N isn't just a bad number; it's the *very smallest* bad number. So any number smaller than N has to behave itself and do as the Survival Theorem says. Okay, now back to our sentence—

—which means that it obeys the Survival Theorem and is a W.

Can you taste the tang of contradiction? N is supposed to be a winning position, but any move you make from N leaves your opponent with a W. So this can't be right.

That leaves the possibility that N isn't a multiple of four, and is an L. But no matter what N is, I can take either one, two, or three flags away and leave the other player a multiple of four; and that, because the new, smaller number of flags can't be bad, has to be an L. If I can leave the other player in a losing spot, my position must have been a W. This is a contradiction, too. There's no way out; no way, that is, except to concede that we must have been wrong in the first place to imagine there were any bad numbers at all. The numbers are good—all of them. And so the Survivor Theorem is proved.

Now there are two ways you might respond to that proof. One is to admire the systematic parade of thoughts that carefully guide us along a winding path to an inescapable conclusion. And the other, which is honestly just as valid, is to say—

"Why did we just spend two pages doing that? I was already completely convinced! I knew what you meant by 'And so on . . . ' and I didn't feel any more explanation was needed. Do you mathematicians *really* spend all day piecing together elaborate arguments to prove what a normal person would already consider established beyond doubt?"

Well . . . *some* days, yes. Not most. Once you've seen a few proofs like this, you don't have to write them down anymore. You see the "And so on . . ." and you take it as a proof, not because it *is* a proof, exactly, but because you have enough experience to know that a careful proof could be constructed to replace the ellipsis.

The game of Nim is a kind of mathematics—or, if you like, this kind of mathematics is a kind of game. It's a game people all over the world enjoy and willingly play. So here's a question: Why don't we teach this in school? Skill at Nim may not be directly relevant to your profession, assuming you're not a reality-show contestant, but if we grant that learning to think mathematically helps us understand everything else better,* doing this analysis has to count as educational. We're always being scolded that the school system is crushing our students' natural sense of play. If we played more games in math class, would students learn more math?

Yes. Also, no. I've been teaching math for more than twenty years now. When I started, I was driven by questions like this: What's the *right way* to teach a mathematical concept? Examples first, then explanation? Explanation followed by examples? Letting students discover principles by examining the examples I present, or stating principles at the blackboard and letting students discover examples? Wait, are blackboards even good?

I've come to feel there's no one right way. (Though there are certainly some wrong ways.) Different students are different and there is no One True Teaching Method that will ring everyone's saliva bell. I myself, I have to admit, do not like games. I hate to lose and they stress me out. I once got in a shouting match with a friend's mom when she shot the moon against me in Hearts. A lesson plan centered on Nim would likely lose me utterly. But it might fascinate the kid sitting next to me! Math teachers, I think, ought to adopt every teaching strategy they can, and shuttle through them in quick succession. That's the way to maximize the chance each student at least sometimes feels that their teacher is finally, after so much boring hoo-hah, talking about things in a way that makes sense.

THE WORLD OF MR. NIMATRON

Did you take my advice and actually play some 2 x 2 Nim? It didn't feel much like play, did it? Once you know the strategy, it's kind of a slog, like carrying out a mindless and purely mechanical process. You're right. It's so mechanical you can make it *literally* mechanical. Here's U.S. Patent 2,215,544, from 1940.

This machine plays Nim, and plays it perfectly. In the electrical

* You're here, with me, quite a ways into this book, so I'm assuming we've granted that?

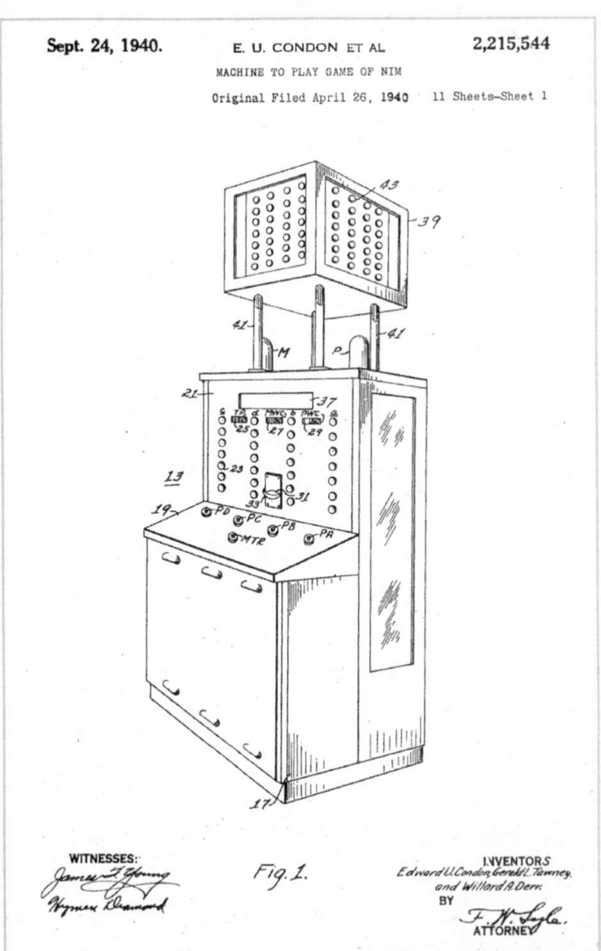

The Nimatron patent

The Nimatron in its finished glory

spirit of the time, it used lighted bulbs in place of stones. A few years later, its co-inventor, the physicist Edward Condon of the Westinghouse Corporation, would become the assistant director of the Manhattan Project (only to quit six weeks later, complaining that the absolute secrecy of the work was "morbidly depressing"). But in 1940, with America at peace, he was showing off the Nimatron at the World's Fair in Flushing Meadows, New York (theme: "The World of Tomorrow"). Nimatron played a hundred thousand games of Nim that summer in Queens. *The New York Times* wrote:

> On the novelty side the Westinghouse Company announced it would introduce a "Mr. Nimatron"—a new electric robot 8 feet tall, 3 feet wide, and a ton in weight, at its exhibit. "Mr. Nimatron," pitting his electrical brain against the human thinking apparatus, will play a variation of an old Chinese* game called "Nim" with all comers. The game will consist of turning out lamps in four rows of lights until the last bulb is extinguished. "Nimatron" usually will win, but if he loses he will present to his opponent a pocket token stamped "Nim Champ," the Westinghouse people promised.

How could the human ever win, if Mr. Nimatron is perfect? Because Mr. Nimatron offered a choice of nine starting configurations, some of which were "W" for the human player; so humans could win, as long as they were perfect, too. Usually they weren't. According to Condon, "Most of its defeats were at the hands of the exhibit attendants as demonstrations to folks who, after numerous trials, became convinced that the machine couldn't be beaten."

In 1951, the British electronics firm Ferranti built its own Nim-playing robot, Nimrod, which drew huge crowds on a world tour. In London a team of psychics attempted to overcome Nimrod's perfect play by means of concentrated telepathic vibrations, with no success. In Berlin the machine took on future West German chancellor Ludwig Erhard and beat him three times in a row. Alan Turing, who worked

* Not actually Chinese, see above!

on Ferranti's Mark One computer, reported that Nimrod so captivated the German public that a free bar down the hall went entirely unpatronized.

That a computer can play Nim as well as a human was seen as amazing, Germans-passing-up-free-beer amazing—but is it? Turing himself expressed some skepticism, writing: "The reader might well ask why we bother to use these complicated and expensive machines in so trivial a pursuit as playing games." Knowing what we now know about Nim, we can see it doesn't require any kind of human-level insight to be a perfect player, only the patience to label the tree, step by step, from leaf to root. If you've played tic-tac-toe, you've probably observed the same thing. That's because tic-tac-toe has the geometry of a tree, too. The first few stages look like this:*

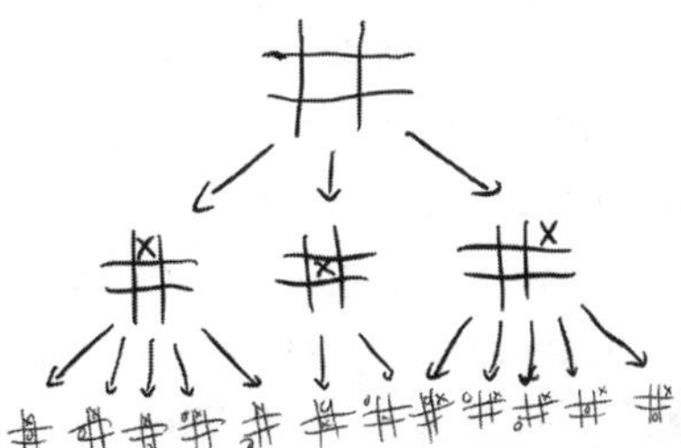

There's a difference, though; a game of tic-tac-toe, unlike a game of Nim, can end in a draw, which for reasons permanently obscure is called a *cat's game*. In fact, when both players are over the age of seven, *most* games end in a draw.

No problem: this just means we need a new letter, "D" for "draw," and Three Rules instead of Two.

* There's some calling different things by the same name in this picture, too; the symmetries of the game mean we can treat all opening corner moves and all opening "mid-edge" moves the same, which allows us to draw only three opening branches instead of nine.

THE THREE RULES:

First Rule: If every move I make leads to a W, my current position is an L.

Second Rule: If some move I can make leads to an L, my current position is a W.

Third Rule: If no move I can make leads to an L, but not every move I make leads to a W, my current position is a D.

The Third Rule is longer, but it captures what it means to be in a drawn position. The first part of the rule says that I haven't won. The second part says that I haven't lost, because there's some move I can make that doesn't present my opponent a path to victory. If I can't win but my opponent can't beat me, it's a draw.

What I want you to notice is that, no matter what the options before us when we play tic-tac-toe, we are always in a situation where one of the Three Rules applies. So, just as with Nim, we can work our way up to the root of the tree, the empty board, which we'll find ourselves marking with a D. Unsurprisingly, there's no secret undiscovered strategy for winning tic-tac-toe. If both players play perfectly, it's a cat's game every time.

Here's something that happens a lot in math. You sit down to solve one problem, and when you finish, the next day or month or year, you realize you've solved a lot more problems at the same time. When a nail requires you to invent a truly new kind of hammer, everything looks like a nail worth hitting with that hammer, and lots of things actually are.

Tic-tac-toe has the geometry of a tree, and so the Three Rules guarantee that tic-tac-toe is either a win for the first player, a win for the second player, or a draw. What's more, a purely mechanical computation can tell us which of those three options is the case, and what a perfect strategy looks like.

By the same reasoning, that's true for *any* game whose geometry is a tree. That means any game where two players take turns, where the outcome of a move is deterministic (no coin flips, spinners, card draws, or other instruments of chance), and where every game ends after some finite number of steps. For such a game, either:

> The first player has a strategy that ensures they always win;
> The second player has a strategy that ensures they always win; or
> Every perfectly played game ends in a draw.

And we can figure out those strategies by labeling a tree, leaves to root, with W, L, and D according to the Three Rules. It might take a long time, but it will always work.

Lots of games are trees. Checkers is a tree. So is Connect Four. So is chess. Yes, even chess! We think of it as a kind of romantic art, a way of distilling the essence of combat onto a little wooden board. It *means* something. There are movies and novels and musicals scored by members of ABBA about it.

But it's a tree. Players take turns, no chance is involved, and a game can last no more than 5,898 moves. That's the theoretical maximum for a legal game, at least, which would never arise in a game the players were actually trying to win. The longest tournament game ever played clocked in at a relatively brisk 269 moves and took a little over twenty hours to complete.

If you don't know chess, you might wonder why the game has a limit. It's not like Nim; you don't lose pieces with every move. Why can't the horsie and the castle just chase each other around the board forever? It's because the masters of chess put in rules to forbid exactly that. If fifty moves go by without anybody capturing anybody or moving a pawn, for instance, the game is over and declared a draw. These "stalemate" rules stem from the same impulse that drives us to exclude 1 from the list of prime numbers. If we declared 1 to be a prime, the process of prime factorization could go on and on endlessly: $15 = 3 \times 5 \times 1$

× 1 × 1. . . . That's not exactly *wrong*, but it's kind of pointless. Stalemate keeps chess from going down the same dull eternal path.*

So chess, for all its lore and mystique, is the same kind of thing as Nim and tic-tac-toe. If two absolutely perfect players faced off, either white would always win, white would always lose, or the game would always end in a draw. And in principle, computing which one is the case is merely a matter of working step by step down the tree to the root. Chess is a hard problem, yes, but it's not a hard problem like writing a poem that captures the intersection of midcentury Atom Age politics with urban renewal, nostalgia for childhood, the endless reverberations of the Civil War, and the replacement of the human spirit by mechanized artifice. It's a hard problem like multiplying two really big numbers together. It might take a long time to do it, but in principle you know how to finish the task, step by step.

"In principle." Those two words, a little straw mat placed daintily over a bottomless abyss of difficulty!

Nim with two piles of two stones is a loss. Connect Four is a win. (Pretty dispiriting, sis!) But we don't know whether chess is a win, a loss, or a draw. We may never know. The tree of chess has many, many leaves. We don't know exactly how many, but it's more than an eight-foot robot can contemplate, that's for sure. Claude Shannon, who we last saw generating faux English text with a Markov chain, also wrote one of the first papers to take machine chess seriously; he thought the number of leaves was on the order of 1 with 120 zeroes after it, a hundred million trillion googols. That's more than the number of . . . okay, actually, it's more than the number of *anything* in the universe, and it is certainly not a number of things you're going to comb through one by one and write little W's, L's, and D's next to. In principle, yes; in reality, no.

This phenomenon of computations we know exactly how to do, but don't have time to do, is a somber minor-key motif that sounds through the whole history of computer-age mathematics. Back to prime factorization for a second. We've already seen that you can carry that out

* If you want to get pedantic, checkers isn't strictly speaking a finite tree, because it doesn't have strict stalemate rules. If the players want to dance their kings around forever, they are technically free to do so. In practice, checkers players agree to a draw when they see neither one can force a win.

without much real thought. If you start with a number like 1,001, you just have to find a number that divides it up evenly, and if you can't find one, 1,001 is prime. Does 2 work? No, 1,001 can't be split in half. 3? No. 4? No. 5? No. 6? No. 7? Yes—1,001 is 7 × 143. (The Thousand and One Nights were the Hundred and Forty-Three Weeks.) Having swung the axe once, we can swing it again at 143, testing division after division until we find that 143 = 11 × 13.

But what if the number we're trying to factor is two hundred digits long? Now the problem is a lot more chess-level. The life span of the universe isn't enough to check every possible divisor. It's just arithmetic, sure. But it's also, as far as we know, completely infeasible.

Which is good, because there's undoubtedly something in the real world you value whose security depends on that problem being hard. What does factorizing numbers have to do with security? For that, we need to go back to Confederate cryptography and the 1914 book of experimental prose poetry *Tender Buttons*, by Gertrude Stein.

ELIMINATING THE NECESSITY FOR *TENDER BUTTONS*, BY GERTRUDE STEIN

Suppose Akbar and Jeff, done with their game, want to communicate in secret. They can do so if they have a secret coding scheme in common. "In common" is the crucial phrase here; they need to be using the same code, and that requires sharing some piece of information, usually called the *key*. Perhaps the key is the text of *Tender Buttons*, by Gertrude Stein. If Akbar wants to privately transmit the message "Nim has grown dreary" to Jeff, here's how he can do it. He writes his message above the opening of the first poem in *Tender Buttons*, by Gertrude Stein ("A CARAFE, THAT IS A BLIND GLASS. A kind in glass and a cousin, a spectacle and nothing strange a single hurt color and an arrangement in a system to pointing") matching letter for letter.

```
NIM HAS GROWN DREARY
ACA RAF ETHAT ISABLI
```

Now we add up each pair of letters. Letters aren't numbers, but they have a position in the alphabet, and that's what we add. It's customary to start with 0, so that A is letter number 0, B is letter number 1, and so on. N is the thirteenth letter of the alphabet and A the zeroth; add those, and get 13, and the thirteenth letter is N. And I + C is 8 + 2 = 10, which goes with K. Keep doing this letter by letter: you get an encoded text that starts NKM YAX K . . .

After that, you hit a small problem: R(17) + T(19) is 36, which is not a letter. But the problem's easily solved: you just wrap around after Z, so that the twenty-sixth letter is A again, the twenty-seventh is B, and so on, until you find that letter 36 is the same as letter 10, or K. Your message ends up looking like this:

```
 NIM HAS GROWN DREARY
+ACA RAF ETHAT ISABLI
---------------------
 NKM YAX KKVWG LJEBCQ
```

Now Jeff, who has the encrypted message, and also, of course, a copy of *Tender Buttons*, by Gertrude Stein, can go backward, by subtracting the letters of the poem instead of adding. N minus A is 13 minus 0 is 13 is N. And so on. When we get to that second K, we find ourselves asked to subtract T (19) from K (10). That leaves –9, but that's okay! The minus-ninth letter is the one nine letters before A(0), and remembering that the letter before A is now understood to be Z, this is the letter eight places before Z, which is R.

If you don't like all the addition and subtraction, you can just keep this table handy:

	A	B	C	D	E	F	G	H	I	J	K	L	M	N	O	P	Q	R	S	T	U	V	W	X	Y	Z
A	a	b	c	d	e	f	g	h	i	j	k	l	m	n	o	p	q	r	s	t	u	v	w	x	y	z
B	b	c	d	e	f	g	h	i	j	k	l	m	n	o	p	q	r	s	t	u	v	w	x	y	z	a
C	c	d	e	f	g	h	i	j	k	l	m	n	o	p	q	r	s	t	u	v	w	x	y	z	a	b
D	d	e	f	g	h	i	j	k	l	m	n	o	p	q	r	s	t	u	v	w	x	y	z	a	b	c
E	e	f	g	h	i	j	k	l	m	n	o	p	q	r	s	t	u	v	w	x	y	z	a	b	c	**d**
F	f	g	h	i	j	k	l	m	n	o	p	q	r	s	t	u	v	w	x	y	z	a	b	c	d	e
G	g	h	i	j	k	l	m	n	o	p	q	r	s	t	u	v	w	x	y	z	a	b	c	d	e	f
H	h	i	j	k	l	m	n	o	p	q	r	s	t	u	v	w	x	y	z	a	b	c	d	e	f	g
I	i	j	k	l	m	n	o	p	q	r	s	t	u	v	w	x	y	z	a	b	c	d	e	f	g	h
J	j	k	l	m	n	o	p	q	r	s	t	u	v	w	x	y	z	a	b	c	d	e	f	g	h	i
K	k	l	m	n	o	p	q	r	s	t	u	v	w	x	y	z	a	b	c	d	e	f	g	h	i	j
L	l	m	n	o	p	q	r	s	t	u	v	w	x	y	z	a	b	c	d	e	f	g	h	i	j	k
M	m	n	o	p	q	r	s	t	u	v	w	x	y	z	a	b	c	d	e	f	g	h	i	j	k	l
N	n	o	p	q	r	s	t	u	v	w	x	y	z	a	b	c	d	e	f	g	h	i	j	k	l	m
O	o	p	q	r	s	t	u	v	w	x	y	z	a	b	c	d	e	f	g	h	i	j	k	l	m	n
P	p	q	r	s	t	u	v	w	x	y	z	a	b	c	d	e	f	g	h	i	j	k	l	m	n	o
Q	q	r	s	t	u	v	w	x	y	z	a	b	c	d	e	f	g	h	i	j	k	l	m	n	o	p
R	r	s	t	u	v	w	x	y	z	a	b	c	d	e	f	g	h	i	j	k	l	m	n	o	p	q
S	s	t	u	v	w	x	y	z	a	b	c	d	e	f	g	h	i	j	k	l	m	n	o	p	q	r
T	t	u	v	w	x	y	z	a	b	c	d	e	f	g	h	i	j	k	l	m	n	o	p	q	r	s
U	u	v	w	x	y	z	a	b	c	d	e	f	g	h	i	j	k	l	m	n	o	p	q	r	s	t
V	v	w	x	y	z	a	b	c	d	e	f	g	h	i	j	k	l	m	n	o	p	q	r	s	t	u
W	w	x	y	z	a	b	c	d	e	f	g	h	i	j	k	l	m	n	o	p	q	r	s	t	u	v
X	x	y	z	a	b	c	d	e	f	g	h	i	j	k	l	m	n	o	p	q	r	s	t	u	v	w
Y	y	z	a	b	c	d	e	f	g	h	i	j	k	l	m	n	o	p	q	r	s	t	u	v	w	x
Z	z	a	b	c	d	e	f	g	h	i	j	k	l	m	n	o	p	q	r	s	t	u	v	w	x	y

This is just like the addition table you learned in elementary school, but for letters! To compute R + T, just look at the R row and the T column (or the T row and the R column) and find K.

Or, better still, you can take advantage of the geometry this code imposes on the alphabet. We've adopted the rule that when you go one letter past Z you don't fall off the edge of the English language; you come back to A. And what that means is that we're thinking of the alphabet, not as a line,

ABCDEFGHIJKLMNOPQRSTUVWXYZ

but as a circle.

Every *A* in *Tender Buttons* by Gertrude Stein is a 0, which means that when the letter in our key is an *A* we leave the corresponding letter of the message alone. Every C is a 2, which means rotate the circle counterclockwise by two jots. From this geometric standpoint it's obvious why this code is easy to decipher, as long as you have the key; all you need to do is rotate the circle by the same amount, but clockwise instead.

This kind of code is called a *Vigenère cipher,* after Blaise de Vigenère, a learned sixteenth-century Frenchman who didn't invent it. Misattributions like this are common in math and science, so common that the statistician and historian Stephen Stigler formalized a law: "No scientific discovery is named after its original discoverer." (Stigler's Law, Stigler observed, was really first formulated by the sociologist Robert Merton.)

Vigenère was a well-connected man of noble birth, writer of many books, a secretary to ambassadors and kings. As such, and especially during his years in Rome, he was exposed to the latest and most intricate in coded messagery. The world of sixteenth-century Roman cryptography was one of rivalries and jealously guarded secrets. Vigenère famously pranked one of those rivals, Paulo Pancatuccio, the pope's personal code breaker, by sending him a message in a childishly easy cipher. Pancatuccio swiftly decoded the message, only to find a stream of insults directed at himself: "O poor wretched slave that you are to your decipherments, on which you waste all your oil and your pains. . . . Come, use your leisure and your work in the future for things more fruitful, and stop uselessly frittering away your time, one lone minute of which cannot be bought back by all the treasures of this world. Put matters to the test now, and see if you can get at the meaning of one little letter of what follows here." At that point, the cipher switched to one of Vigenère's own high-test home brews, which, as Vigenère knew very well, was beyond Pancatuccio's ability to break. We know all this from Vigenère's book *Traicté des Chiffres ou Secrètes Manières d'Escrire* ("Treatise on codes and secret writing"), which became a standard reference on cryptography while the rest of Vigenère's belletristic work was forgotten. The book contains many of Vigenère's own complex codes, and also the essential ideas of the simpler Vigenère cipher described here, which in fact were laid out in 1553 by Giovan Battista Bellaso, who had devised the cipher while working as secretary and cryptographer to Cardinal Durante Duranti in Camerino. (How far down the ecclesiastical ladder did you have to go in those days before you didn't get your own cryptographer?)

Bellaso had a high opinion of his code, advertising it as being "of such marvelous excellence that all the world may use it, and notwithstanding no one will be able to understand what the other writes, save only those who possess a very brief key, as is taught in this booklet, together with his explanation and method of use." The world largely agreed with his assessment; the so-called Vigenère cipher became widely known as *le chiffre indechiffrable*, the undecodable code. No reliable way to unravel the Vigenère existed until the development of the "Kasiski examination," which, as Stigler might have predicted, was actually invented two decades before Friedrich Kasiski by Charles Babbage. But even that method doesn't work well if the key is something as long as *Tender Buttons*, by Gertrude Stein.

COMPLETE VICTORY

A code, of course, is only as good as the work ethic of the people who deploy it. For instance: you probably know that the Confederacy was a rebellious splinter nation that waged war against the United States of America in a desperate effort to preserve a monstrous system of mass enslavement, but were you aware they were also really terrible at cryptography? The Confederates used a Vigenère with a short key repeated throughout the length of the message, and they only bothered to encode the words of the message they deemed strategically important. So a dispatch sent by Jefferson Davis to General Edmund Kirby Smith on September 30, 1864, reads, in part:

> By which you may effect O—TPQGEXYK—above that part HJ—OPG—KWMCT—patrolled by the ZMGRIK—GGIUL—CW—EWBNDLXL.

Not only did the Confederate cryptographers leave much of the message in plain text, they left the spaces between the encrypted words intact, making it quite natural for the Union soldiers who intercepted the message to guess that the phrase after "above that part" should be "OF THE RIVER." And once you have a passage of decoded text to work with, you can go backward and work out the key. Look back at that alphabetic square on page 131. To send O to *H* requires the corresponding letter in the key to be *T*. To send *F* to *J* you need an *E* in the key. In the arithmetic language we used earlier, we're subtracting: H – O = T and J – F = E. Keep going and you get

```
 OF THE RI VER
-HJ OPG KWMCT
_____________
 TE VIC TORYC
```

From this one small phrase the Union codebreaker has already revealed more than half the key the Confederates were using, which is—a little ironically considering what was about to happen to the Confederacy—"Complete Victory." And once the key is known, decrypting the rest of the message is just a few minutes' work.

The Vigenère code with a long key retains its status as a more or less undecodable code. But there's a problem, a big one. Someone else besides Akbar and Jeff might have a copy of *Tender Buttons,* by Gertrude Stein. And anyone who does can decode their messages with ease. If Akbar or Jeff wants to include Sheba as another trusted communicant, they need to get Sheba a copy of *Tender Buttons,* by Gertrude Stein. If you want to send someone your key, you can't encrypt it, because they don't yet have the key required to decode it; but if you send it unencrypted and the message is intercepted, then the eavesdropper has your key, and you might as well not encrypt at all.

This used to be thought of as a basic structural problem of cryptography, one that you couldn't solve, but just had to live with. Sheba and the enemy eavesdropper, after all, are in the same position; neither one knows the key. You can't get the key to Sheba without sending her a message, but without the key you can't protect that message from enemy eyes. How can you send a message that Sheba can read but the eavesdropper can't? That's when—quite unexpectedly and wonderfully, like a delivery of flowers from an intriguing new acquaintance—prime factorization arrives at the door.

Multiplication of large numbers, it turns out, is what mathematicians call a *trapdoor function.* A trapdoor is a door it's easy to go through in one direction and really, really hard in the other. Multiplying two one-thousand-digit numbers is something your phone can do before you can blink. Separating that product into the original two multiplicands is a problem no known algorithm can solve within a million million lifetimes. And you can use this asymmetry to get your key to Sheba without your adversary listening in. There's a wonderful algorithm to do this called RSA, after Ron Rivest, Adi Shamir, and Leonard Adleman, the people who invented it in 1977. At least, they were the people who invented it and *told* everyone about it. The real story is a little more interesting. As Stigler's Law demands, the people RSA is named after aren't the people who actually first created it. The system was really invented in the early 1970s, by Clifford Cocks and James Ellis. At least this time there's a very good reason for the misattribution: Cocks and Ellis were working at GCHQ, the top-secret British intelligence shop. Until the 1990s, nobody outside the classified circle was allowed to know that RSA predated R, S, and A.

The details of the RSA algorithm involve a bit more number theory than I want to pack into this space, but here's the key feature. Sheba has in mind two very large prime numbers, p and q. Nobody but Sheba knows them: not Akbar, not Jeff, nobody. Those numbers are the keys. The RSA algorithm can be used to decode messages by anyone who knows those two big primes.

But to encipher the message in the first place, you don't need to know p and q, but only their product, an even huger number we call N.* So it's not like the Vigenère cipher, where decoding is just encoding run backward, using the same key. In RSA, encoding and decoding are completely different processes, and thanks to the trapdoor, the former is much easier than the latter.

That big number N is called the *public key*, because Sheba can tell it to everybody. She can tape it to the front door of her house if she wants. When Akbar sends Sheba a message, he only needs to know the product N; using that number, he can encode a message, which Sheba, at home with her secret p and q, can turn back into readable text. *Anybody* can send Sheba a coded message using N; they can even post those messages publicly. Everybody can see all those messages, but nobody except Sheba, holder of the private keys, can read them.

The advent of public-key cryptography has made everything easier and simpler. You (or your computer, or your phone, or your fridge) can send messages very securely to tons of people at once without having to find a way to share privileged information. But it all rests on the trap-

* My hardworking copy editor asked: Why are p and q in lowercase, but N is in uppercase? This reflects a mathematical habit of using small letters for numbers we think of as small and capital letters for those we think of as big. In this case, p and q might be three hundred digits long, which may not sound small, but compared to their product N, they are indeed itsy-bitsy.

door actually *being* a trapdoor. If someone builds a ladder under it, making two-way travel easy, the whole edifice falls apart. That is: if someone devised a way to separate a large number N into its component prime factors p and q, that person would have instant access to every formerly private message encoded with that N.

If the problem of factoring primes turns out, like the problem of winning a game of chess, to be easier for a computer program than we thought, the transfer of information suddenly becomes a much more dangerous enterprise. That's why you get thriller novels—I saw this in the airport, it's real—with breathless back cover copy like:

> Teenager Bernie Weber is a math genius. Washington, the CIA, and Yale invade Milwaukee to kidnap him. They need to know his secret for factoring prime numbers.

(If you didn't laugh at the last sentence, take a second and think carefully about the mathematical task Bernie is said to have mastered.)

MINE WAS THE LORD

Chinook played checkers better than anyone alive, dead, or neither. But that didn't mean it couldn't *in principle* be beaten. Maybe there was some superior strategy hiding in the depths of the checkers tree, not yet dreamed of by human or machine, which would fell the champ. The only way to rule that out entirely was to analyze checkers down to the very bottom, to say for certain how to label the root. Which of the three types of game is checkers? A first-player win, a second-player win, or a draw?

Let me not keep you in artificial suspense. It's a draw. Mathematically, checkers is more or less a big bicolored version of tic-tac-toe. Two players who never make a mistake will never win and never lose; they'll draw every time. This might not come as a big surprise for followers of Marion Tinsley, who, as you'll recall, made very few mistakes. His competition didn't slip up much more. And when two of those near-perfect players faced off against each other, they mostly played to draws. In 1863, James Wyllie, the Scottish champion known as the "Herd Laddie,"* faced the Cornishman Robert Martins in Glasgow for a world championship match. They played fifty games, of which every single one ended in a draw. And twenty-eight of those games were exactly, move-for-move, the same. Boring! The Glasgow debacle led checkers to adopt a "restriction" system, with the first two moves of the game drawn at random from a deck of permissible openings; the idea was to keep players from plodding down the same well-worn paths of the tree and ending up, again and again, at the same old leaf. But after Samuel Gonotsky and Mike Lieber† played to forty straight draws in a 1928 match for a $1,000 purse at the Garden City Hotel in Long Island, New York, the system moved on to the current "three-move restriction," with the first three moves drawn from 156 opening options, with names like "Dreaded Edinburgh," "The Henderson," "The Wilderness," "Fraser's Inferno," "Waterloo," and "Oliver's Twister."‡ Even with the three-move restriction, modern championship checkers has many more draws than wins or losses.

But that's just a mass of evidence; an actual *proof* that there's no winning strategy for either player that generations of checkers masters had somehow missed would be a different matter.

Chinook was just five years old when it took the checkers crown from Marion Tinsley in 1994. It would be thirteen more years before Jonathan Schaeffer and the rest of the Chinook team could prove Tinsley couldn't *possibly* have beaten it. And neither can anyone else. Certainly not you.

You can try, though! Chinook runs all day and night on a server at its home in Edmonton, Alberta, taking on all comers. As you play, Chinook calmly assesses its position. "Chinook has a small advantage," it

---

* Because he was a country kid who would lead herds of cattle into Edinburgh and hustle the city slickers when he got there, betting opponents he could beat them ten times for every game they took from him. Indeed he could.

† Lieber was a high school classmate of Asa Long's from Toledo, which was apparently the San Pedro de Macoris of competitive checkers.

‡ "Checkers opening or moderately difficult ski slope?" would be a fun parlor game.

reports at first. Then "Chinook has a big advantage." And then—after seven moves in the game I'm playing as I write this: "You lose." That means you've arrived at a position that Chinook knows, from its panoptic point of view, to be a W. Doesn't mean you have to stop playing! Chinook is patient. Chinook has no place else to be. You can make another move. Chinook moves its own piece, and remarks, again, "You lose." Keep it up as long as you can stand it.

Playing against Chinook is unsettling but also somehow soothing. It's very different from the experience of playing a game against a very, very skilled human who's trying to beat you, which is unsettling and not soothing at all. I once played a game of Go against my cousin Zachary, who at the time was fifteen years old, the drummer in a thrash metal band called Sinister Mustard, and one of the top junior chess players in Arizona. Zachary had never played Go before, and at first I was able to gain a fair amount of advantage. But about a quarter of the way into our game, something locked in for him—he grasped the logic of the game, as he had grasped chess long before, and with a real vigor wiped me off the table. Playing against Tinsley was said to be much the same: "Terrible Tinsley," the invariably courteous and kind math professor was called, just because the experience of sitting across the checkerboard from him was a near guarantee of being bulldozed. Tinsley, like the 1994 iteration of Chinook, was essentially perfect at checkers. But unlike Chinook, he *cared* if he won. "I'm basically an insecure individual," he said in an interview. "I hate to lose." To Tinsley's way of thinking, even though he and Chinook were carrying out the same task, they were fundamentally different kinds of beings. "I have a better programmer than Chinook," he told a newspaper before the two met in the 1992 tournament. "His was Jonathan, mine was the Lord."

THE AFRICAN GLASGOW

Checkers, according to Schaeffer, has 500,995,484,682,338,672,639 possible positions, though many of these can't ever be arrived at in a legal game. Because checkers is a tree,* we can work our way backward from the end of the game, assigning each of those positions a W, L, or D.

But even this set of positions, tiny compared to what chess or Go has to offer, is beyond our capacity to work out exhaustively. Fortunately, you can get by with a lot less, thanks to the power of the Three Rules.

The most popular of the seven possible opening moves in checkers is the one denoted "11-15," but which is so fervently loved by expert players that it's usually called "Old Faithful." Suppose Black starts with Old Faithful and White responds with the move called "22-18," the beginning of an opening called the "26-17 Double Corner." Now it's Black's turn to play again. At this point, Schaeffer proves that Black might have an L or might have a D, but definitely can't force the win. So we mark this position LD, to show we haven't quite finished computing it.

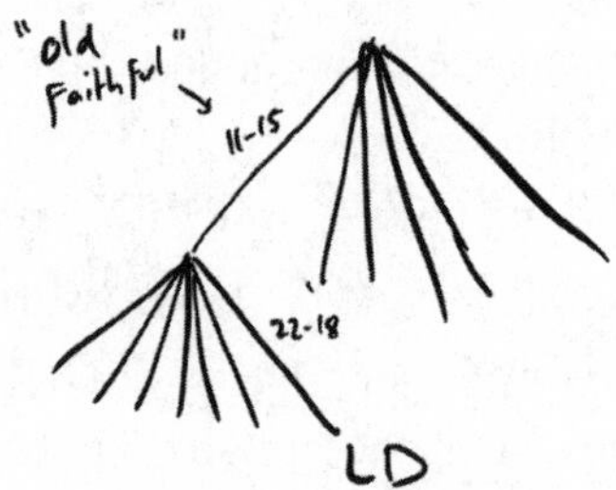

But that already tells us something about Old Faithful! By the Three Rules, a position is an L *only* if every position descending from it in the tree is a W. That's not true for Old Faithful, because White has a choice, 22-18, which leads to either L or D. So we know Old Faithful is either D or W. And we know this without bothering to study any of the many other possible responses to Old Faithful that White might have made, or pinning down the exact label we're supposed to assign to 22-18. In computer science and also arborist lingo, we have "pruned" the branches we can get away with not considering. This is a tremendously important technique. Often people think of developments in computation as arising when we make our computers more blazingly fast, so they can compute *more stuff, bigger data*. It's actually just as important to prune away

* Return of the pedantic footnote: it's a finite tree as long as you impose the chesslike rule that a position repeated three times ends the game in a draw, which is what Schaeffer does.

big parts of the data that aren't relevant to the problem at hand! The fastest computation is the one you don't do.

In fact, all of the seven opening moves can be shown to lead to either D or W in the same efficient way. Only for one of them, 9-13, does Schaeffer need to dig deeper, and show that it's D.

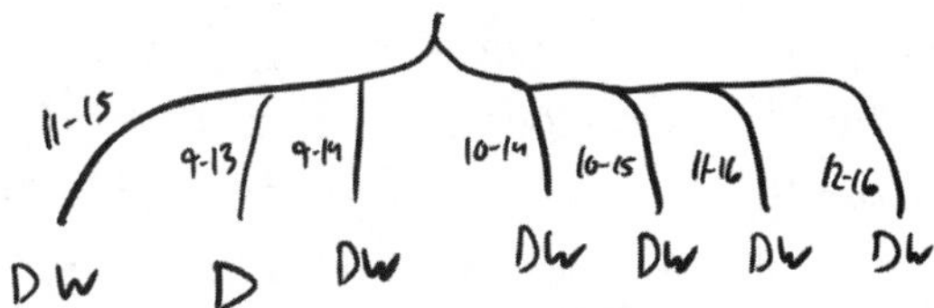

And that's enough to solve checkers! We know that Black, starting the game, has an option that doesn't give White a winning position—namely, 9-13—so the initial position can't be an L. But we also know that none of Black's options give White an L, so the initial position isn't W either. That only leaves D; checkers is a draw.

We don't have an analysis like this for chess, not yet and maybe not ever. The chess tree is a redwood to checkers's shrub, and we don't know whether the root should be marked W, L, or D.

But what if we did? Would people still give their lives to chess if they knew a perfect game always ended in a tie, that there was no winning by magnificence, only losing by screwing up? Or would it feel empty? Lee Se-dol, one of the best Go players alive, quit the game after losing a match to AlphaGo, a machine player developed by the AI firm DeepMind. "Even if I become the number one," he said, "there is an entity that cannot be defeated." And Go isn't even solved! Compared to the redwood that's chess, Go is—well, if there were a tree somewhat bigger than a googol redwoods it would be that tree. Read chess and Go forums and you'll see a lot of people grappling with the same anxieties Lee expressed. If a game is just a tree with letters written on it, is it still really a game? Should we just quit when Chinook tells us, with calmness and infinite patience, that we have lost?

The International Checker Hall of Fame used to be the biggest attraction in Petal, Mississippi, a city of about ten thousand people just outside the college town of Hattiesburg. The Hall was a thirty-two-thousand-square-foot mansion that featured a bust of Marion Tinsley, and the largest checkerboard in the world, and also the second-largest checkerboard in the world. It closed down in 2006 after its founder was sentenced to five years in federal prison for money laundering. In 2007—the same year Schaeffer proved checkers was a draw—it burned to the ground.

And yet people are still playing checkers, all around the world, still vying to be the human champion. (As I write, that title is held by the Italian grandmaster Sergio Scarpetta.) It's not as popular as it once was, sure, but the decline started long before Schaeffer's proof, and new players keep joining the pool. Amangul Berdieva of Turkmenistan, one of the world's top players, was a seven-year-old girl when Chinook took Tinsley's crown. The current world champion of "go-as-you-please" checkers (where you choose your own opening) is Lubabalo Kondlo, from South Africa, who's forty-nine. Kondlo has pioneered a variant of the very opening Wyllie and Martins played to forty draws in a row in Scotland in 1863; Kondlo's version, in that match's honor, is now known as the African Glasgow.

If the point of playing checkers is to be the best at winning, there's no point anymore in playing checkers. But the point of playing checkers isn't to be the best at winning. No human was better at winning than Marion Tinsley, and Tinsley knew that wasn't what mattered. "I surely have an intense dislike for losing," he told an interviewer in 1985, "but if we play a lot of beautiful games, that will be my reward. Checkers is such a beautiful game that I don't mind losing." Chess is no different. The current world champion, Magnus Carlsen, told an interviewer, "I don't look at computers as opponents. For me it is much more interesting to beat humans." Garry Kasparov, the longtime champion, dismisses the idea that human chess playing is obsolete, because for him, the computation carried out by the machine and the game play executed by a human are fundamentally different things. "Human chess," he says, "is a form of psychological warfare." It's not a tree; it's a battle that takes place in a tree. Reflecting on a game he'd played against Veselin Topalov twenty years earlier, Kasparov says, "I was amazed by the beauty of this geometry." The tree geometry tells you how to win; it doesn't tell you what makes a game beautiful. That's a subtler geometry, and for now it's not one a machine can compute step by step with a short list of rules.

Perfection isn't beauty. We have absolute proof that perfect players will never win and never lose. Whatever interest we can have in the game is there only because human beings are imperfect. And maybe that's not bad. Perfect play isn't play at all, not in the plain English sense of that word. To the extent that we're personally present in our game playing, it's by virtue of our imperfectness. We *feel* something when our own imperfections scrape up against the imperfections of another.

Chapter 6

The Mysterious Power of Trial and Error

We don't know how to fully label the tree of chess with W's, L's, and D's, and when I say we may never know, I don't mean because we aren't very clever, I mean because the number of positions of the tree to be labeled is so very great, more than any physical process could label with letters before the universe winks out. It is, strictly speaking, possible there's some way to bypass the recursive rigamarole of starting at the leaves (*so many leaves*) and labeling our way back. That's what happened with the "subtraction game" the Survivors played. When the game starts with 100 million flags, you *could* work laboriously backward from the end of the game and fill in all the W's and L's, or you could use the Survivor Theorem we proved a few pages ago. Since 100 million is evenly divisible by four, the theorem tells us, the second player can always win. And we even know how: if the first player takes one flag, you take three. If they take two, you take two. If they take three, you take one. Repeat 24,999,999 more times and enjoy your victory.

I cannot prove there's no simple-minded winning strategy like this for chess. But it doesn't seem likely.

And yet computers *do* play chess. They play chess really well. Better than me, better than you, better than Garry Kasparov, better than my cousin Zachary, better than anybody. How can they do that if they can't possibly compute the labels on all the states of the game?

They do it because the machines of the new wave of artificial intelligence don't even try to be perfect. They are going for something entirely different. To explain what that is, we have to go back to the prime numbers.

Remember: the apparatus of public-key cryptography that so much rests on depends on being able to come up with two large prime numbers to use as your private key, where "large" means three hundred digits or so. Where do you get these? They don't have a prime number store at the mall. Even if they did, you wouldn't want to use store-bought primes, because unless you're a Confederate cryptography reenactor the whole *point* of your secret key is that it isn't publicly available.

So you have to brew your own. Which seems hard at first. If I want a three-hundred-digit number that's *not* prime, I know what to do: just multiply together a bunch of smaller numbers until I get up to three hundred digits. But the prime numbers are precisely those that aren't made up of smaller building blocks; how to even start?

This is one of the questions I hear most as a math teacher: "How do I even start this?" I'm always happy to hear it, no matter how stricken the student looks as they ask, because the question is an opportunity to teach a lesson. The lesson is that it matters much less how you start than *that* you start. Try something. It might not work. If it doesn't, try something else. Students often grow up in a world where you solve a math problem by executing a fixed algorithm. You're asked to multiply two three-digit numbers and the first thing you do is multiply the first number by the last digit of the second number and you write that down and you're off.

Real math (like real life) is nothing like this. There's a lot of trial and error. That method gets looked down on a lot, probably because it has the word "error" in it. In math we're not afraid of errors. Errors are great! An error is just an opportunity to run another trial.

So you need a three-hundred-digit prime. "How do I even start this?" You start by picking a random three-hundred-digit number. "How do I know which one?" Seriously, doesn't matter. "Okay, how about 1 with three hundred zeroes after it?" Okay, maybe not that one because it's obviously even, and an even number other than 2 can't be prime because you can factor it as 2 times something else. That's an error, on to the next trial. Pick another three-hundred-digit number, one that is odd this time.

So at this point you have come up with a number that, as far as you can see, is prime. At least, you can't see any obvious reason it *isn't* prime. But how can you be sure? You could try swinging your factoring axe at your number and seeing what happens. Is it divisible by 2? No. Is it divisible by 3? No. Is it divisible by 5? No. You're making progress, but again, you have much more than the age of the universe left to go. You can't check primeness this way, in practice, any more than you can solve chess by labeling tree branches one by one.

There's a better way, but we have to enlist a different geometry: the geometry of the circle.

OPALS AND PEARLS

This is a *bracelet*: seven stones arranged in a circle, some opals, some pearls.

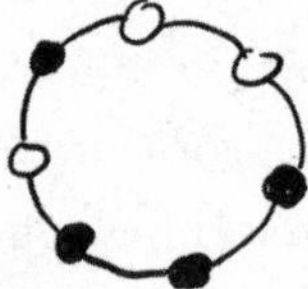

Here are some more bracelets:

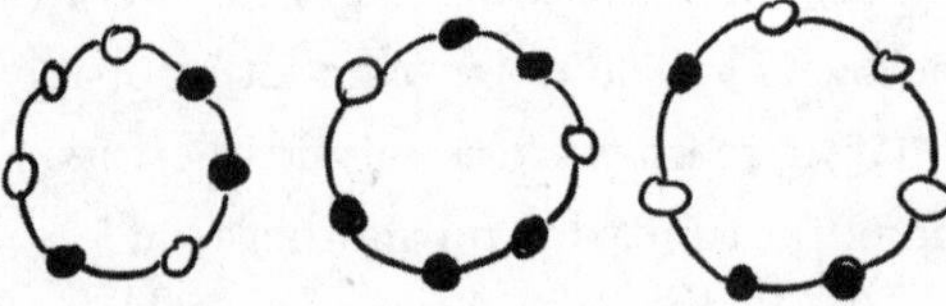

Here are all the bracelets with four stones:

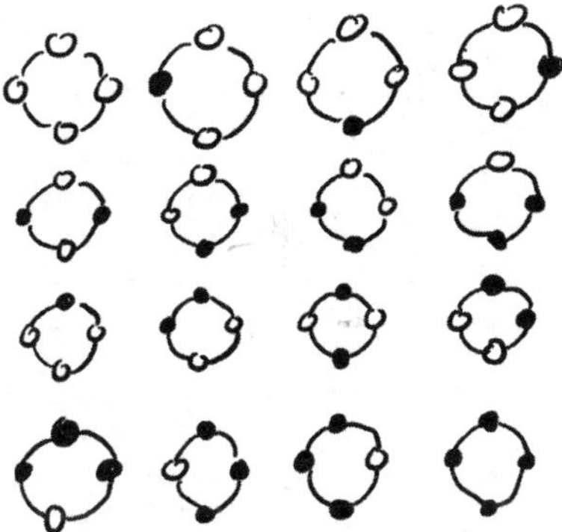

There are sixteen of them. You can just count the bracelets in the picture and satisfy yourself I didn't miss any, but there's also a more fancy-pants way. Starting from the top and working clockwise, the first stone is either an opal or a pearl: two choices. For *each* of these two choices, there are two choices for the next stone; that's four choices in all for the disposition of the first two stones. And for each of those four choices, two choices of the third stone, and we're up to eight choices in all. For each of *these* eight choices, you have one bracelet where the final stone is an opal and one where the final stone is a pearl; so you end up with twice eight, or 2 × 2 × 2 × 2, or 16.

Or you could just have counted! But the advantage of wearing the fancy pants is that we can export this reasoning to bigger bracelets, like the seven-stone number on the previous page. The number of ways to make a seven-stone bracelet is 2 × 2 × 2 × 2 × 2 × 2 × 2, or 128; my Sharpie isn't thin enough to draw all those on the page.

But maybe, I hear you suggest, I'm drawing more bracelets than I need to. Look at those three bracelets above—the third bracelet is what you get when you rotate the first one two clicks to the right. Is that really a *different* bracelet, or just the same one viewed from a different angle?

For now, let's stick with the convention that we're counting bracelets as different if they look different on the page. But let's not forget about this idea of rotating. We might call two bracelets *congruent* if the first one can be rotated to yield the second (which also means the second can be rotated to yield the first).*

Maybe it makes for a nicer jewelry drawer to organize the bracelets by congruence. Each bracelet can be rotated in seven ways, so we're grouping our 128 bracelets into piles of seven. How many piles? Just divide 128 by 7 and you get the answer: 18.2857142. . . .

Hurray, another error! Something has gone wrong, because 128 is not a multiple of seven.

The problem is with some of the bracelets I didn't draw. Like the all-opal version:

The seven rotations of this bracelet are all the same bracelet! So it's not a group of seven; it's a group of one. The all-pearl bracelet is its own group, too.

Should we worry about other too-small groups? Definitely. These two four-stone bracelets

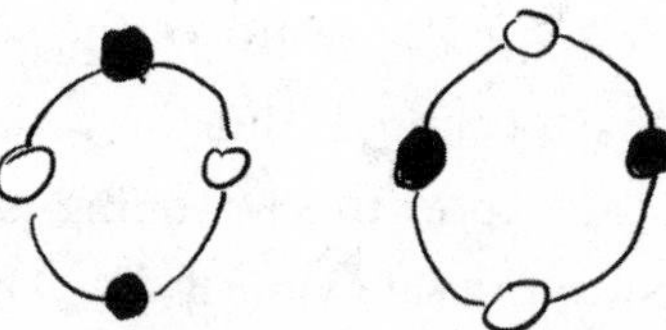

are a group of their own. That's because the alternating opal-pearl pattern repeats itself every two places. So you don't have make four full rotations to get back the original bracelet; just two will do.

But with seven stones on the bracelet, this kind of thing doesn't happen. Fire up your imagination and suppose you had a bracelet you could rotate, say, three places and get back the same bracelet you started with. Then you'd have a group of three: the original bracelet, the bracelet

* Which matches perfectly the notion of congruence we encountered in chapter 1, where two figures in the plane are called congruent if you can turn one into the other by a rotation or some other rigid motion.

rotated once, and the bracelet rotated twice. Wait, what if some of these were the same? To push that obnoxious possibility away, let's say three is the *smallest* number of turns* that returns the bracelet to its initial form.

If a triple turn brings us back to the same bracelet, then so does turning it six times, and so does turning it nine times; but now we have a problem, because seven turns of the bracelet *definitely* brings the bracelet back where it started, so nine turns is the same as two turns, but two turns can't bring the bracelet back where it started because we *just decided* no fewer than three turns could do that.

And the tang of contradiction fills our nostrils once again.

Maybe starting with three was a bad idea. What if you have a group of five, so that five is the smallest number of rotations that restores the bracelet? Then ten rotations also restores the bracelet, and ten turns is the same as three, and we are contradicted again. How about two turns? That worked for the four-stone bracelet. If two turns keep the bracelet the same, then the same applies to four, and six, and eight, and uh-oh—eight turns is the same as one.

We didn't have this problem when there were only four stones. You turn the bracelet twice, you get the same bracelet. You turn it four times, you get the same bracelet; but now there's no contradiction because you already knew four turns would bring you back to the beginning. What makes it work is that four is a multiple of two. And what caused all our problems with seven stones is that seven is *not* a multiple of three, or of five, or of two. Seven is not a multiple of *anything*, because seven is a prime.

Did you remember we were originally talking about primes?

This same principle, by the way, has a lot to tell us about cicadas. Every seventeen years, my home state of Maryland is visited by the Great Eastern Brood, a population of hundreds of billions of insects that emerge from the earth and cover the landscape of the whole mid-Atlantic like a chirping carpet. For a while you try to avoid crunching

* Smallest number greater than zero, you pedant.

them underfoot as you walk and then you just give up, there are too many.

But why seventeen? A lot of cicada specialists believe—though I'm going to be honest with you here, there is serious disputation on this point among cicada specialists, of whom there are more than you would probably think, and who are surprisingly entertaining in their salty takedowns of each other's cicada periodicity hypotheses—that the cicadas count to seventeen underground because seventeen is a prime number. If it were sixteen, instead, you could imagine a similarly periodic predator evolving to emerge every eight years, or four, or two; and every time they came out, there would be heaps of cicadas to consume. But no hungry lizard or bird can sync up with the Great Eastern Brood unless it itself evolves a period seventeen years long.

When I say 7 (like 5, and 17, and 2) isn't a multiple of anything, I overstate the case; it's a multiple of 1, of course, and it's a multiple of 7. So there are two kinds of groups of bracelets: groups of one and groups of seven. And a group of one has to have all its stones the same, because *any* rotation leaves it unchanged.

So the all-opal and all-pearl bracelets are the lone groups of one; and the other 126 bracelets fall into groups of seven. *Now* the division works; there are 126/7 = 14 of those groups.

What if we went up to eleven stones? The total number of bracelets is two multiplied by itself eleven times, a quantity we call 2^{11}, or 2,048. Again, there are two monochromatic bracelets, and the other 2,046 have to fall into groups of 11; 186 of those groups, to be exact. You can just keep going:

$$2^{13} = 8{,}192 = 2 + 630 \times 13$$
$$2^{17} = 131{,}072 = 2 + 7710 \times 17$$
$$2^{19} = 524{,}288 = 2 + 27594 \times 19$$

Did you notice I skipped 15? I skipped it because it's not prime, being 3 times 5, but I also skipped it because it doesn't work! $2^{15} - 2$ is 32,766, which doesn't divide evenly into 15s. (The bracelet rotation enthusiasts among you are encouraged to verify on their own time that the

32,768 bracelets break down into 2 groups of one, 2 groups of three, 6 groups of five, and 2,182 groups of fifteen.)

We thought we were screwing around with spinning bracelets, but in fact we were using the geometry of the circle and its rotations to prove a fact about prime numbers, which on the face of it you'd never have thought geometric at all. Geometry is hiding everywhere, deep in the gears of things.

Our observation about primes isn't just a fact, it's a fact with a name: it's called Fermat's Little Theorem, after Pierre de Fermat, the first person to write it down.* No matter which prime number n you take, however large it may be, 2 raised to the nth power is 2 more than a multiple of n.

Fermat wasn't a professional mathematician (in seventeenth-century France there were hardly any such people) but a provincial lawyer, a comfortable member of the Toulouse bourgeoisie. Fermat, far from the center of things in Paris, participated in the scientific life of his times largely though correspondence with his mathematical contemporaries. He first stated the Little Theorem in a 1640 letter to Bernard Frénicle de Bessy, with whom Fermat was engaged in a vigorous exchange on the subject of perfect numbers.† Fermat stated the theorem but didn't set down a proof; he *had* one, he told Frénicle, which he would definitely have included in the letter "if he did not fear being too long." This move is *classic* Pierre de Fermat. If you've heard his name at all, it's not because of Fermat's Little Theorem, but the other FLT, Fermat's *Last* Theorem, which was neither his theorem nor the last thing he did; it was a conjecture about numbers Fermat jotted down in the margin of his copy of Diophantus's *Arithmetic*, sometime in the 1630s. Fermat noted that he'd come up with a really handsome proof that the margin of the book wasn't big enough to contain. Fermat's Last Theorem did turn out to be a theorem after all, but only centuries later, in the 1990s, when Andrew Wiles and Richard Taylor finally finished off the proof.

---

* This is actually just one case of Fermat's theorem; in fact, for *any* number m, not just 2, m^n is m more than a multiple of n.

† Perfect numbers are numbers equal to the sum of all their smaller factors, like 28 = 1 + 2 + 4 + 7 + 14. To a modern mathematician their charm is somewhat obscure, but Euclid loved them, and that gave them a certain cachet for early number theorists. It feels good to outdo Euclid.

One way to read this is that Fermat was a kind of visionary, able to reliably infer the correctness of mathematical statements without proving them, the way a master checkers player can feel the soundness of a move without nailing down the winning sequence all the way to the end. A better take is that Fermat was an ordinary person who wasn't always careful! Fermat certainly realized quickly that he didn't have a proof of his so-called Last Theorem, since he later wrote about special cases of the theorem without ever again claiming he knew a proof of the whole shebang. The French number theorist André Weil* wrote of Fermat's premature assertion that "there can hardly remain any doubt that this was due to some misapprehension on his part, even though, by a curious twist of fate, his reputation in the eyes of the ignorant came to rest chiefly upon it."

At the end of Fermat's letter to Frénicle he expresses his belief that all numbers of the form $2^{2^n} + 1$ are prime. In Fermat fashion, he didn't offer a proof, but said, "I am almost persuaded," having checked that his conjecture held when n was 0, 1, 2, 3, 4, and 5. But Fermat was wrong. His statement wasn't true for all numbers. It wasn't even true for 5! He had neglected to notice that 4,294,967,297, which he thought he'd checked was prime, was actually 641 × 6,700,417. Frénicle didn't notice Fermat's mistake (too bad, since the tone of the letters suggests he was really eager to get one over on his more famous correspondent) and neither did Fermat, who stood by this conjecture for the rest of his life, apparently never bothering to check the arithmetic he'd done in his initial exploration. Sometimes things just *feel* right; but even when you're a mathematician of Fermat's stature, not everything that feels right is right.

THE CHINESE HYPOTHESIS

The theorem of the bracelets allows us to check a purported prime's credentials, like the bouncer at the door of a tony club. If the number

* Simone's brother, though in mathematical circles she's André's sister.

1,020,304,050,607, standing at the door in its shiniest suit, tries to get past the rope, it might take me a while to test numbers, one by one, to see if any of them divides evenly into 1,020,304,050,607. Much easier to multiply 2 by itself 1,020,304,050,607 times and see if the result is 2 more than a multiple of 1,020,304,050,607.* It isn't—which means 1,020,304,050,607 is for sure not prime, and I can shoo it away with one muscular arm.

Here's what's strange: we have proven, without a doubt, that 1,020,304,050,607 factors into smaller numbers, but this proof gives us no clue what the smaller numbers are! (Which is good; remember that the whole apparatus of public-key cryptography depends on it being difficult to find the factors . . .) This kind of "non-constructive proof" takes some getting used to, but it's ubiquitous in mathematics. You can think of such a proof as being akin to a car that's damp inside every time it rains.† You know from the water and the smell that there's a leak. But the proof, annoyingly enough, doesn't tell you where the leak is, just that it exists.

There's another important feature of this proof we need to dig into. If your floor mats are wet when it rains, there's a leak; but that doesn't mean that if your floor mats are dry, there's no leak! The leak might be somewhere else, or your mats might be very quick-drying. There are two different assertions one might make:

If your floor mats are wet, there's a leak.
If your floor mats are dry, there's no leak.

The second one, in logical terms, is called the *inverse* of the first. There are more variants, too:

---

* Why is that easier? It sounds like we have to multiply by 2 a trillion times or so, which sounds time-consuming. There's a clever technique called binary exponentiation that allows us to do this really fast, but this margin is too small for me to explain it.

† For example, the used Chevy Cavalier I drove from 1998 to 2002 until it broke down for good at the Delaware Memorial Bridge toll plaza on the New Jersey Turnpike. In my mind I can still smell the floor mats. I never did find the leak.

Converse: If there's a leak in your car, your floor mats will get wet.

Contrapositive: If there's no leak in your car, your floor mats will stay dry.

The original statement is equivalent to its contrapositive; they are just two different sets of words that express the same idea, like "1/2" and "3/6" or "the greatest shortstop to play in my lifetime" and "Cal Ripken Jr." You don't have to agree with either one, but if you agree with one, you have to agree with the other. But a statement and its converse are just two different things: both might be true, or one but not the other, or neither.

Fermat showed that, if n is prime, 2^n is 2 more than a multiple of n. The converse would say that if 2^n is 2 more than a multiple of n, then n is prime. That converse, which would make Fermat's test a perfectly reliable one, is sometimes called "The Chinese Hypothesis." Is it true? No. Is it Chinese? Also no. The name comes from a persistent and wrong belief that Fermat's Little Theorem was actually known to mathematicians in China around the time of Confucius. As with Nim, Western mathematicians found themselves strangely attracted to the idea that a mathematical concept lacking a clear origin should be presumed old and Chinese. The claim that the antique Chinese mathematicians asserted the false converse to Fermat's Little Theorem, which appears to originate in a short 1898 note written by the British astrophysicist James Jeans* as an undergraduate, just adds insult to misattribution.

The converse to Fermat's Little Theorem isn't true, because, just as a youth with a fake ID can fool the sternest bouncer, some non-primes pass Fermat's prime test. The smallest such is 341. (Though this example seems not to have been discovered until 1819!) Sneaky 4,294,967,297, the one that fooled Fermat, is another. And there are infinitely many more.

But that doesn't make the test useless; it's merely imperfect. The broader world often thinks of mathematics as the science of the flawless or the certain, but we like imperfect things, too, especially when we have some bounds on how imperfect they are. Here's how to generate

* Same person who seven years later was fighting over quantum physics in the letters page of *Nature*, next to Karl Pearson's question about the random walk.

large very-probably-primes by trial and error. Write down a three-hundred-digit number. Apply the Fermat test (or, better, its modern improvement, the Miller-Rabin test). If it fails, pick another number and try again. Keep going until you arrive at a number that passes the test.

DRUNK DRUNK GO

Which brings us back to computer Go. The game of Go is much older than checkers or chess—in fact, just for a change of pace, it *actually is* ancient and Chinese. Machines that play Go, on the other hand, came later than machines for other games. In 1912, the Spanish mathematician Leonardo Torres y Quevedo built a machine, El Ajedrecista, to play out certain chess endgames, and Alan Turing laid out the plan for a functional chess computer in the 1950s. The *idea* of a chess-playing robot is even older, dating back to Wolfgang von Kempelen's "Chess Turk," a wildly popular chess-playing automaton of the eighteenth and nineteenth centuries, which inspired Charles Babbage, baffled Edgar Allan Poe, and checkmated Napoleon, but which was in fact controlled by a diminutive human operator concealed inside the works.

The first computer program that played Go didn't come until the late 1960s, when Albert Zobrist wrote one as part of his University of Wisconsin PhD thesis in computer science. In 1994, while Chinook was matching Marion Tinsley blow for blow, Go machines were helpless against professional human players. Things have changed fast, as Lee Se-dol found out.

What does a top-tier Go machine like AlphaGo, without a small human crouched inside it to move the pieces, actually do? It doesn't label each node of the Go tree with a W or an L (we don't need D, since there aren't draws in standard Go). The tree of Go is deep and bushy; *no one* can solve the damn thing. But as with Fermat's test, we can be content with an approximation, a function that assigns each position of the board a score in some readily computable way. The score should be high if the position is a good one for the person about to move, low if the board favors the opponent. A score suggests a strategy; among all the moves available to you, choose the one yielding a board with the *lowest* score, since you want to put the other player in the maximally disadvantageous position. It's useful to imagine yourself into the inner life of an algorithm like that. You're going about your daily business, and every time you face a decision—chocolate croissant or almond croissant, or do I want a bagel?—you briskly riffle through all possible choices available to you, each one near-instantly flashing a superimposed numerical score that registers your best estimate of the net benefit to you of each available baked good, deliciousness plus satiation minus cost minus processed carbohydrate intake, etc. It sounds sort of awesome and science-fictionally horrifying at once.

There's a kind of trade-off here which is fundamental to everything we do in artificial intelligence. The more accurate the scoring function is, the more time it typically takes to compute; the simpler it is, the less accurately it captures the thing it's supposed to be measuring. Most accurate of all would be to assign each winning position a 1 and each losing position a 0; that would yield absolutely perfect play, but we have no earthly way of actually computing this function. At the other extreme, we could just assign each position the same value. ("I dunno, all those pastries seem fine.") That would be very simple to compute and would offer no useful advice about game play at all.

The right place to be is somewhere in between. You want a way of roughly judging the worth of a course of action without laboriously reckoning your way through all its consequences. That could be "do what feels good in the moment, you only live once" or "give away that copy of *Bleak House* from college unless it sparks joy" or "obey the instructions of your local religious official." None of these strategies is perfect, but all are probably better for you than completely unconsidered action (apart from certain exceptional cases concerning the local religious official).

It's hard to see how this applies to a game like Go. If you're not an expert in the game, or if you're a computer, no position of the stones on the board sparks joy or misery. By contrast with checkers or chess, where the player with the most pieces is usually in some sense "ahead," Go doesn't have an obvious notion of material advantage. Whether a position is a winner or a dog is a subtle matter of positioning.

Important math tactic: when you have no idea what to try, try something that seems very dumb. Here's what you do. Starting from a given position, you imagine that Akbar and Jeff begin to drink heavily, so heavily they lose all sense of strategy and desire to win, yet remembering in some dim cranny of their consciousness the rules of the game. They are, in other words, like the inebriated wanderer in an open field Karl Pearson had imagined. Each player at each turn selects a legal move at random, and they play until the game ends and they collapse under the table, spent. The players are carrying out a random walk on the tree of Go.

Drunk Go is easy to simulate on the computer because it requires no careful judgment, just knowledge of the rules and a carefree random spin of the wheel to choose one of the available moves at each turn. You can simulate the game, and then, when it's done, simulate it again: once, twice, a million times, always starting from the same position. Sometimes Akbar wins, sometimes Jeff wins. And the score you assign to the game, measuring how much you think it favors Akbar, is the proportion of the simulations where Akbar ends up the winner.

As coarse as this measure is, it's not totally useless. Consider this metaphor. Akbar is alone in a long hallway, with one exit at the front and one at the back. He's still dead drunk. Akbar wanders back and forth aimlessly until he finds his way to one of the doors. It seems reasonable to guess that the closer Akbar starts to the front, the more likely he is to emerge from the front door, even though he's not *trying* to get to the front door, or anywhere in particular. And we can use this reasoning in reverse; if Akbar comes out the front door, that's evidence (though not, of course, *proof*) that his starting position was nearer to the front.

This kind of reckoning was part of the theory of random walks centuries before Pearson gave the random walk a name. Arguably it goes back as far as the book of Genesis, where Noah, sick of being cooped up in the ark with a few hundred animal couples, sends out a raven to fly "to and fro" looking for land left exposed by the receding water. The raven found nothing. Next Noah sent out a dove, which also came back from its wandering having encountered no sign of land. But the *next* time the dove went out on a random flight, it came back with an olive leaf in its beak, and by this Noah was able to infer that the ark was close to the water's edge.*

And random walks have appeared in the study of games for centuries, especially in games of chance, where the walk through the tree is *always* random, at least in part. Pierre de Fermat, when he wasn't writing letters about prime numbers, was corresponding with the mathematician and mystic Blaise Pascal about the problem of the gambler's ruin. In this game, Akbar and Jeff go head to head at dice, each starting with a stake of twelve coins and rolling three dice at each turn. Every time Akbar rolls an 11, he gets one of Jeff's coins; every time Jeff rolls a 14, he takes one from Akbar. The game ends when one player runs out of coins and is "ruined." What's the chance that Akbar wins?

This is just a question about a random walk, which starts with the players evenly well funded and ends when one player has hit their number twelve more times than the other. An 11 is about twice as likely as a 14 to come up on a roll of three dice, simply because there are only fifteen ways three dice can add up to 14, and twenty-seven ways three dice can add up to 11. So it seems reasonable to guess Jeff is at a disadvantage in this game. But *how much* of a disadvantage? That was the question Pascal set to Fermat. It turns out (as Fermat immediately wrote back to Pascal, and Pascal huffily made clear he had already worked out) that Jeff is more than a thousand times as likely as Akbar to wind up ruined! A modest bias in the random walk is magnified into a massive one in the gambler's ruin game. Jeff might get lucky and hit 14 once or twice before Akbar rolls 11, but his lead is unlikely to be long-lasting, let alone ever make it to twelve.

The easiest way to see how this works is to replace the problem with a much simpler one, what mathematicians like to call a "baby example." Suppose Akbar and Jeff play a game in which Akbar has a 60% chance of winning each point, and the first player to two points wins. The

* For completeness, I should concede that the text is somewhat obscure about the raven, and interpreters have had to flesh out the story. According to the third-century bandit-turned-rabbi Reish Lakish (Talmud Sanhedrin 108b), the raven didn't go in search of dry land at all. Rather, his "to and fro" motion was confined to a tight circle around the ark so he could keep an eye on Noah, who the suspicious bird was grimly certain had sent him off the boat as a pretext so Noah could have sex with Mrs. Raven. Always read the commentary, there's wild stuff in there.

chance that Akbar wins the first two points, thus winning the game, is 0.6 × 0.6 = 0.36. And the chance that Jeff wins two points in a row and thereby ruins Akbar is just 0.16. Putting those two options aside, what's left is a 48% chance that the first two points are split 1–1, and the game goes on. In 60% of those cases, or 28.8% of all games, Akbar wins the next point and the game; in the other 40% of the 1–1 scenarios, or 19.2% of all games, Jeff wins and ends up with a 2–1 victory. So Akbar has, overall, a 36% + 28.8% = 64.8% chance of winning, a little better than the chance he has of winning each individual point. If the game is played to three points instead of two, you can check in similar fashion that Akbar's chance of winning goes up to about 68.3%. The longer the game, the better the odds that the slightly better player is going to win.*

The gambler's ruin principle underlies the design of tournaments in sports. Why don't we determine the world championship of baseball or the winner of a tennis tournament by the outcome of a single game? Because that would be too uncertain; in any given game of tennis, the better player might well lose, and the point of a tournament is to find out who's really best.

Instead, a set of tennis continues until one of the two players has won six games *and* is two games ahead. That's hard to parse in words, so here's a picture:

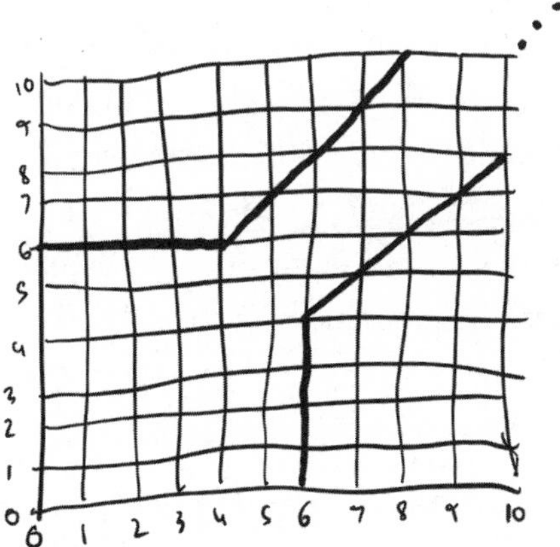

* You might note that this game is a bit different from the original gambler's ruin; in the problem studied by Pascal and Fermat, you have to pull ahead by 12 points to win, not just be the first to get to a score of 12. The baby example is easier to analyze on paper.

You can think of a tennis set as a random walk on this picture; each time a game is played, you walk either up or to the right, and you stop when you strike one of the two boundaries, "ruining" one of the players. If player A is even a little bit better than player B—that is, if an upward step is more likely than a rightward one—ending up on the upper boundary is much more likely than hitting the lower one.* Because the long diagonal corridor in the diagram is infinite, there's no definite bound to how long a tennis set can go. Unless the players are very well matched indeed, it's extremely unlikely the walk will move very far along the corridor without hitting one wall. But it can happen. It happened to John Isner and Nicolas Mahut, who met at Wimbledon on June 23, 2010. The two players kept matching each other game for game. Hours went by. The sun started to set. The scoreboard on the court hit its preprogrammed maximum number and shut itself down. At about nine o'clock, with the set tied 59–59, it got too dark to play. Isner and Mahut resumed again the next afternoon, and kept alternating victories. Finally, late in the afternoon of June 24, Isner hit a backhand past Mahut to win the set's 138th game and claim a 70–68 victory. "Nothing like this will ever happen again," Isner said. "Ever."

But it could! It might sound bizarre to design a sport this way, but to me it's part of tennis's charm. No clock, no buzzer, no limit to the number of games. The only way out is for someone to win.

Most sports championships work differently. When two baseball teams compete in the World Series, the champion is the first team to win four games. It can't go longer than seven games, and if both teams have won three, the next game decides the championship. There's no possibility of the series stretching into a 138-game ultramarathon like the Isner-Mahut set.† The geometry of the World Series boundary is different.

---

* Tennis fans will observe that the alternation of service makes the random walk a little more complicated; it's true, but it doesn't materially affect the nature of the math involved.

† Though an individual baseball *game* can in principle extend indefinitely, as long as the score's tied at the end of each inning; if that possibility intrigues you I highly recommend W. P. Kinsella's novel *The Iowa Baseball Confederacy*.

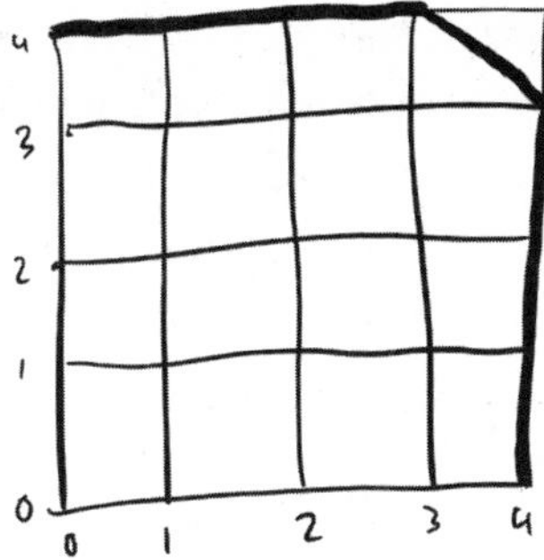

We've come again to the trade-off between accuracy and speed. You can think of a set of tennis as an algorithm, whose purpose is to figure out which player is better at tennis, just as the World Series is an algorithm to figure out which team is better at baseball. (A sporting event isn't *just* an algorithm; it may also be intended to provide entertainment, generate tax revenue, narcotize a seething populace, etc.—but an algorithm is one of the things it is.) A set of tennis spends longer computing its output, and is more accurate at teasing out fine distinctions between players; the World Series is cruder and gets the job done faster. The difference comes from the geometry of the boundary; is it squared off and blunt, like the World Series, or long and pointy, like a tennis set? And these aren't the only two choices. You can position yourself anywhere you like along the accuracy-speed trade-off by your choice of shape. I've always liked this one:

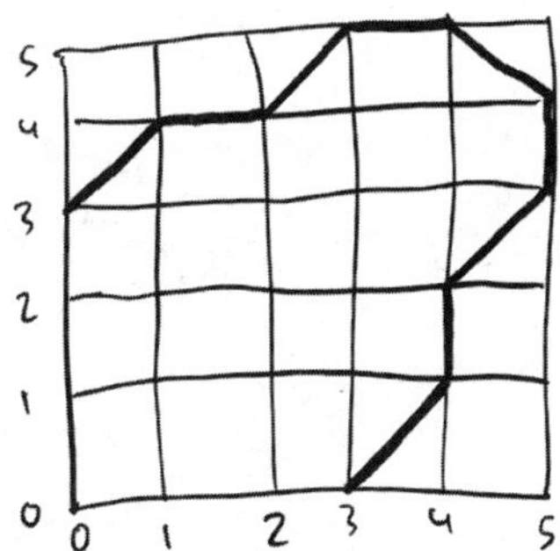

This system has a "mercy rule"; you lose if you go down 3–0. On the other hand, if both teams win three games, suggesting they're evenly matched, you have to get a *fifth* win to be crowned the champ. Yes, you'd lose those rare but thrilling moments where a team like the 2004 Red Sox comes back from a 3–0 deficit to win the American League Championship Series; but that hardly ever happens. And would that be too high a price to pay for all the Game 8s and winner-take-all Game 9s we'd have between closely matched teams?

THE SPACE OF STRATEGIES

Back to Go. We've seen that the outcome of a random walk can give you clues about where your starting position was; it's reasonable to guess that a position from which Akbar is likely to accidentally win is also one where he's well set up to win if he actually tries. You can test this by playing Go using this as your strategy; at each stage, move to the position with the best Drunk Go score. If you adopt this rule, it turns out, you won't beat a player with any skill, but you'd play better than a total novice.

Better still is to blend drunken stumbling with the kind of tree analysis we used for Nim. It goes something like this.

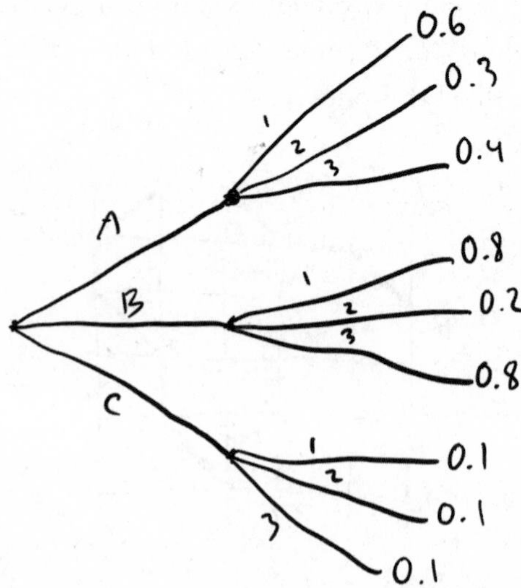

The time has come for me to reveal something about myself. I don't know how to play Go. The game my cousin Zachary crushed me in was the last one I ever played. I don't even remember the rules. But that doesn't matter; I can write this section about Go anyway, because the tree tells you what to do whether or not you know the rules. The tree could be a Go tree or a checkers tree or a Nim tree; you analyze it in exactly the same way. Everything relevant to choosing your strategy is contained in the patterns of its branches and the numbers on the leaves. The geometry of the tree is all that matters.

The numbers on the leaves indicate the Drunk Go score for the corresponding sequence of moves; if Akbar plays move A and Jeff follows up with move 1, and the players move randomly from then on, Akbar ends up winning 60% of the games. So A1 has Drunk Go score 0.6.

But the Drunk Go score of move A on its own is not quite as good; assuming Jeff gets drunk and plays randomly from there, there is a 1/3 chance the game moves to A1, a 1/3 chance the game moves to A2, and a 1/3 chance the game moves to A3. If we're playing 300 drunken trials, 100* will end up at A1, of which Akbar will win 40. And Akbar wins 50 of the A2 games and 60 of the A3 games, for a total of 150 out of 300, exactly half. So A has Drunk Go score 0.5. In a similar way we can find that position B scores 0.4 and position C scores 0.9. (Remember, the Drunk Go score of a position where it's Jeff's turn to play is the chance that drunk Jeff beats drunk Akbar, not the reverse.)

How Akbar plays this game depends on when the drinking starts. If he looks just one branch down the tree, figuring it's random from there onward, he'll pick move B, the one with the lowest Drunk Go score. But if he works his way farther down the tree, he can reason as follows. What's actually likely to happen if he picks move B? Jeff, sober as a pope, will choose B2, giving Akbar a 20% chance of winning. That beats the lousy move C, where Akbar's chances are just 10% whatever Jeff

* Or rather "it is very likely that if we run this experiment many times, the average number of times the game moves to A1 will approach 100," but I am not going to type this phrase before every single number. You're welcome.

does next. But move A actually gives Jeff less to work with; his best choice is to move to A1, which gives Akbar a 40% chance. So Akbar, contemplating two steps down the tree instead of one before giving way to drunk analysis, can see that A, not B, is the better move.

And, of course, a deeper analysis might do better still. B2 is a position that ends very poorly for Akbar when played out at random. That might be because it's just an objectively Akbar-unfavorable scenario. Or it could be that from that position Akbar has one devastatingly good move and many lousy ones. For a random Akbar, this is a bad position to be in, since the odds of choosing the one good move are very low. For an Akbar who can look another move ahead, it's great.

A blended strategy like this still relies heavily on the semi-ridiculous method of Drunk Go. So it might be surprising that Go-playing computer programs with methods like this at their heart were at the top of the field just a few years ago, playing competitively at an advanced amateur level.

But this strategy isn't what powers the new generation of machines, the ones that drove Lee Se-dol into early retirement. They still use a scoring function that rates a position as "good for Akbar" or "bad for Akbar" on a numerical scale, and use that score to decide on a move. But the scoring mechanism a program like AlphaGo uses is much, much better than any you can get from a random walk. How do you build a mechanism like that? The answer, as you surely knew I'd say, is geometry. But it's a geometry of a higher order.

Whether grappling with tic-tac-toe, checkers, chess, or Go, you start with the geometry of the board. From there, and from the rules of the game, you go one level up and develop the geometry of the tree, which in principle contains everything about the perfect strategy for playing the game. But when the perfect strategy is too computationally hard to find, you'll settle for finding a strategy *close enough* to the perfect one to provide high-quality game play.

To locate a strategy that's close to the unknown and practically unknowable perfect one, you have to navigate a new geometry: the geometry of the space of strategies, a terrain much harder to draw than a tree.

And we're trying to locate in that infinite-dimensional abstract haystack a decision-making protocol better than anything Marion Tinsley's or Lee Se-dol's experience-sharpened intuition could devise.

That sounds hard. How do we proceed? It all comes down to that crudest and most powerful of methods, trial and error. Let's see how it works.

Chapter 7

Artificial Intelligence as Mountaineering

My friend Meredith Broussard, an NYU professor with an expertise in machine learning and the societal impact of same, was on television not long ago, tasked with explaining to a national audience, in two minutes or so, what artificial intelligence is and how it works.

It's not killer robots, she explained to the anchors, or passionless androids whose mental powers dwarf our own. "The really important thing to remember," she told the anchors interviewing her, "is that this is just math—it doesn't have to be scary!"

The stricken expression of the anchors suggested they'd prefer killer robots.

But Meredith's answer was a good one. And I have more than two minutes to talk about it. So let me take the baton and explain the kind of math that machine learning is, because the big idea is easier than you think.

Suppose, to start with, that instead of a machine you are a mountaineer. As such, you're trying to achieve the summit. But you're a mountaineer without a map. You're surrounded by trees and brush and there's no vantage point from which the broader landscape is visible. How do you get to the top?

Here's one strategy. Assess the slope around your feet. Perhaps the ground grades slightly upward as you step northward, and slightly downward to the south. Turn your feet to the northeast and you notice an even steeper upward incline. Shuffling in a little circle, you survey all the directions you can possibly go; among them all, there is one that offers you the steepest upward slope.* Walk a few steps that way. Then make a new circle, pick the steepest ascent among all the directions available to you, and continue.

Now you know how machine learning works!

Okay, maybe there's a little bit more to it than that. But this idea, called *gradient descent,* is at the core of it all. It's really a form of trial and error; you try a bunch of possible moves and pick the one that helps you out the most. The "gradient" associated to a certain direction is math for "how much does the height change when you take one tiny step in that direction"—in other words, it's the slope of the ground as you walk that way. If you're into calculus, it's the same thing as the "derivative," but nothing we say here is going to require you to be into calculus. Gradient descent is an algorithm, which is math for "an explicit rule that tells you what to do in any situation you might encounter." And the rule is just this:

Consider all the small moves you can make, figure out which one offers you the biggest gradient, and do that. Repeat.

Your path to the peak, charted on a topographical map, would look something like this:

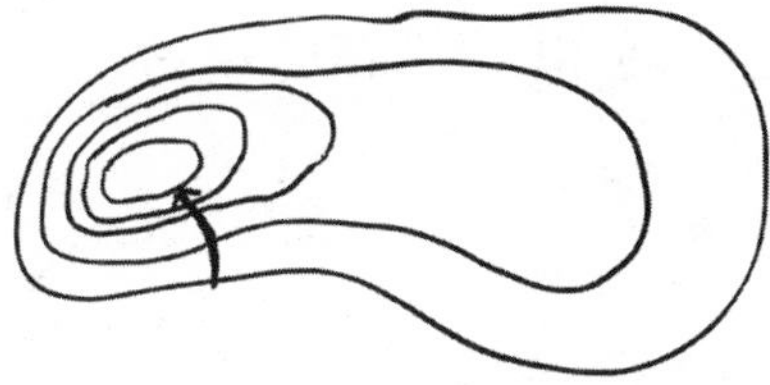

(Another nice piece of geometry: when you navigate by gradient descent, your path on a topo map will always cross the lines of equal elevation *at a right angle*. Explanation in the endnotes.)

* What if there's a tie? Then you just pick whichever of the steepest options you like.

You can see that this might be a good idea for mountain climbing (though not always—we'll come back to this), but what does it have to do with machine learning?

Let's say I'm not a mountaineer after all, but a computer trying to learn something. It might be one of the machines we've already encountered, like AlphaGo, the machine that learns to play Go better than a master, or GPT-3, the machine that produces long strings of discomfitingly plausible English text. But to start with, let's stick with the classics, and suppose I'm a computer trying to learn what a cat is.

How am I supposed to do that? The same way a baby does, more or less. The baby lives in a world where every so often some large person points at something in their visual field and says "Cat!" You can provide a computer that kind of training, too: supply it with a thousand images of cats, in various positions, lightings, and moods. "All these are cats," you tell the computer. Indeed, if you really want to be helpful, you throw in an equal number of images of things that *aren't* cats, and tell the computer which ones are which.

The machine's task is to develop a strategy, so that it can make that distinction between cats and non-cats on its own. It is wandering around the landscape of all possible strategies, trying to find the *best* one, the apex of accuracy in feline identification. It is a would-be mountain climber. And so the way to proceed is by gradient descent! You pick some strategy, thus placing yourself in the landscape, and then proceed, as the rule of gradient descent demands:

Consider all the small changes you can make to your current strategy, figure out which one offers you the biggest gradient, and do that. Repeat.

GREED IS PRETTY GOOD

That sounds good, until you realize you have no idea what it means. What, for instance, is a strategy? It has to be something a computer can carry out, which means it has to be expressed in mathematical terms. A picture, to a computer, is a long list of numbers. (*Everything* to a computer is a long list of numbers, except things that are a short list of numbers.) If the picture is a grid of 600 x 600 pixels, then each pixel has a brightness, given by a number between 0 (pure black) and 1 (pure white), and to know those 600 × 600 = 360,000 numbers is to know what the picture is. (Or at least what it is in black and white.)

A strategy is just a way to take in a list of 360,000 numbers and turn it into either "cat" or "not-cat," which, in the language of computers, is "1" or "0." It is, in mathematical terms, a *function*. In fact, to make this more psychologically realistic, the output of the strategy might be a number *between* 0 and 1; that represents the uncertainty the machine might reasonably want to express when presented with an ambiguous image, like a lynx, or a Garfield pillow. An output of 0.8 should be interpreted as "I'm pretty sure this is a cat, but doubts linger."

Your strategy could be, for instance, the function "output the average of all 360,000 numbers in your input." That would give 1 if the image was all white and 0 if the image was all black, and in general measures the overall average brightness of the image on the screen. What does that have to do with whether it's a cat? Nothing. I didn't say it was a *good* strategy.

How do we measure the success of a strategy? The simplest way is to see how it does on the two thousand images Kittytron has already seen. For each one of those images, we can assign our strategy a "wrongness score."* If the image is a cat and the strategy says 1, that's zero wrongness; it got the answer right. If the image is a cat and the strategy says 0, that's a wrongness of 1, the worst possible. If the image is a cat and the strategy says 0.8, that's like getting the answer right but tentatively; it gets a wrongness of 0.2.†

You add up those scores for all two thousand images in your training set, and you get an overall total wrongness, which is the measure of your strategy. Your goal is to find a strategy with as little total wrongness as possible. How do we get our strategy not to be wrong? That's where gradient descent comes in. Because now you know what it means for a

---

* Among actual computer scientists, usually called an *error* or a *loss*.

† There are lots of different ways to measure wrongness; this one isn't the most popular in practice but is simple to describe. We're not going to sweat details at this level of fineness.

strategy to get better or worse as you modify it. The gradient measures how much wrongness changes when you change your strategy a little. And of all the different little ways you could change it, you pick the one that decreases wrongness the most. (This, by the way, is why it's called gradient descent, not ascent! Often our goal in machine learning is to minimize something bad like wrongness, not to maximize something awesome like height above the plain.)

This method of gradient descent doesn't just apply to cats; you can apply it any time you want a machine to learn a strategy from experience. Maybe you want a strategy that takes someone's ratings of a hundred movies and predicts their rating for a movie they haven't seen. Maybe you want a strategy that takes a checkers or Go position and returns a move that places your opponent in a losing situation. Maybe you want a strategy that takes video input from cameras mounted on a car and returns a steering column motion that doesn't crash the car into a dumpster. Whatever! In any of these cases, you can start with a proposed strategy, assess which small changes would decrease wrongness the most in the examples you've already observed, make those changes, and repeat.

I don't want to understate the computational challenges here. Kittytron is more likely to be training itself on millions of images than a thousand. So computing that total wrongness might involve adding up a million individual wrongness scores. Even if you have a very fancy processor, that takes time! So in practice, we often use a variation, called *stochastic gradient descent*. There are untold different flavors and tweaks and complications of the method, but here's the basic idea: Instead of adding up all the wrongnesses, you pick *one* picture out of your training set at random, just one Angora kitten or fish tank, then take whatever step most decreases how wrong you are about that image. At the next step, you pick a new random image, and continue. Over time—because this process is going to take a lot of steps—you're likely to eventually take all the different images into account.

What I like about stochastic gradient descent is how nuts it sounds. Imagine, for instance, that the president of the United States made decisions without any kind of global strategy; rather, the nation's chief executive is surrounded by a crowd of shouting subordinates, each hollering for policy to be tweaked in a way that suits their own particular interest. And the president, every day, chooses one of those people at random to listen to, and changes course accordingly.* That would be a ridiculous way for a person to run a major world government, but it works pretty well in machine learning!

Our description so far is missing something important: How do you know when to stop? Well, that's easy; you stop when no small change you can make creates any improvement at all. But there's a big problem: you might not actually be at the top!

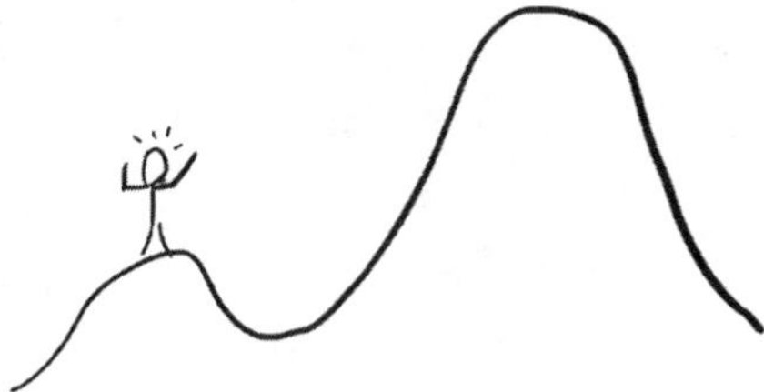

If you're the happy mountaineer in this picture, you can take a step to your left or take a step to your right and see that neither direction offers you an upward slope. That's why you're happy! You're at the summit!

But no. The real summit is far away, and gradient descent can't get you there. You're caught in what mathematicians call a *local optimum*,† a point from which no small change can generate improvement but which is far from the actual best possible place to be standing. I like to think of the local optimum as a mathematical model of procrastination. Suppose you're faced with a task you find aversive; say, organizing a huge teetering stack of files, most of them related to goals you've been meaning to get to for years, and whose disposal would represent a final concession that you're never going to go down those paths. On any given day, gradient descent would counsel you to take whatever small step

* A slightly more accurate analogy to stochastic gradient descent would be to place the advisers in a random order and have the president cycle through them, one adviser per day; that, at least, would guarantee that everyone gets heard their fair share of the time.

† Also often known as a *local maximum* or a *local minimum*, depending on whether you're thinking of your goal as reaching the top or hitting bottom.

most increases your happiness that day. Will starting in on the stack do that? No, quite the contrary. Starting the stack is going to feel *terrible.* Putting it off another day is what gradient descent demands of you. And the algorithm tells you the same thing tomorrow, and the day after that, and the day after that. You're caught in the local optimum, the low summit. To get to the higher peak you'll have to endure a walk through the valley, maybe quite a long one—you have to go down to get all the way up. Gradient descent is a "greedy algorithm," so-called because at each moment it takes the step that maximizes short-term advantage. Greed is one of the chief fruits on the tree of deadly sins, but then again, according to a popular capitalist saying, greed is good. In machine learning it would be more accurate to say "greed is *pretty* good." Gradient descent *can* get caught in a local optimum, but in practice this seems not to happen as much as it theoretically might.

And there are ways around a local optimum; you just have to momentarily suspend your greed. Every good rule has some exceptions! You might, for instance, instead of stopping when you get to a summit, choose another random location and start the gradient descent all over again. If you keep ending up at the same place, you start to gain more confidence it really is the best place to be. But in the picture on the previous page, gradient descent from a random starting point will more likely end up at the big peak than get stuck atop the small one.

In real life, it's pretty hard to reset yourself in a completely random life location! It's more realistic to take a random large step from your current position instead of greedily choosing a small one; often this is enough to kick you into a new place from which the best summit is reachable. This is what we do when we ask for life advice from a stranger outside our usual circle, or draw cards from a deck like Oblique Strategies, whose dicta ("Use an unacceptable color," "The most important thing is the thing most easily forgotten," "Infinitesimal gradations,"* "Discard an axiom"†) are meant to knock us out of whatever local optimum we're stuck on, allowing us to make moves that don't immediately

* This could almost be a description of gradient descent!
† And this could almost be a description of non-Euclidean geometry!

"work." The very name suggests a trajectory aslant to what you'd ordinarily do.

AM I RIGHT? AM I WRONG?

There's one more snag, a big one. We blithely resolved to consider all small changes we could make and figure out which one provides the best gradient. When you're a mountaineer in a landscape, that's a clearly defined problem; you're in a two-dimensional space, and choosing a step to take is just choosing a point on the circle of compass directions; your goal is to find the point on the circle with the best gradient.

But the space of all possible strategies for assigning cat scores to images? That's a much huger space, *infinite*-dimensional, in fact. There's no meaningful way to consider all your options. This is obvious if you consider it in human rather than machine terms. Suppose I were writing a self-help book about gradient descent, and I said, "The way to improve your life choices is really simple; just think of every possible way you could change your life, and then pick the one which would have improved your previous choices the most." You'd be paralyzed! The space of all possible behavior modifications is just too big to search.

And what if, through some kind of superhuman feat of introspection, you *could* search it? Then you hit yet another problem. Because here is a strategy for your life that absolutely minimizes wrongness on all your past experiences.

> *Strategy*: If a decision you have to make is exactly identical with one you've made before, make the decision you now consider, in retrospect, the right one. Otherwise, flip a coin.

In the Kittytron scenario, the analogue of this rule would be

> *Strategy:* For any image identified for you in training as a cat, say "cat." For any image identified for you as a non-cat, say "non-cat." For all other images, flip a coin.

That strategy has zero wrongness! It gets the right answer for every single image in the group you trained it on. But it stinks. If I present Kittytron with a picture of a cat it hasn't seen before, it flips a coin. If I present a picture I've already told it was a cat, but rotate it a hundredth of a degree, it flips a coin. If I present a picture of a refrigerator, it flips a coin. All it can do is reproduce the exact finite list of cats and non-cats I've told it about. It's not learning; it's just remembering.

We've seen two ways strategies can be ineffective, which are in some sense opposites.

- The strategy is wrong a lot in situations you've already encountered.
- The strategy is so precisely tailored to situations you've already encountered that it's useless for new situations.

The former problem is called *underfitting*—you haven't used your experience enough when forming your strategy. The latter is *overfitting*—we've relied on our experience *too much*. How can we find a happy medium between these two extremes of uselessness? We can do that by making the problem more like the mountain hike. The climber is searching a very limited space of options, and so can we, if we decide to restrict our options in advance. Let's go back to my gradient descent self-help book. What if, instead of instructing my readers to sort through *every* possible intervention they could undertake, I told them to think about just one dimension; say, for working parents, how much weight they put on the needs of their job relative to how much weight they put on the needs of their kids. That's one dimension of choice, one knob on your life apparatus you can turn. And you may ask yourself—looking back on how things have gone so far, would I rather have had that knob turned more toward my job, or more toward my kids?

Instinctively, we know this. When we think about assessing our own life strategies, the metaphor we use is typically a choice of direction on the surface of the Earth, not a wander through infinite-dimensional space. Robert Frost frames it as "two roads diverged." The Talking Heads song "Once in a Lifetime,"* a sort of sequel to Frost's "The Road Not Taken," is almost a depiction of gradient descent, if you squint:

You may ask yourself
Where does that highway go to?
And you may ask yourself
Am I right? Am I wrong?
And you may say to yourself
"My God! What have I done?"

You don't have to restrict your controls to a single knob. A typical self-help book might provide multiple questionnaires, assessing: Do you want to turn the knob toward your kids and away from your job, or the reverse? Toward your kids or toward your spouse? Toward ambition or toward ease of life? But no self-help book, no matter how authoritative, has *infinitely* many questionnaires. Somehow, from the infinite list of possible knobs you could twiddle on your life, the book chooses a finite set of directions you might consider stepping.

Whether or not it's a good self-help book depends on whether it chooses good knobs. If the questionnaires were about whether you should read more Jane Austen and less Anthony Trollope, or whether you should watch more hockey and less volleyball, they probably wouldn't help most people with their highest-priority problems.

One of the most common ways to choose knobs is called *linear regression*. It's the workhorse tool statisticians reach for as a first resort whenever they're looking for a strategy to predict one variable given the value of another. A miserly baseball team owner might, for instance, want to know how much the winning percentage of the team affects the number of tickets sold. You don't want to put too much talent on the field unless that translates into rears in the seats! You'd make a chart like this:

* Produced and cowritten by Brian Eno, who also cocreated the Oblique Strategies cards!

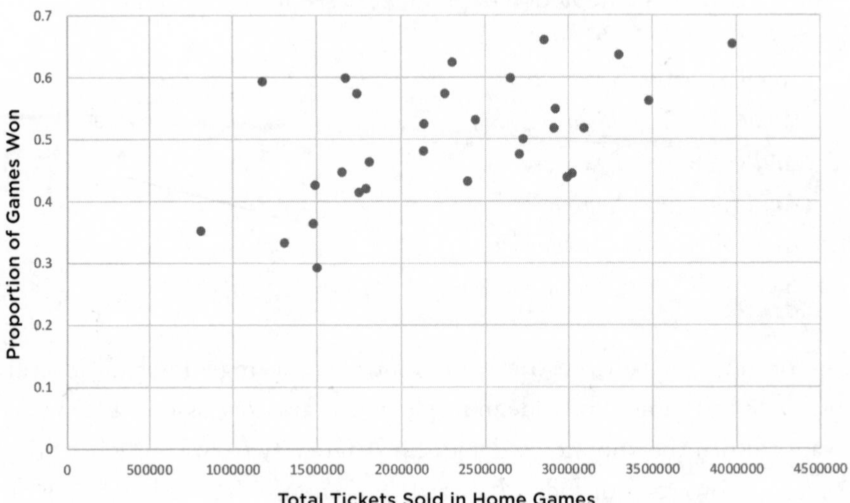

Each dot on the chart is a baseball team, its vertical position determined by the proportion of games the team won in 2019, the horizontal position by the overall attendance for the year. Your goal is to have a strategy for predicting the attendance in terms of the winning percentage, and the small space of prediction strategies you allow yourself to consider consists of those which are *linear*:

> attendance = mystery number 1 × winning percentage + mystery number 2

Any strategy like this corresponds to a line drawn over the graph, and you want that line to match your data points as well as possible. The two mystery numbers are the two knobs; and you can do gradient descent by turning them up and down, nudging the numbers until your strategy's total wrongness can't be improved by any little tweak.*

* The notion of wrongness that works best here is, for reasons we will have to go into another time, the *square* of the difference between the linear strategy's prediction and the truth, summed over every baseball team;

The line you wind up with looks like this:

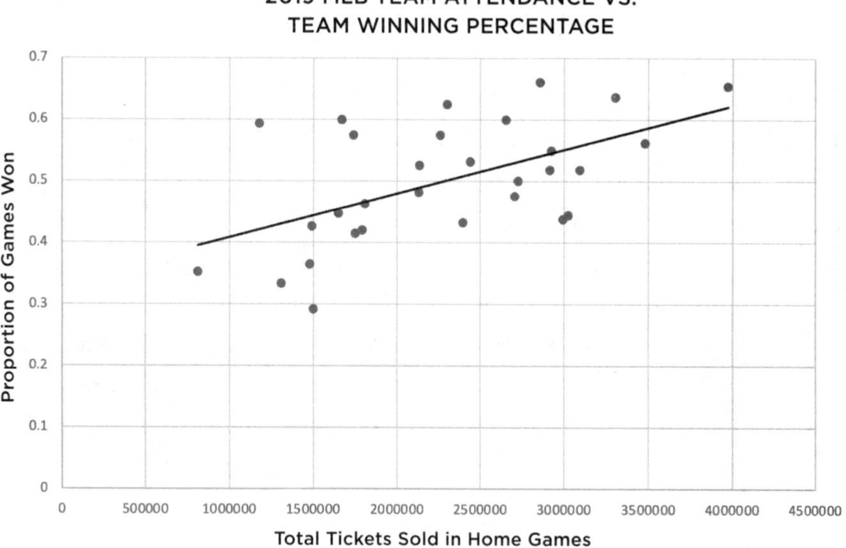

You may notice that the least wrong line is still pretty wrong! Most relationships in the real world are *not* strictly linear. We could try to solve the problem by taking in more variables as input (you have to figure the size of the team's stadium would be relevant, for instance), but in the end, linear strategies only get you so far. That class of strategies is just not big enough to, for instance, tell you which images are cats. For that, you have to venture into the wild world of the nonlinear.

DX21

The biggest thing going on right now in machine learning is the technique called *deep learning*. It powers AlphaGo, the computer that beat Lee Se-dol, it powers Tesla's fleet of sort-of-self-driving cars, and it

so this method is often called *least squares*. The method of least squares being so old and perfected by now, there are much faster ways to find the optimal line than gradient descent; but gradient descent will work.

powers Google Translate. It is sometimes presented as a kind of oracle, offering superhuman insight automatically and at scale. Another name for the technique, *neural networks,* makes it sound as if the method is somehow capturing the workings of the human brain itself.

But no. As Broussard said, it's just math. It's not even new math; the basic idea has been around since the late 1950s. You can already see something kind of like neural network architecture in my bar mitzvah present from 1985. Along with checks and several kiddush cups and more than two dozen Cross pens, I got, from my parents, the present I most fervently desired, a Yamaha DX21 synthesizer. It's in my home office right now. I was extremely proud, in 1985, that I had a *synthesizer,* not a *keyboard.* What that meant is that the DX21 didn't just play preset fake piano, fake trumpet, fake violin sounds installed at the factory. You could program your own sounds, so long as you could master the somewhat impenetrable seventy-page manual, which featured a lot of pictures like this:

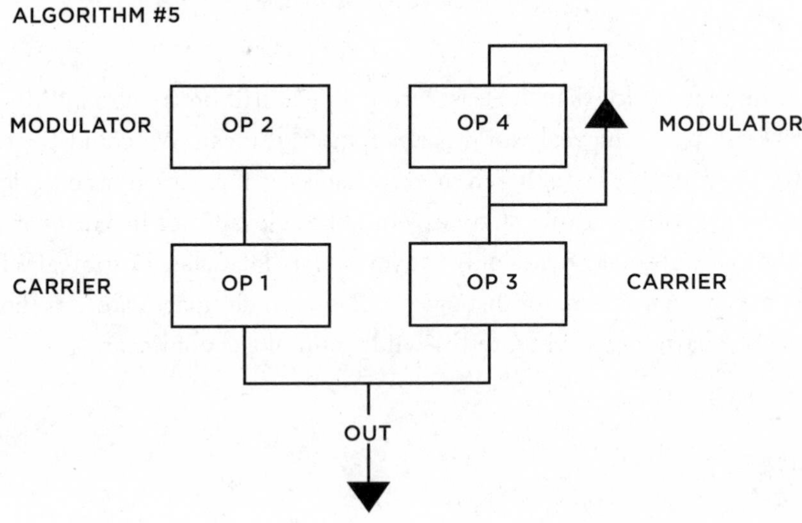

Each one of those "Op" boxes represents a wave, which has a handful of knobs you can turn, if you want to make it louder, softer, fade out or fade in with time, whatever. That's all standard. The real genius of the DX21 is the *connection* between the operators, expressed in the diagram above. There's a kind of Rube Goldberg process, where the wave that comes out of Op 1 doesn't just depend on the knobs you turn on that box, but on the output of Op 2, which feeds into it. Waves can even modify themselves; that's the "feedback" arrow attached to Op 4.

In this way, by turning a few knobs on each box, you can get a remarkably broad range of outputs, which afforded me endless opportunities to create new homemade sounds, like "ELECTRIC DEATH" and "SPACE FART."*

A neural network is a lot like my synthesizer. It's a network of little boxes, like this:

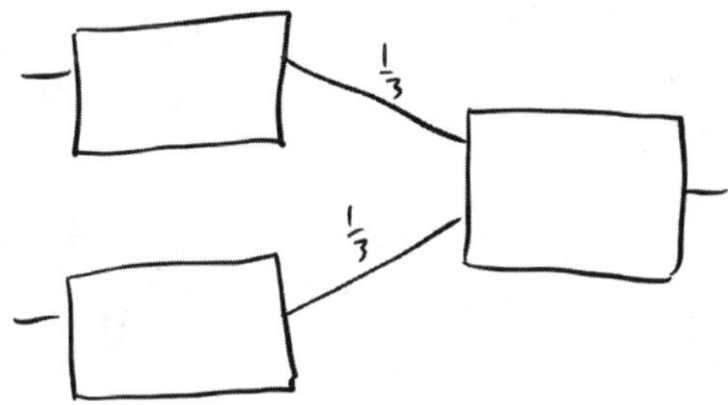

Each of these boxes does the same thing: it takes as its input a single number, and outputs either 1, if its input is bigger than or equal to 0.5, or 0, if its input is smaller. The idea of using this kind of box as the basic element of a learning machine was devised in 1957 or 1958 by Frank Rosenblatt, a psychologist, as a simple model of the way a neuron works; it sits, quiescent, until the stimulation it receives exceeds a certain threshold, at which point it fires off a signal. He called his machines *perceptrons*. In honor of the history we still call these networks of fake neurons "neural networks," though most people no longer think of them as emulating our own cerebral hardware.

Once the box emits its output, that number goes out along any arrow heading rightward from the box. Each of those arrows has a number written over it, called the *weight*, and the output gets multiplied by that weight as it zooms along that arrow. Each box takes as input the sum of all the numbers entering it from the left.

Each column is called a *layer*; so the network above has two layers,

* Perhaps relevant here, for those who don't know, is that a bar mitzvah present is something you get when you're thirteen.

with two boxes in the first layer and one in the second. You start with two inputs, one for each of the two boxes. Here's what can happen:

- Both inputs are at least 0.5. Then both boxes in the first column put out 1, each of which turns into 1/3 as it moves along its arrow, so the box in the second column receives 2/3 and spits out 1.
- One input is at least 0.5 and the other is smaller. Then the two outputs are 1 and 0, the box in the second column receives 1/3 as input, and it outputs 0.
- Both inputs are less than 0.5. Then the boxes in the first column both output 0 and so does the final box.

In other words, this neural network is a machine that takes two numbers and tells you whether or not they're *both* bigger than 0.5.

Here's another neural net that's a little more complicated.

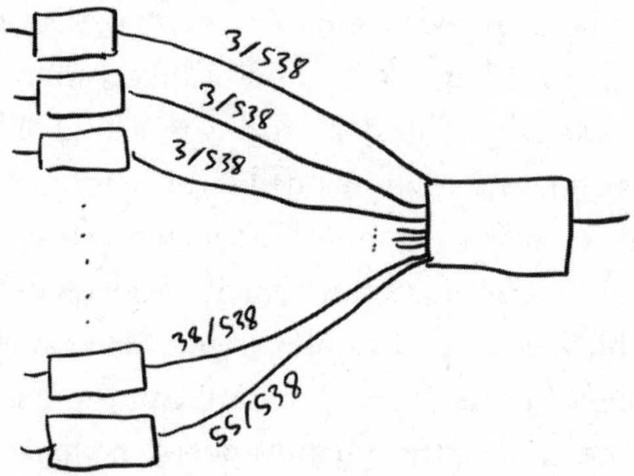

Now there are fifty-one boxes in the first column, all feeding into the single box in the second column, with different weights on the arrows. Some of the weights are as small as 3/538; the biggest is 55/538. What does this machine do? It takes as input fifty-one different numbers, and activates each box whose input is bigger than 50%. Then it adds up the weights attached to each of those boxes and checks whether the sum is bigger than 1/2. If so, it outputs 1; if not, it outputs 0.

We could call this a two-layer Rosenblatt perceptron. But it's more commonly called the Electoral College. The fifty-one boxes represent the fifty states and Washington, D.C. A state's box is activated if the Republican candidate wins there. Then you add up the electoral votes for all those states, divide by 538, and if the answer is more than 1/2, the Republican candidate is the winner.*

Here's a more contemporary example. It's not as easy to describe in words as the Electoral College, but it's a little closer to the neural networks that are driving modern progress in machine learning.

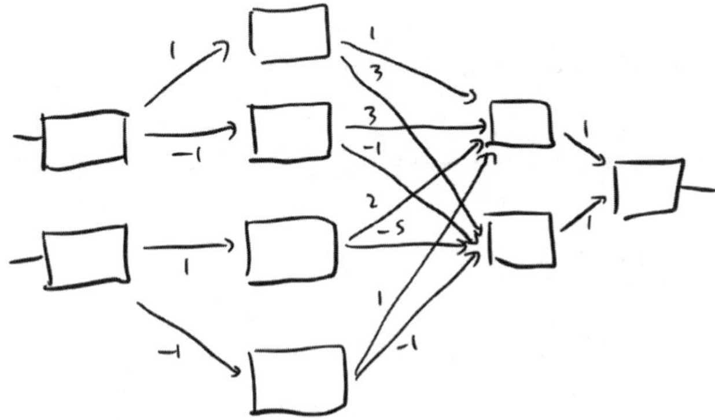

The boxes here are a little more refined than Rosenblatt's; a box takes a number as input, and outputs that number or zero, whichever is larger. In other words, if the box gets a positive input, it just passes along whatever it got; but if it gets a negative input, it outputs zero.

Let's take the device for a spin. Suppose I start with inputs 1 and 1 on the far left. Both those numbers are positive, so both boxes in the first column will output 1. Now the top box in the second column receives 1 × 1 = 1, and the second box receives –1 × 1 = –1. The other two boxes in the second column get 1 and –1, similarly. Since 1 is positive, the top box outputs 1. But the box below it, which gets a negative input, fails to fire and outputs 0. Similarly, the third box puts out a 1 and the fourth box 0.

* The Electoral College departs from Rosenblatt's definition in one small way; the final box outputs 1 if its input is bigger than 0.5 and 0 if its input is less than 0.5, but if the input is *exactly* 0.5, the box passes responsibility for deciding the election to the House of Representatives.

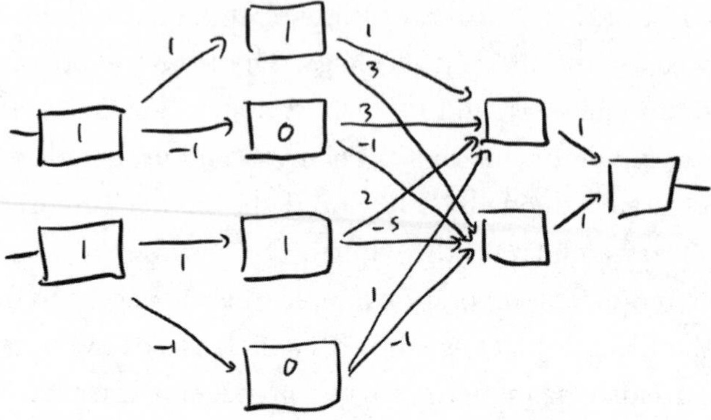

Now on to the third column: the top box receives

$$1 \times 1 + 3 \times 0 + 2 \times 1 + 1 \times 0 = 3$$

and the bottom box

$$3 \times 1 - 1 \times 0 - 5 \times 1 - 1 \times 0 = -2$$

So the top box outputs 3, and the bottom box fails to fire and outputs 0. Finally, the lone box in the fourth column receives the sum of its two inputs, which is 3.

It's okay if you didn't follow that in every detail. What's important is that the neural network is a *strategy*; it takes two numbers as input and returns one as output. And if you change the weights on the lines—that is, if you turn the fourteen knobs—you change the strategy. The picture gives you a fourteen-dimensional landscape you can explore, looking for a strategy that fits best whatever data you already have. If you're finding it hard to imagine what a fourteen-dimensional landscape looks like, I recommend following the advice of Geoffrey Hinton, one of the founders of the modern theory of neural nets: "Visualize a 3-space and say 'fourteen' to yourself very loudly. Everyone does it." Hinton comes from a lineage of high-dimension enthusiasts: his great-grandfather Charles wrote an entire book in 1904 about how to visualize four-dimensional cubes, and invented the word "tesseract" to describe them.* If you've seen Dalí's painting *Crucifixion (Corpus Hypercubus)*, that's one of Hinton's visualizations.

This network, with the given weights, assigns a point (x,y) in the plane a value of 3 or less whenever it lies inside the following shape:

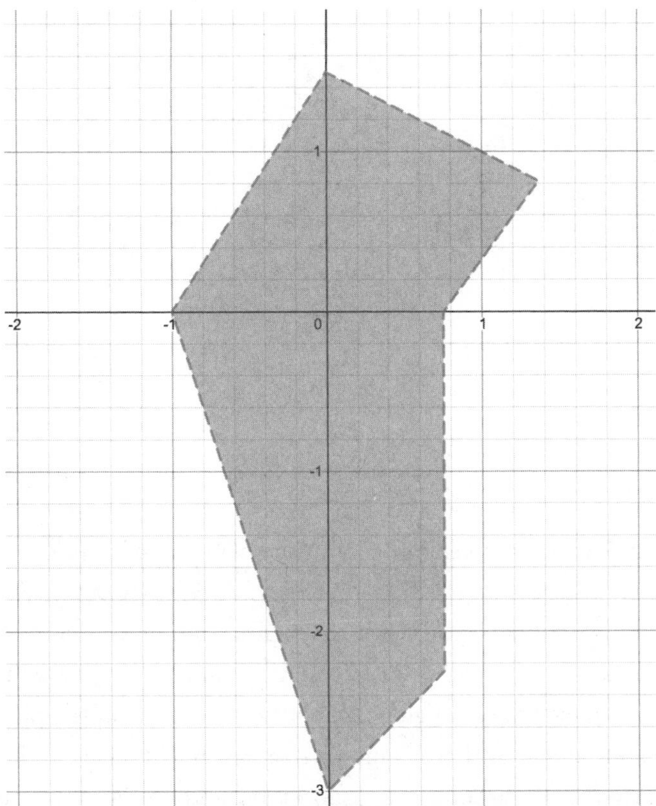

(Notice how the point (1,1), where our strategy returns exactly 3, is on the boundary of the shape.) Different values of the weights yield different shapes; though not *any* shape. The nature of the perceptron means

* The elder Hinton also wrote numerous science fiction novels, then called "scientific romances," was convicted of bigamy and had to leave England for Japan, and ended up teaching math at Princeton, where he developed a gunpowder-driven pitching machine for the baseball team, which won him great publicity but which was retired after it injured several players.

the shape will always be a polygon, a shape whose boundary is made out of line segments.*

Suppose I have a picture like this:

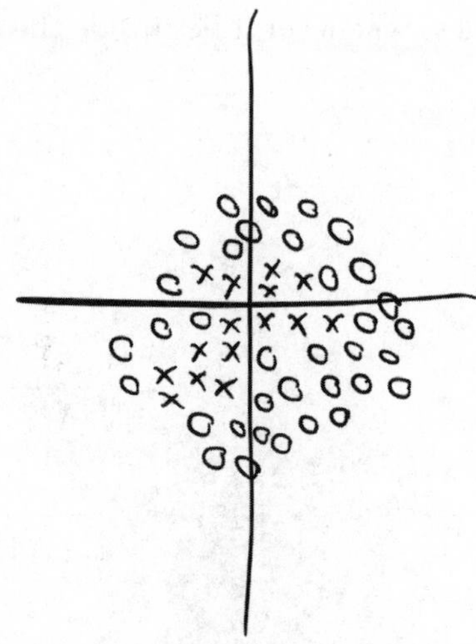

I've marked some points on the plane with an X, and some others with an O. My goal for the machine is that it learn a strategy for assigning an X or an O to the other, unlabeled points on the plane, just based on the labels I've given it. Maybe—hopefully—there is some strategy obtainable by setting the fourteen knobs just right that will assign large values to all the points with an X and small values to all the points with an O, and thereby allow me to make educated guesses about other points in the plane I haven't yet labeled. And if there *is* such a strategy, hopefully I can learn it by gradient descent, twiddling each knob a bit and seeing how much that diminishes my strategy's wrongness about the examples it's already been given. Find the best small twiddle you can make, make it, repeat.

The "deep" in deep learning just means the network has a lot of columns. The number of boxes in each column is called the *width*, and that number can get pretty big, too, in practice, but "wide learning" just doesn't have the same terminological zing.

Today's deep networks are more complicated than the ones in these

* Didn't I say this was supposed to be nonlinear? Yes, but the perceptron is *piecewise linear*, which means it's linear in different ways in different regions of space. More general neural nets can yield curvier results.

pictures, to be sure. What goes in the boxes can be more complicated than the simple functions we've talked about. In a so-called recurrent neural network you can have feedback boxes that take their own output as input, like "Op 4" on my DX21. And they're just plain faster. The idea of neural nets, as we've seen, has been around for a long time; I remember a time not long ago when the idea was seen as a dead end. But it turned out to be a good idea that just needed the hardware to catch up with the concept. Chips called GPUs, designed to render graphics really fast for gaming, turned out to be the ideal tool for training really big neural networks really fast. That allowed experimenters to jack up the depth and the width of their networks. With modern processors, you don't have to settle for fourteen knobs; you can have thousands of knobs, or millions, or more. The neural net GPT-3 uses to generate plausible English text has 175 billion knobs.

A 175-billion-dimensional space sounds big, sure; but 175 billion is really puny compared to infinity. We are still exploring only a tiny subspace of the space of all possible strategies. And yet this seems to be enough, in practice, to get text that looks like a human could have written it, just as the tiny network available to the DX21 is enough to enable a plausible imitation of a trumpet, a cello, and a space fart.

That's surprising enough, but there's a further mystery. The idea of gradient descent, remember, is to turn your knobs until you do as well as you possibly can on the data points you trained on. Today's networks have so many knobs that they can often attain *perfect* performance on the training set, calling each of the thousand cats a cat and each of the thousand other images a non-cat. In fact, with so many knobs to play with, there's a colossal space of possible strategies that *all* get the training data 100% correct. Most of these strategies, it turns out, perform terribly when presented with images the net hasn't seen. But the dumb, greedy process of gradient descent lands at some strategies much more often than others, and the strategies that gradient descent prefers seem to be in practice much more able to generalize to new examples.

Why? What is it about this particular form of network that makes it so good at such a wide variety of learning problems? Why does *this* tiny region of strategy space we're searching happen to contain a good strategy?

As far as I know, it's a mystery. Though let me be honest, there's a lot of controversy about whether it's a mystery! I have asked a lot of researchers in AI this question, famous, important people, and each one has happily talked my ear off about it. Some of them have very confident accounts to offer of why it all works. No two of the accounts I heard were the same.

But I can tell you, at least, *why* the landscape of neural networks is the one we've chosen to explore.

CAR KEYS EVERYWHERE

Famous old story: Man walking home late at night sees a friend of his, despondent, on his hands and knees under a lamppost. "What's wrong?" the man says. "I lost my car keys," says his friend. "Bummer," the man says, "let me help you," and he gets down on his knees, too, and they both scrabble around in the grass companionably for a time. After a while, the man says to his friend, "I dunno, are you sure it's here? We've been looking for a while," and the friend says, "Oh, no, I have no idea, I've been all over town since the last time I'm sure I had them," and the man says, "Then why have we been looking under this lamppost for the last twenty minutes?" and the friend says, "Because everywhere else it's too dark to look!"

The friend is much like a contemporary practitioner of machine learning. Why do we look to neural networks among the vast sea of strategies we could be searching? It's because neural networks are exquisitely well adapted to gradient descent, the only way of searching we really know. The effect of turning one knob is easily isolable; it affects the output from that box in an understandable way, and from there you can follow the lines and see how the change in that output affects the boxes that take input from that box's output, and how each of *those* boxes affects the boxes downstream from it, and so on.* The reason we

* For calculus fans: the real reason this is easy is that the function the neural net computes is built up by adding functions and composing them, and both of those operations play very nicely with computing derivatives, thanks to the chain rule.

choose this particular part of space to search for good strategies is that it's the part where it's easiest to see where we're going. Everywhere else it's too dark!

The car-keys story is supposed to cast the friend as a fool. But in a slightly alternate universe, the friend is not so foolish. Suppose car keys are actually strewn around all over the place—in the street, in the woods, and, very probably, somewhere in the circle of light under the lamppost. In fact, there are probably multiple car keys lying around in the grass there. Maybe the friend has found, in practice, that previous searches of that area have turned up keys to much better cars than he'd expected! The keys to the very nicest car in the whole city may well be elsewhere, true. But given enough time searching under the lamppost, abandoning each set of keys every time you see the keys to a more deluxe car nearby, you can do pretty well.

Chapter 8

You Are Your Own Negative-First Cousin, and Other Maps

What is a circle? Here's the official definition:

> *A circle is the set of points in the plane at a given distance from a fixed point, which is called the center.*

Okay, what's distance?

Already we are faced with a subtle problem. The distance between two points might be the crow-flying distance between them. But if, in reality, someone asks you how far away you are from their house, you might say "Oh, it's just about fifteen minutes away." That's a notion of distance, too! And if distance is understood in this way, "time it takes to travel," circles might look like this:

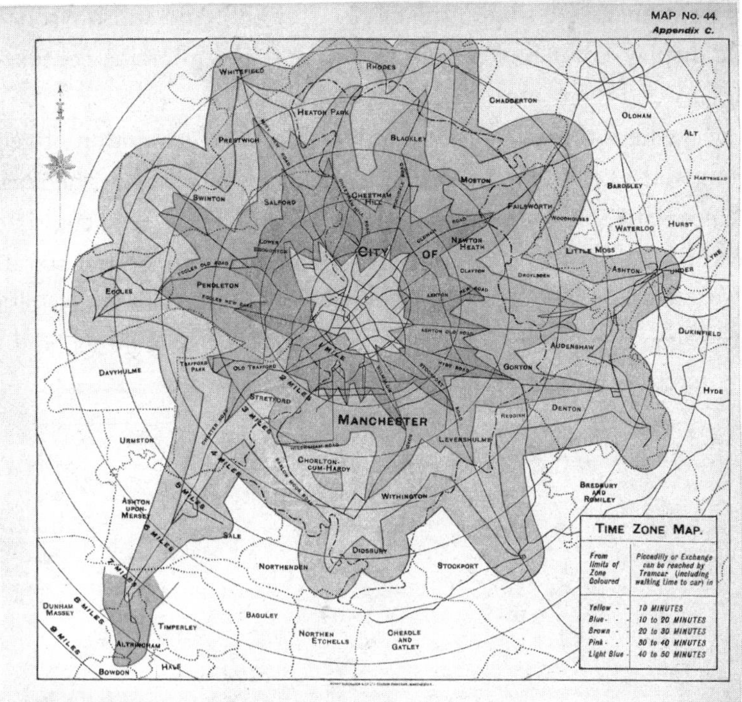

These pointy starfish guys are concentric circles, representing points exactly ten, twenty, thirty, forty, and fifty minutes distance by tram from the circles' common center, Piccadilly Gardens, in downtown Manchester, England. This kind of map is called an *isochrone.*

Different urban geometries yield different types of circles. In Manhattan (motto: "I'm walkin' here!") people travel on foot, and if someone asks you how far you are from home, you answer in blocks. The circle of points four blocks away from a given center will look like a square set on its corner:

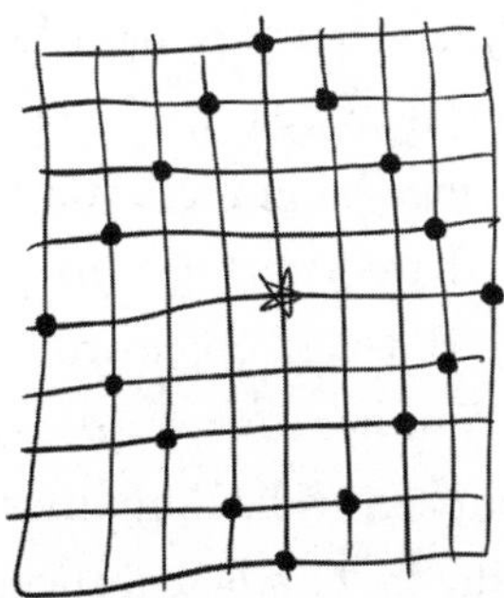

(See, we managed to square the circle after all!) And an isochrone map would display a bunch of concentric squares-which-in-this-context-are-circles around the central point.

Everywhere there's a notion of distance, there's a notion of geometry, too, and a concomitant idea of a circle. We are used to the idea of a "distant relative," and that's the very notion of distance we could derive from the geometry of the family tree. You and your sibling are at distance two from each other, because to get from you to your sibling in the tree you need to go up one limb to one of your parents, and then down one to get to the sibling.

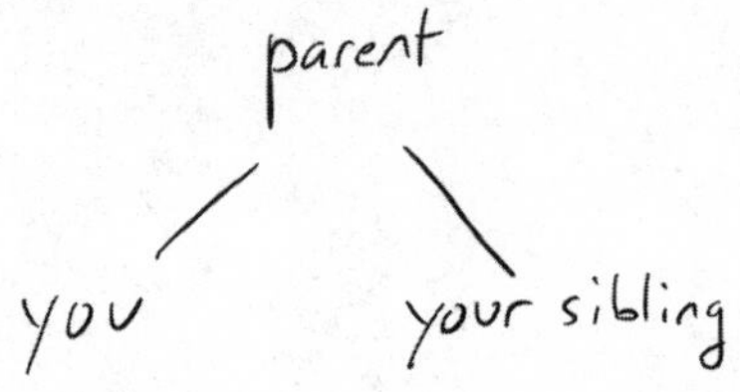

parent

you your sibling

Your distance from your uncle is three (up one step to your parent, who is at distance two from their sibling). Your distance from a first cousin is four: up two to Grandma, back down two to your cousin. You can do this for any level of cousindom, and get a nice algebraic formula:

Distance from your nth cousin = (n + 1) × 2

since an nth cousin is a person with whom you share an ancestor n + 1 levels up from you.

You, yourself, are your own negative-first cousin, because the relative you and yourself share is you, zero steps up! (And the formula still works: your distance from yourself is twice (–1 + 1), or zero.) As for your parents, they don't share a known ancestor (unless you're from a truly aristocratic clan), but they *do* have a shared relative—namely, you—one level *down* the tree, which is to say, –1 levels up; so your parents are each other's negative-second cousins. Your negative-third cousin is someone you share a grandchild with, say, your son-in-law's mother. That sometimes fraught relationship, called samdhi in Hindi, consuegro in Spanish, athoni in

Kikamba, and machatunim in Hebrew and Yiddish, has no name in English, which tends to be a little impoverished in the kin-word department.

If you think of the people in your generation of your family as a "plane," a disc of radius 2 around me in that plane consists of me and my siblings; a disc of radius 4 is me, my siblings, and my first cousins; a disc of radius 6 includes my second cousins, too. Here we can see a charmingly weird feature of cousin geometry. What does the disc of radius 4 around my first cousin Daphne look like? It consists of Daphne, her siblings, and her first cousins, or, in other words, all the grandchildren of the grandparents Daphne and I share. But that's the same as the disc of radius 4 around me! So who's the center, me or Daphne? No way around it: we both are. In this geometry, *every* point in a disc is its center.

Triangles in the cousin plane are also a little different than the ones you're used to. My sister and I are at distance 2 from each other, and each of us is 4 away from Daphne, so the triangle we form is isosceles. Guess what: *every* triangle in the cousin plane is isosceles. I'll leave it to you to satisfy yourself that's true. Weird geometries like this one, which are called *non-Archimedean,* might seem like misshapen scientific curiosities, but no: geometries like this show up all over mathematics. For instance, there's a "2-adic" geometry of whole numbers in which the distance between two numbers is the reciprocal of the largest power of two dividing their difference. Seriously, this turns out to be a good idea.

There's almost no context so abstract we can't invent a notion of distance, and with it a notion of geometry. Dmitri Tymoczko, a music theorist at Princeton, writes whole books about the geometry of chords and the way composers instinctively try to find short paths from one musical location to another. Even the language we speak can be said to have a geometry. Charting that geometry leads us to the map of all words.

THE MAP OF ALL WORDS

Imagine somebody tried to describe what Wisconsin looked like by giving you a list of towns and telling you the distance between any two of them. Yes, that would in principle tell you what shape Wisconsin was and where all the towns were within that shape. But in practice, a human, even a number-loving human like me, can't do anything with that long list of names and numbers. Our eyes and brains take in geometry in the form of maps.

It's not completely obvious, by the way, that the distances tell you the shape of the map! If there were only three towns in Wisconsin, to know the distance between each pair is to know the lengths of all three sides of the triangle they form, and it's a proposition of Euclid, gestured at in chapter 1, that if you know all three side lengths you know the shape of the triangle. It's yet more work to prove the fact that you can reconstruct the shape formed by *any* set of points if you know the distance between each pair; you and I, given that data, might make different maps, but mine would be related to yours by a rigid motion, moving and rotating the map without changing its shape.*

Why *would* you present the shape of Wisconsin in this hard-to-grasp tabular form, when maps of Wisconsin already exist? You wouldn't. But for other, non-geographic sorts of entities, we can define a notion of distance, and use that to create new kinds of maps. You could, for instance, make a map of personality traits. What could we mean by the distance between two traits? One simple way is to ask people. In 1968, the psychologists Seymour Rosenberg, Carnot Nelson, and P. S. Vivekananthan handed out packets of sixty-four cards to college students, each card labeled with a personality trait, and asked the students to group the cards into clusters of traits they thought likely to be common to a single person. The distance between two traits is then determined by the frequency with which the students grouped those two cards together. "Reliable" and "honest" were found together a lot, so they should be close; "good-natured" and "irritable" not so much, so they should be further apart.†

* For four points, this amounts to saying that the shape of a quadrilateral is determined if you know all four side lengths and the lengths of both diagonals. It's fun to try to convince yourself of this fact by contemplation. Would the length of one diagonal be enough?

† Actually, this turned out to not quite be good enough, since so many pairs of traits were essentially never grouped together; to get a more refined picture, you rate "reliable" and "honest" as close not just because they're often grouped together, but because third words like "finicky" tend to be grouped about equally often with both "reliable" and "honest."

Once you have those numbers, you can try to put the personality traits on a map so that the distances between traits on the page match the distances found in your experiment.

You might not be able to! What if, for instance, you find that each pair of traits among "reliable," "finicky," "sentimental," and "irritable" are at the same distance from each other? You can try and try to draw four points on the page in such a way that each pair of points delineates the same distance; you will fail. (I highly recommend *actually trying* this, to get your geometric intuition locked in on why it's impossible.) Some sets of distances are possible to achieve in the plane; others are not. A method called *multidimensional scaling* still allows you to make the map, though, as long as you're willing to allow the distances on your map to only *approximately* match the distances you're looking for. (And you should be willing; college students doing psychology experiments for beer money aren't exactly providing electron-microscope-level precision.) You get the following picture:

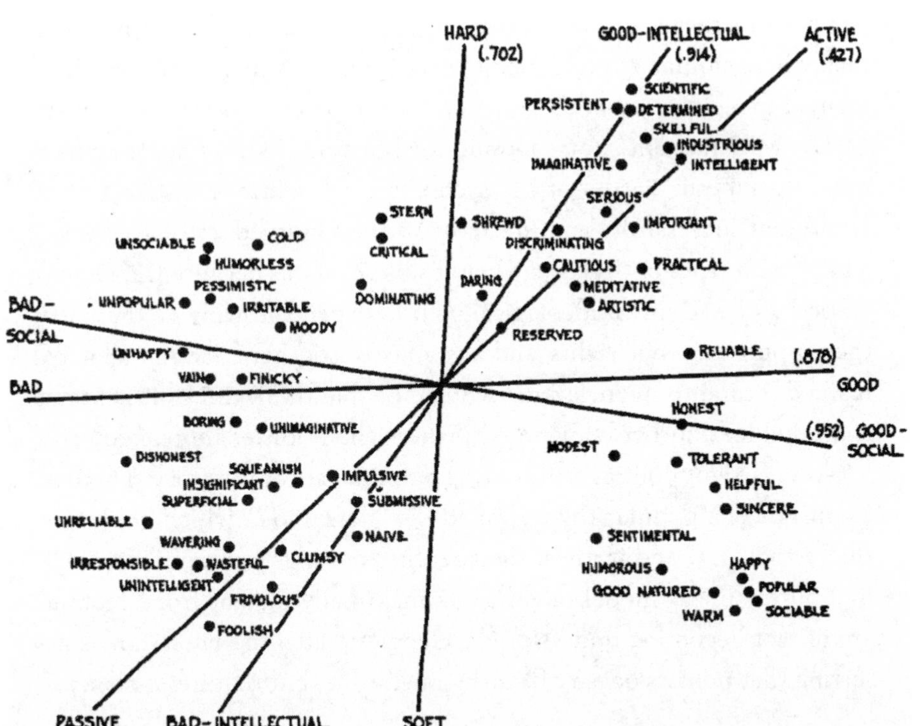

which I think you'll agree captures something of the geometry of personality. (The "axes" in the picture were drawn in by the researchers, and are their interpretation of what the directions in this map really mean.)

In three dimensions, by the way, it's easy to make the distances between four points all the same; you place the four points at the corners of a shape called a regular tetrahedron:

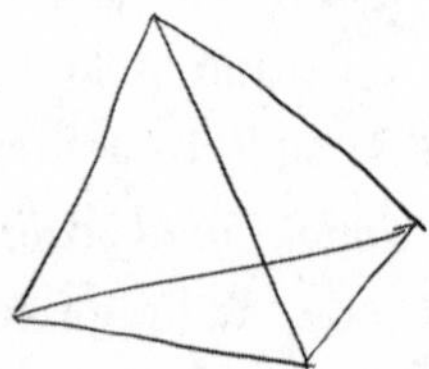

The more dimensions you allow yourself, the better you can get the distances between points on your map to match the ones you've measured. Which means the data can *tell* you which dimension it "wants" to be in. Political scientists measure the similarity between members of Congress by means of their votes; you can then put representatives on a map where similarly voting members are close together. You know how many dimensions you need to match the voting data in the U.S. Senate pretty well? Just one. You can rank the senators along a line, from leftmost (Elizabeth Warren of Massachusetts) to rightmost (Mike Lee of Utah) and successfully capture most of the observed voting behavior. That's been true for decades; when it wasn't, it was because the Democratic Party had a real ideological split between the wing of the party that supported civil rights and the mostly southern faction that remained militantly segregationist. Some people think the United States is headed for another realignment, where the traditional breakdown of left-versus-right politics will once again miss part of the story. There is, for instance, a popular theory called the "horseshoe," which holds that the farthest left and farthest right reaches of American politics, which in a purely linear model ought to be maximally distant from another, are in fact becoming quite similar. Geometrically, the horseshoe is asserting that politics doesn't fit in a straight line, but requires a plane:

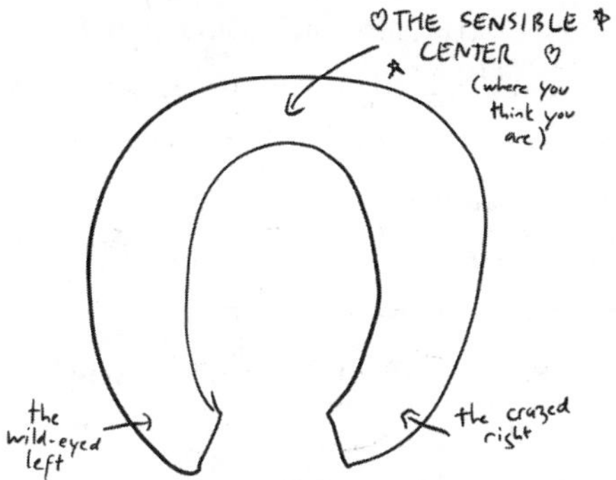

If that's true, and if the opposite ends of the horseshoe have enough of a constituency to get elected to Congress, we'll see that in the voting data; the one-dimensional model of Congress will start to get less and less accurate. That hasn't happened yet.

For bigger data sets, two dimensions will rarely be enough. A team of Google researchers led by Tomas Mikolov developed an ingenious mathematical device called Word2vec, which one might call a *map of all words*. We no longer need to rely on college students and index cards to gather numerical information on which words go together. Word2vec, trained on a body of text from Google News six billion words long, assigns to each English word a point in three-hundred-dimensional space. That's hard to picture, but remember, just as a point in two-dimensional space can be rendered as a pair of numbers, a longitude and a latitude, a point in three-hundred-dimensional space is nothing more than a list of three hundred numbers, a longitude, a latitude, a platitude, an amplitude, an attitude, a turpitude, etc., etc., as far as your rhyming dictionary will take you. There is a notion of distance in three-hundred-dimensional space that is not so different from the two-dimensional one you know.* And Word2vec's goal is to place similar words at points that are not too far from one another.

* The distance between two points is computed as follows: Compute the difference between the two longitudes, the two latitudes, the two platitudes, and so on. Now you have three hundred numbers. Square them all, add up those squares, and take the square root, and that's your distance. This is the three-hundred-dimensional version of the Pythagorean Theorem, though Pythagoras himself might well have rejected this characterization of something so far from physical geometry.

What makes two words "similar"? You can think of each word as having a "neighbor cloud" of words that often appear near it in the Google News text corpus. To a first approximation, Word2vec rates two words as similar when their neighbor clouds have a lot of overlap. In a chunk of text that contains the words "glamour" or "runway" or "jewel" you might expect to find words like "stunning" or "breathtaking," but not "trigonometry." So "stunning" and "breathtaking," which share "glamour," "runway," and "jewel" in their clouds, would be rated as similar, reflecting the fact that these two nearly synonymous words often appear in the same contexts. Word2vec places them at a distance of 0.675 from each other. In fact, "breathtaking" is the very *closest* word to "stunning" out of the 1 million words Word2vec knows how to encode. The distance from "stunning" to "trigonometry," on the other hand, is 1.403.

Once we have the idea of distance, we can start talking circles and discs. (Though maybe, being in three hundred dimensions instead of two, it would be better to talk about spheres and balls, their higher-dimensional analogues.) A disc around "stunning" of radius 1 has forty-three words in it, including "spectacular," "astonishing," "jaw-dropping," and "exquisite." The machine is clearly capturing something about the word, including that it can be used to indicate either great beauty or surprise. What it is not doing, I need to point out, is numerically distilling the *meaning* of the word. That would be quite a feat. But it's not what the strategy is built to do. "Hideous" is just 1.12 away from "stunning"—even though the two are nearly opposite in meaning, you can imagine those two words frequently appearing with the same neighbors, as in "That sweater is truly _____." The disc of words at distance at most 0.9 from "teh" consists of "ther," "hte," "fo," "tha," "te," "ot," and "thats"—those aren't even words, let alone synonyms, but Word2vec correctly recognizes that all these words are likely to show up in contexts with a lot of typos.

We need to talk about vectors. That's a technical term whose formal definition looks forbidding, but its meaning comes down to this. A point is a noun. It represents a thing: a location, a name, a word. A vector is a *verb.* It tells you what to do to a point. Milwaukee, Wisconsin, is a point. "Move thirty miles west and two miles north" is a vector. If you apply that vector to Milwaukee, you get Oconomowoc.

How would you describe that vector, the one that gets you from Milwaukee to Oconomowoc? You could call it the "due-west outer-ring suburb vector." Apply it to New York City* and you get Morristown, New Jersey, or, more precisely, the Dismal Harmony Natural Area, a state park just west of town.

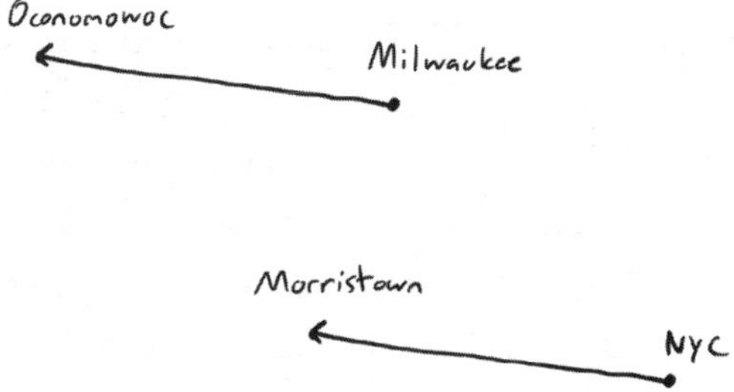

You might phrase this as an analogy: Morristown is to New York as Oconomowoc is to Milwaukee. And as Boinville-en-Mantois is to Paris, and as San Jerónimo Ixtapantongo is to Mexico City, and as the Farallon archipelago, an uninhabited former nuclear waste dump now said to be the most rodent-dense island chain on earth, is to San Francisco.

Which brings us back to "stunning." The developers of Word2vec noticed an interesting vector: the one that tells you how to get from the word "he" to the word "she." You might think of it as the "feminization" vector. If you apply it to "he" you get "she." What if you apply it to "king"? You get a point that, like the Dismal Harmony Natural Area, doesn't land squarely on a place you have a word for. But the *nearest* word—the Morristown, New Jersey, of this scenario—is "queen." "Queen" is to "king" as "she" is to "he." This works well for other words, too: the feminized version of "actor" is "actress," and of "waiter" is "waitress."

* The official municipal boundaries of New York are pretty broad, so let's stipulate that the precise geographic location of "New York" meant here is the Strand Book Store in the East Village.

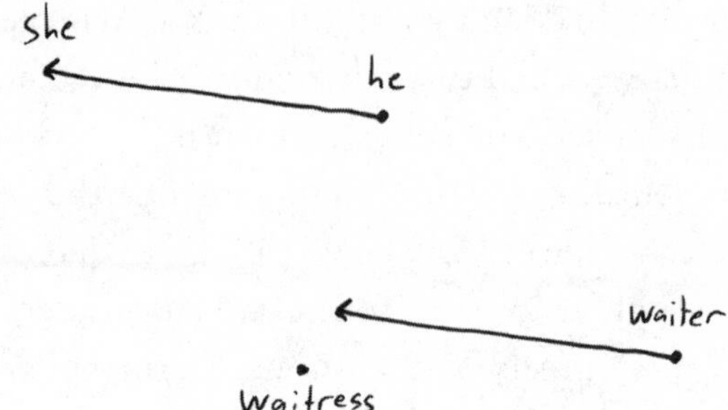

And what about "stunning"? Guess what: you get "gorgeous." "Gorgeous" is to "she" as "stunning" is to "he." Apply the vector in the other direction, asking Word2vec how to "masculinize" the word "stunning," and you get "spectacular." Because these analogies represent only approximate numerical equalities, not exact ones, they're not always symmetric: female for "spectacular" is indeed "stunning" but male for "gorgeous" is "magnificent."

What does this mean? That, in some mathematical and universal and utterly objective sense, gorgeousness is the feminine version of stunningness? Certainly not. Word2vec doesn't know what words mean, and has no way to know. All Word2vec knows is the massive corpus of English text it trained itself on, chewed down into numerical pulp from decades of transcribed newspapers and magazines. When English speakers want to talk about stunningness and we're talking about a woman, we have a statistically detectable habit of saying "gorgeous." And when we're talking about a man, we don't. The geometry teased out by Word2vec might look at first like a geometry of meaning, but it's actually a geometry of our way of speaking, from which we can learn as much about ourselves and our gendered biases as we can about our language.

Messing around with Word2vec is like putting the collected writings of the Anglophone world on the psychoanalyst's couch and peeking into its grotty unconscious. The "feminized" version of "swagger" is "sassiness." The feminized version of "obnoxious" is "bitchy." The feminized version of "brilliant" is "fabulous." The feminized version of "wise" is "motherly." A female "goofball" is a "ditz," with, no lie, "perky blonde" as the number-two choice.* And a female "genius" is a "minx." Non-symmetry again: a male "minx" is a "scallywag." Male for "teacher" is "headmaster." Male for "Karen" is "Steve."

A lady bagel is a "muffin." And a Hindu bagel—that is, what you get if you apply the vector that takes "Jewish" to "Hindu" to the point representing "bagel"—is a "vada pav," a popular Mumbai street snack. A Catholic bagel is a "sandwich"; "meatball sub" is number two.

Word2vec knows the names of cities, too. If you use its conceptual vector analysis in place of plain longitude and latitude, the Oconomowoc of New York is not Morristown but Saratoga Springs. I have no idea why.

Playing with this is deeply fun and in some ways enlightening. But I've indulged myself in a bad behavior that's endemic to writing about machine learning, and I'd better cop to it: I have been vigorously cherry-picking. It's fun to share the punchiest and most impressive examples! This can mislead; Word2vec is not a magic meaning machine. More often than not, its proposed "analogy" is no more than a synonym (female "boring" is "uninteresting," female "mathematics" is "math," female "amazing" is "incredible") or a misspelling (female "vicious" is "viscious") or just wrong: male "duchess" is "prince," female "pig" is "piglet," female "cow" is "cows," female "earl" is "Georgiana Spencer" (the right answer is "countess," which, to be fair, Spencer was). When you read about the latest advance in AI, don't be dismissive; progress really has been rapid and exciting. But the odds are that what you're seeing in the press release are the very shiniest results out of many, many attempts. So be skeptical, too.

* Word2vec actually works with "lexical tokens," which are usually words but are sometimes names or short phrases.

Chapter 9

Three Years of Sundays

A really important and in some ways underpublicized fact about math is that math is very hard. We sometimes conceal this fact from our students, with the idea that we're doing them a service. It's just the opposite. Here's a plain fact I learned as an apprentice instructor from master teacher Robin Gottlieb. When we say the lesson at hand is "easy" or "simple," and it manifestly isn't, we are telling the student that the difficulty isn't with the mathematics, it's with them. And they will believe us. Students, for better or worse, trust their teachers. "If I didn't even get this and it was easy," they'll say, "why bother trying to understand something hard?"

Our students are afraid to ask questions in class because they're afraid of "looking stupid." If we were honest about how difficult and deep mathematics is, even the mathematics that appears in a high school geometry classroom, this would surely be less of a problem; we could move toward a classroom where asking a question meant not "looking stupid" but "looking like someone who came here to learn something." And this doesn't just apply to students who find themselves struggling. Yes, some have no trouble picking up the basic rules of algebraic manipulation or geometric constructions. Those students should still be asking questions, of their teachers and of themselves. For example: I have done what the teacher asked, but what if I'd tried to do this other thing that the teacher didn't ask of me, and, for that matter, why did the teacher ask for one thing and not the other? There's no intellectual vantage from which you can't easily sight a zone of ignorance, and that's where your eyes should be pointed, if you want to learn. If math class is easy, you're doing it wrong.

What is difficulty, anyway? It's one of those words we feel we know well, but which falls apart into related but distinct concepts when you try to circumscribe it. I like this story the number theorist Andrew Granville tells about the algebraist Frank Nelson Cole:

> At the 1903 meeting of the American Mathematical Society, F. N. Cole came to the blackboard and, without saying a word, wrote down
> $2^{67} - 1 = 147573952589676412927 = 193707721 \times 761838257287$,
> long-multiplying the numbers out on the right side of the equation to prove that he was indeed correct. Afterwards he said that figuring this out had taken him "three years of Sundays." The moral of this tale is that although it took Cole a great deal of work and perseverance to find these factors, it did not take him long to justify his result to a room full of mathematicians (and, indeed, to give a proof that he was correct). Thus we see that one can provide a short proof, even if finding that proof takes a long time.

There's the difficulty of recognizing a statement as true, and the difficulty, which is not the same, of *coming up with* the statements whose truth is to be recognized. That's the achievement Cole's audience clapped for. We have already seen that finding the prime factors of a large number is a problem recognized as difficult; but 147573952589676412927, by the standards of modern computing machinery, is not a large number. I just factored it on my laptop and it took not even a single Sunday, but an amount of time so small as to be imperceptible. So is this problem difficult or not?

Or consider the problem of computing hundreds of digits of π, a practice that once would have counted as research mathematics, but now is a mere computation. This presents yet another kind of difficulty, the difficulty of motivation. I don't doubt that my technical ability to compute is sufficient to allow me to work out seven or eight digits of π by hand. But it would be hard for me to *make* myself do that—because it would be dull, because my computer could do it for me, and, perhaps most of all, because there is no reason to know very many digits of π. There are real-world contexts where you'd want to know seven or eight digits, sure. But the hundredth digit? It's hard to imagine what you'd need that for. Forty digits is already enough to compute the circumference of a circle the size of the Milky Way to within the size of a proton.

To know one hundred digits of π is not to know more about circles than other people. What's important about π is not so much what its value is, but *that* it has a value. The meaningful fact is that the ratio of a circle's circumference to its diameter doesn't depend on what circle it is. That's a fact about the symmetries of the plane. Any circle can be made into any other by means of the so-called *similarities*, composed of translations, rotations, and changes of scale. A similarity might modify distances, but it changes them by means of multiplying by a fixed constant; maybe it doubles every distance, maybe it shrinks every distance by a factor of ten, but in any event it leaves the *ratio* between any two distances—say, the distance around the circumference of a circle and the distance straight across the diameter—the same. If you deem two figures to be the same if one can be transformed into the other via those symmetries, calling different things by the same name à la Poincaré, then there is really only *one* circle, which is why there's only one π. Similarly, there's only one square, and so there's only one answer to the question "What's the ratio of the perimeter of a square to its diagonal?,"* and that answer is twice the square root of 2, about 2.828 . . . , which you could say is the π of the square. There is only one regular hexagon, and its π is 3.

* Why the diagonal? I take that to be a good analogue of "diameter" since it's the greatest distance between any two points in the figure.

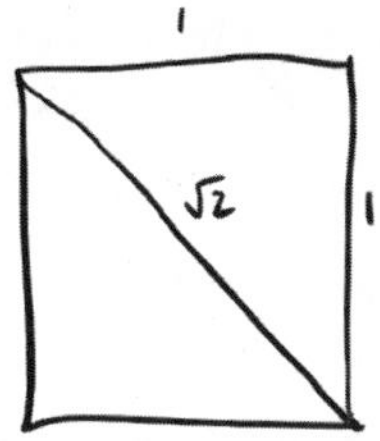

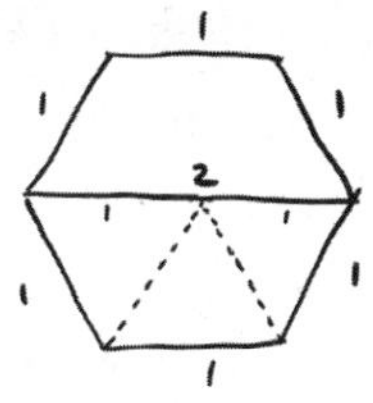

But there's no π of the rectangle, because there is not just one rectangle but many different rectangles, distinguished by the ratio between their long side and their short one.

Is it difficult to play a perfect game of checkers? For a person, yes—but the computer program Chinook can do it. (Is the right question here about the difficulty Chinook faces playing checkers, or the difficulty scientists faced in building Chinook?) As we've seen, the problem of playing perfect checkers, or perfect chess, or perfect Go is no different in principle from multiplying two very large numbers—isn't there a sense in which this is therefore conceptually easy? We know exactly what steps we'd have to take to analyze the tree of the game, even if there's not enough time in the universe to actually do it.

One easy answer would be to say that some problems, like factoring numbers and playing Go, are easy for computers and hard for us, because computers are better and smarter than us. This answer implicitly models difficulty as a point on a line, a line on which humans and computers are situated, too, able to handle any problem we're greater than or equal to:

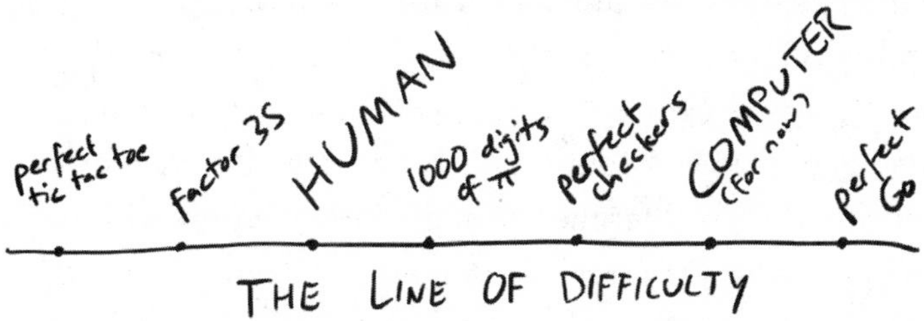

But this is wrong; the geometry of difficulty isn't one-dimensional. There are problems, like factoring large numbers or playing perfect checkers or storing billions of words of text with perfect fidelity, at which computers are much, much better than we are. (For one thing, computers don't face the difficulty of motivation; they do—for the moment, at any rate—what we tell them to do.) But there are problems that are hard for computers and easy for us. The *parity problem* is a famous one; standard neural network architectures do a terrible job at learning whether a string of *X*'s and O's has an even or odd number of *X*'s. Extrapolation is hard, too. If you give a person a bunch of examples like

input: 2.2	output: 2.2
input: 3.4	output: 3.4
input: 1.0	output: 1.0
input: 4.1	output: 4.1
input: 5.0	output: 5.0

and then ask, What is the output when the input is 3.2?, a human would say 3.2, and so would a neural net trained on this data. What if the input is 10.0? The human would say 10.0. But the neural net could say just about anything. There are all kinds of crazy rules that agree with "output = input" between 1 and 5 but which look completely different outside that range. A human knows that "output = input" is the simplest, most natural way to extend that rule to a larger class of possible inputs, but a machine-learning algorithm may not. It has processing power but not taste.

I cannot, of course, rule out that machines will eventually (or even imminently!) surpass the human cognitive capacity in every respect. It's a possibility that researchers in artificial intelligence, and their promoters, have always recognized. AI pioneer Oliver Selfridge, in a television interview from the early 1960s, said, "I am convinced that machines can and will think in our lifetime," though with the proviso "I don't think my daughter will ever marry a computer." (There is no technical advance so abstract that people can't feel sexual anxiety about it.) The multidimensional geometry of difficulty should remind us that it's very hard to know which competencies machines are on the verge of acquiring. An autonomous vehicle may be able to make the right choice 95%

of the time, but that doesn't mean it's 95% of the way to making the right choice *all* the time; that last 5%, those outlier cases, might well be a problem our sloppy brains are better equipped to solve than any current or near-future machine.

And of course there's the question, naturally of interest to me, of whether machine learning can replace mathematicians. I won't presume to predict. But my hope is that the mathematicians and the machines will continue to be partners, just as we are now. Many computations that would have taken mathematicians years of Sundays to work out can now be delegated to our mechanical colleagues, leaving us to specialize in what *we're* especially good at.

A couple of years ago, Lisa Piccirillo, then a PhD student at the University of Texas, solved a long-standing geometry problem about a shape called the Conway knot. She proved the knot was "non-slice"—this is a fact about what the knot looks like from the perspective of four-dimensional beings, but for this story it doesn't matter exactly what it means. This was a famously difficult problem. Though even here, the meaning of the word gets complicated; is the problem difficult, because many mathematicians worked on it and failed, or easy, because Piccirillo found a crisp solution that took just nine pages, of which two pages are pictures? One of my own most cited theorems is of the same nature, settling in a six-page paper a problem I and many other people had been wrestling with for twenty years. Maybe we need a new word that conveys not "it's easy" or "it's difficult," but "it's difficult to realize that it's easy."

A few years before Piccirillo's breakthrough, a topologist named Mark Hughes at Brigham Young had tried to get a neural network to make good guesses about which knots were slice. He gave it a long list of knots where the answer was known, just as an image-processing neural net would be given a long list of pictures of cats and pictures of non-cats. Hughes's neural net learned to assign a number to every knot; if the knot, in reality, were slice, the number was supposed to be 0, while if the knot were non-slice the net was supposed to return a whole number bigger than 0. In fact, the neural net predicted a value very close to 1—that is, predicted the knot was non-slice—for every one of the knots

Hughes tested, except for one. That was the Conway knot. Hughes's neural net returned a number very close to 1/2: its way of saying it was deeply unsure whether to answer 0 or 1. This is fascinating! The neural net correctly identified the knot that posed a really hard and mathematically rich problem (in this case, reproducing an intuition topologists had already arrived at). Some people imagine a world where computers give us all the answers. I dream bigger. I want them to ask good questions.

Chapter 10

What Happened Today Will Happen Tomorrow

I'm writing this chapter from inside a pandemic. COVID-19 has been ravaging the world for months, and no one is quite sure what the course of the disease's spread will look like. That's not a math question, but it's a question with math in it—how many people will get sick, and where, and when? The whole world has been getting a crash course in the mathematics of disease. And that subject, in its modern form, brings us back to the mosquito man, Ronald Ross. His lecture on the mosquito's random walk at the 1904 St. Louis exposition was part of a larger project: bringing disease into the realm of the quantifiable. Historically, plagues were like comets, appearing unexpectedly and terrifyingly, then vanishing again, on no fixed schedule. Newton and Halley had tamed the theory of comets, binding them to fixed elliptical orbits via the laws of motion. Why shouldn't epidemics be just as subject to universal laws?

Ross's lecture was not a success. "I was really to have opened the whole discussion on Pathology," he later wrote, "but was led to believe that I might choose my own subject, and hence I read the mathematical paper . . . to hundreds of disappointed doctors who did not understand a word I said!"

The quote captures him well. Ross was truly devoted to bringing a mathematical outlook to medicine, not always to the acclaim of his fellow doctors. "[S]ome members of this profession," the editor of the *British Medical Journal* wrote, "will learn with surprise, possibly mingled with regret, that this distinguished exponent of the experimental method is an enthusiast for the application of quantitative processes to the problems of epidemiology and pathology."

He was also a little full of himself. An appreciation in the *Journal of the Royal Society of Medicine* concedes:

> Sir Ronald Ross left behind him the reputation of being conceited, quick to take offence and greedy for fame and money. He was, to a certain degree, all of those things, but they were not his only characteristics nor his most dominant ones.

He was known, for instance, to be generous and supportive to younger scientists. In any hierarchical organization, you can find people who are pleasant to the people at or above their status, and treat the people beneath them like garbage; and you can find people who see established big shots as rivals and enemies, while showing nothing but kindness to the new people coming up. Ross was of the latter type, which is, on the whole, to be preferred.

Ross spent the years around 1900 in a vicious academic battle with the Italian parasitologist Giovanni Grassi over credit for the malaria breakthrough, and even after Ross won the Nobel Prize and Grassi was shut out, Ross never seemed to feel the recognition he got was quite what it should have been. His dispute with Grassi evolved into a generalized sense of grievance against the Italians who had taken Grassi's side. His lecture in St. Louis almost never happened, because when Ross learned his panel was to include the Roman physician Angelo Celli, he immediately canceled his trip, to be coaxed back only after being assured by telegram that Celli had been persuaded to withdraw.

Ross was knighted, he was given the directorship of a scientific institute named after him, he collected scientific honors like they were vintage Pez dispensers, but the hole was never filled. He spent years, though under no financial strain, publicly campaigning for Parliament to award him a monetary prize for his contribution to public health. Edward Jenner had gotten one in 1807 for developing the smallpox vaccine, and Ross felt he deserved no less.

Possibly his lifelong peevishness stemmed from a lurking feeling he wasn't following his true life's path. Astonishingly for a doctor so distinguished, Ross says he entered the medical profession "merely and purely as a duty," putting aside the two pursuits that truly sang to his heart. One was poetry, which he wrote throughout his career. The verse he spontaneously composed upon obtaining experimental proof of his malaria theory ("With tears and toiling breath / I find thy cunning seeds / O million-murdering Death") was, at the time, a well-known part of his legend. Twenty years later, very much in character, he wrote a follow-up poem, "The Anniversary," to complain about being underappreciated. ("What we with endless wonder won / the thick world scorned . . .") At some point he adopted a phonetic alphabet he thought best-suited for reproducing the Latin virtues in English verse:

Aa hwydhr dúst dhou flot swit sælent star
Yn yóndr flúdz ov ivcnyngz dæyng læt ?

("Ah, whither dost thou float, sweet silent star / In yonder floods of evening's dying light?")

The other thing he cared about was math. He recalls his early geometric education: "Regarding mathematics, Euclid was amazingly incomprehensible to me until I came to Book I, Prop 36, when his meaning suddenly flashed upon me, with the result that he no longer presented any difficulty whatever. I became very good at geometry and liked to solve problems for myself, and remember that I solved one in my sleep during the early morning." As a young doctor in Madras, he picked off his shelf a book on celestial mechanics, unlooked at since his student years, and experienced what he refers to as "the great calamity"—a sudden plunge into mathematical obsession. He bought every math book the local bookstore offered and read them all in a month: "up to the end of the Calculus of Variations, though I had not advanced beyond Quadratic Equations at school." He was startled by how easy he found it all now, and ascribed this to the fact that nobody was *making* him do it: "education must be chiefly self-education, during or after school, or it will never approach completion at all."

On this point, no one who teaches math can really disagree. I *wish* my explanations at the blackboard were so magisterially clear and my path through the material so efficient and direct that students could walk out of their fifty-minute hour with me in a condition of complete mastery. Not so. Education, as Ross understood, is self-education. Our job as teachers is, yes, to explain—but our job is just as much a species of marketing. We have to sell students on the idea that it's *worth it* to spend the time outside of class it takes to truly learn this stuff. And the best way to do this is to let our own hot feelings for the math spill out into the way we talk and comport ourselves.

Ross, looking back from middle age, summons those hot feelings in a typically poetical manner:

> It was an aesthetic as well as an intellectual enthusiasm. A proved proposition was like a perfectly balanced picture. An infinite series died away into the future like the long-drawn variations of a sonata. . . . The aesthetic sense is indeed largely intellectual satisfaction at perfection achieved; but I saw also future perfection to be achieved by the potent weapon of pure reason. The stars of the evening and of the dawn . . . were now doubly beautiful since they had been caught in the net of analysis. I soon began to read the applications of mathematics to motion, heat, electricity, and the atomic theory of gases; and remember thinking from the first of their possible application for explaining why epidemics of disease exist. . . . But I was always impatient of reading mathematics, and felt that I should like to have created the propositions myself; and indeed new propositions suggested themselves while I was reading the old ones.

This disinclination to learn from his forebears ran deep into his character. Writing about an admired uncle with a chemistry hobby (but of course really writing about himself), he says: "Nearly all the ideas in science are provided by amateurs, such as my uncle Ross; the other gentlemen write the textbooks and obtain the professorships." And as a mathematician, he was never more than an amateur, though this didn't stop him from publishing papers in pure mathematics with rather grand titles ("The Algebra of Space") which more or less recapitulated ideas already existing in the literature, and nursing frustration that full-time mathematicians didn't engage more with his work.

NOT THE MOST IMPORTANT OF GOD'S THOUGHTS

By the middle of the 1910s, Ross was ready to attack in earnest the problem that had burst into his mind back in Madras: the creation of a mathematical theory that would do for epidemics what Newton's had done for the celestial bodies. Actually, that wasn't quite ambitious enough for Ross; he wanted to develop a theory that would govern the quantitative spread of *any* change of condition through a population—conversion between religions, elections to professional societies, military enlistments, and of course infections by epidemic disease. He called it "The Theory of Happenings." In 1911, Ross wrote to a protégé of his, Anderson McKendrick: "We shall end by establishing a new science. But first let you and me unlock the door and then anybody can go in who likes."

And despite his high assessment of his own abilities and his love for amateurism, he did what he had to do to get that door unlocked; he hired an actual mathematician to help him. Her name was Hilda Hudson. Hudson was by far the deeper of the two mathematically. Her first publication was a short new proof of a proposition of Euclid's, obtained by a clever dissection of squares into smaller geometric figures. She was ten. (It helped that both her parents were mathematicians, too.)

Hudson was a specialist in a field that intermingled geometry and algebra, called (we don't *always* come up with inventive names) algebraic geometry. René Descartes was the first to make really systematic use of the idea that points on the plane can be thought of as pairs of numbers, an x-coordinate and a y-coordinate, allowing us to transform a geometric object like a circle (the set of points at fixed distance from a given center) into an algebraic one (say, the set of pairs (x,y) such that

$x^2 + (y - 5)^2 = 25$). By Hudson's time, this melding of algebra and geometry had become a subject in its own right, applying not just to curves in the plane but figures of any dimension. Hudson was a leading figure in the area of so-called Cremona transformations of two- and three-dimensional bodies, and in 1912 was the first woman to lecture at the International Congress of Mathematicians.

If I were to tell you that a Cremona transformation is "a birational automorphism of projective space," I would just be flinging phonemes at you, so let me put it a different way: What is 0/0? You have probably learned at some point that you're supposed to say "undefined," and that is, in a way, correct, but it's also the coward's way out. It really depends which zeroes you're dividing! What is the ratio of the area of a size-zero square to its perimeter? Sure, you could say it's undefined, but why not be bold and define it? If the square has a side of length 1, that ratio is 1/4 or 0.25. When the side length shrinks to 1/2, the area is 1/4 and the perimeter is 2, so the ratio has gone down to 1/8. If the side length is 0.1, the ratio is 0.01/0.4 or 0.025. The ratio gets smaller and smaller as the square does, which means there's only one good answer for what happens when the square shrinks to a point: in *this* case, 0/0 = 0. On the other hand, what if we ask about the ratio of a line segment measured in centimeters to its length measured in inches? That ratio is 2.54 for a long segment and it's 2.54 for a short segment, and if the segment shrinks down to a point, *that* 0/0 should be 2.54.

Geometrically, you can follow Descartes's example and think of a pair of numbers as a point in the plane. The point (1,2) is the point 1 unit to the right and 2 units up from the center. And the ratio 2/1 is the slope of the line joining the center to (1,2). When the point is (0,0), the center itself, there *is* no line segment, so there's no slope. The very simplest type of Cremona transformation swaps out the plane for a very similar geometry, in which (0,0) is replaced with multiple points—infinitely many, in fact! Each one remembers not just its location at (0,0) but a slope as well, as if you were keeping track not just of where you were but the direction of the path you took to get there.* This kind

* This might sound a little bit like the coordinates Poincaré used for the three-body problem, where he needed to track not only the position of each planet but the direction of its motion; yep, same deal.

of transformation, in which one point explodes into infinitely many, is called a *blow-up.* The higher-dimensional Cremona transformations Hudson studied are decidedly more involved; you might call them a general geometric theory of assigning values to those "undefined" ratios a more timid calculator would back away from.

In 1916, just as she was beginning her work with Ross, Hudson published an entire book about straightedge and compass constructions in the style of Euclid, the sort of thing Abraham Lincoln had been vainly struggling with in his attempts to square the circle. Her geometric intuition was so strong that her writings were sometimes criticized for being light on proofs; things were obvious to her that ought to have been justified in writing for those of us less able to trace out a geometric surface in our mind's eye. There's no evidence that Ross, despite his affection for geometry, had any interaction with or interest in Hudson's work in pure mathematics. Perhaps for the best, since algebraic geometry was loaded with Italians.

Ross's first paper with Hudson begins with a substantial list of errata from Ross's previous paper. Ross blames the mistakes on the fact that he was overseas when the page proofs for his paper arrived for checking; I like to imagine that Hudson, immediately upon joining the collaboration, started by gently informing Ross about the errors in the work carried out before her arrival. Very little is recorded about the interaction between the two—Ross mentions Hudson exactly once in his memoirs—but it's fascinating to imagine the relationship between these very different scientists. Ross had the limitless ambition, Hudson the mathematical depth and know-how. Ross had the titles and the prizes, while Hudson, in an era of all-male faculties, was a mere lecturer. If Ross had religious feelings, he didn't make much of them; Hudson's devout Christianity was a central fact of her life. After publishing a treatise on Cremona transformations in 1927, she seems to have left mathematics behind, working for years as an official of the Student Christian Movement. Her 1925 essay, "Mathematics and Eternity," is a remarkable document of an intellectual world in which faith and science each felt some need to justify themselves to the other. "We can practice the presence of God in an algebra class," she writes, "better than in Brother Lawrence's Kitchen; and in the utter loneliness of an unfashionable corner of research work, better than on a mountain top." Every mathematician, religious or not, will understand what she means in this should-be-famous epigram:

> [T]he thoughts of pure mathematics are true, not approximate or doubtful; they may not be the most interesting or important of God's thoughts, but they are the only ones that we know exactly.

NOT TOO REASSURING

Ross's ideas about epidemic growth were governed by an underlying principle, which is in fact the single principle underlying all mathematical predictions: *what happened today will happen tomorrow.* All the grotty details are in figuring out what that means in practice.

Here's the simplest thing it could mean. Suppose people carrying a contagious virus infect, on average, two other people during the period of their contagiousness, which lasts, let's say, ten days. If we start with one thousand infected people, then, ten days later, roughly two thousand new people will be infected. The original thousand are now recovered and infect no more, but the two thousand new people are going to infect four thousand over the next week, and a week after that, another eight thousand or so will catch the bug. So over the first four weeks the number of infections is

Day 0: 1,000
Day 10: 2,000
Day 20: 4,000
Day 30: 8,000

This kind of sequence is called a *geometric progression,* although the connection to geometry is a little obscure. Here's where it comes from: each term is the *geometric mean* of the one before it and the one after it. And what does "mean" mean, and what does it mean for a mean to be geometric?

A mean is a kind of average. The average you're probably used to is the one you get by drawing a point midway between two numbers on the number line. The average of 1 and 9 is 5, because 5 is 4 away from 1, and it's 4 away from 9. That's called the *arithmetic mean,* I suppose because it arises from the arithmetic operations of addition and subtraction, and a sequence where each term is the arithmetic mean of its predecessor and its successor is an arithmetic progression.

The geometric mean is a different kind of average. To take the geometric mean of 1 and 9, you make a rectangle whose sides have lengths 1 and 9.

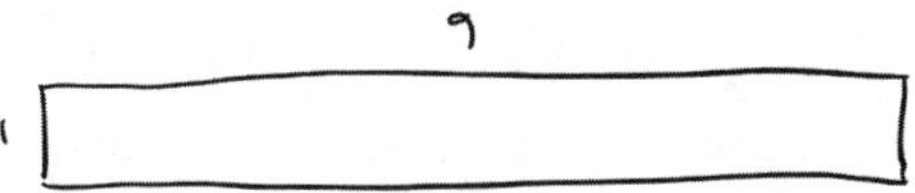

The geometric mean is the length of the side of a square whose area is the same as that of the rectangle. (The Greeks were very big on thinking of areas in terms of squares; this was one reason they kept trying, and failing, to square the circle.) The geometric mean was a favorite of Plato, who by some accounts considered the geometric mean the *truest* mean. The rectangle has area 1 × 9 = 9; if a square has the same area, the length of its side is a number that yields 9 when multiplied by itself. That is a long-winded way of saying "3." So 3 is the geometric mean of 1 and 9, and

1, 3, 9

is a geometric progression.

Nowadays we're more apt to define the geometric mean in a different but equivalent way; the geometric mean of numbers x and z is that number y such that

$$y/x = z/y.$$*

* If you like algebra, multiply both sides of this equation by xy, to get $y^2 = xz$; x and z are the sides of the rectangle, and there in the middle is y, squared, just as the geometry asks of us.

Compare this crisp formula with the verbal knots Plato had to twist himself into when stumping for the geometric mean:

> Now the best bond is one that really and truly makes a unity of itself together with the things bonded by it, and this in the nature of things is best accomplished by proportion. For whenever of three numbers which are either solids or squares the middle term between any two of them is such that what the first term is to it, it is to the last, and, conversely, what the last term is to the middle, it is to the first, since the middle term turns out to be both first and last, and the last and the first likewise both turn out to be middle terms, they will all of necessity turn out to have the same relationship to each other, and, given this, will all be unified.

and you really start to appreciate the virtues of algebraic notation!

Viruses don't spread by geometric progression because they like computing the area of rectangles, or because they've read Plato; they do it because the mechanics of viral spread demand that the ratio between last week's infections and this week's infections is the same as that between this week's infections and next week's infections. What happens today will happen tomorrow, and what happens in our running example is that every ten days the number of new cases gets multiplied by 2. When a sequence of numbers goes up in a geometric progression, we say it's growing exponentially. People often use "growing exponentially" as a synonym for "growing really fast," but the former is much more specific. Every math teacher has longed for an example that will really bring home to students how exponential growth behaves. At the moment, unfortunately, we have one close at hand.

Our factory-standard intuition is poorly adapted for grasping exponential growth. We are used to physical objects moving at roughly constant speed. Drive 60 miles an hour and your hourly distance progress looks like

60 miles, 120 miles, 180 miles, 240 miles . . .

That's an arithmetic progression—the difference between each term and the next never changes, and the numbers grow at a constant rate.

Geometric progressions are a different story; our minds interpret them as slow, steady, manageable growth followed by an abrupt and terrifying steepness. In the geometric sense, though, the speed of increase never changes. This week is like last week but twice as bad. The disaster is entirely predictable, but we are somehow unable to fully expect it. Heed the words of John Ashbery, probably the only major American poet to have addressed this issue, in his 1966 poem "Soonest Mended":

> like the friendly beginning of a geometrical progression
> Not too reassuring . . .

In Italy, one of the hardest-hit countries in the early days of the COVID-19 outbreak, it took almost a month for the disease to kill a thousand people. The next thousand deaths happened in four days. On March 9, 2020, after the disease had already started to spread worldwide, one U.S. government official* aggressively downplayed the threat, comparing it with the thousands of Americans who succumb to flu each year: "At this moment there are 546 confirmed cases of CoronaVirus, with 22 deaths. Think about that!" A week later, twenty-two Americans were dying of COVID-19 every day. A week after that, it was almost ten times that many.

The thing about geometric progressions is, there are good ones and there are bad ones. Suppose the carriers of a disease pass their little friend along to 0.8 people on average instead of 2. Then the geometric progression of infections looks like this:

Day 0: 1000
Day 10: 800

* Okay, it was the president of the United States, but that's not important right now.

Day 20: 640
Day 30: 512

and the next four numbers are even better:

Day 40: 410
Day 50: 328
Day 60: 262
Day 70: 210

This is exponential *decay,* the mathematical signature of an epidemic that's been licked.

That one number—the ratio of each term of the geometric progression to the previous one—means a lot. When it's bigger than 1, the virus rapidly spreads to a sizable proportion of the population. If it's smaller, the epidemic dwindles and dies. In epidemiology circles it's called R_0.* In the spring 1918 wave of the Spanish flu, R_0 is estimated to have been 1.5. For the mosquito-borne Zika virus in 2015–16, it was around 2. For measles, measured in Ghana in the 1960s, it was 14.5!

An epidemic with a small R_0 looks like this:

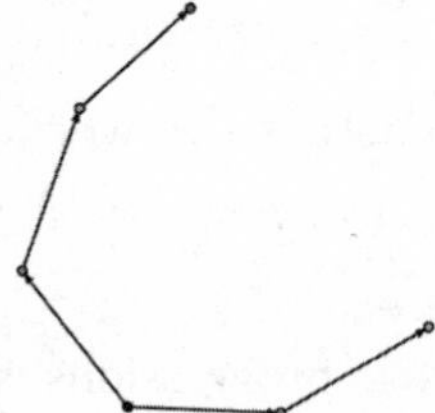

Most people, if they infect anyone at all, infect just one other person, and the chain of infection typically dies out before it spreads too much. When R_0 is a little bigger than 1, you start to see some branching out:

* Pronounced "R nought," as in "You R nought worried enough about the next pandemic."

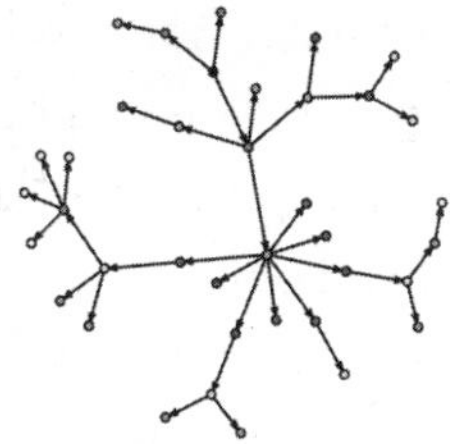

And when R_0 is substantially larger than 1, you see rapid exponential growth, an outbreak constantly splitting off new branches and extending ever further into the population.

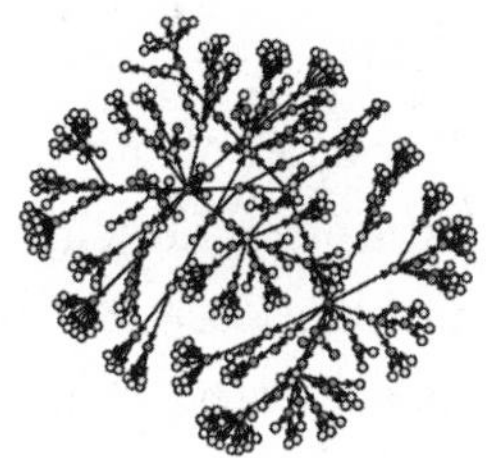

If the disease confers immunity once you've caught it, those branches never cycle back and attach to a person who's already been sick, which makes the epidemic network a kind of geometry we've already seen: a tree.

The existence of this fundamental threshold at $R_0 = 1$ was central to Ross's ideas about epidemics. Ross's discovery that mosquitoes were carrying malaria was a tremendous advance, but it also created a certain amount of pessimism. It's easy to kill mosquitoes, but it's hard to kill *all* the mosquitoes. So you might think it's hopeless to stop the spread of malaria. Not so, Ross insisted. As long as there are anopheles mosquitoes around, some of them are going to drink from a malaria-infected human, and then, having flitted around a bit, bite someone else who doesn't have malaria yet. So the disease keeps transmitting itself. But if the density of mosquitoes is low enough, that magic number R_0 drops below 1, which means there are fewer and fewer new cases each week, and the epidemic exponentially decays away. You don't have to stop all transmission; you just have to stop *enough* transmission.

This was the idea Ross was promoting at the St. Louis exposition in

1904. His argument on the random walk was meant to show that, after the number of mosquitoes in a region had been reduced, it would take quite some time for enough mosquitoes to wander into the area to push it back over the epidemic threshold.

That's a key idea for the battle against COVID-19, too. We don't need to eliminate every transmission of the disease, which is a good thing, since that's impossible. Epidemic control is not about perfectionism.

77 TRILLION PEOPLE WILL CATCH SMALLPOX NEXT YEAR

In the spring of 2020, at the outset of the COVID-19 pandemic in the United States, the disease was clearly tracing out the kind of geometric progression you don't want to see. Cases of COVID-19 were growing by about 7% every day. That meant every week the cases were getting multiplied by 1.07 seven times, which amounts to a 60% increase. If that's how things were going, 20,000 confirmed cases per day at the end of March would turn into 32,000 in the first week of April, 420,000 in the middle of May. A hundred days later, in early July, there would be 17 million new cases every single day.

You see the problem here. You can't keep up a pace of 17 million new infections per day, because in less than three weeks that adds up to more infected Americans than there are Americans. It was reasoning of this too-casual kind that led an intrepid group of post-9/11 modelers headed by the CDC's Martin Meltzer to project in 2001 that an intentional release of smallpox in the United States could lead, in just a year, to 77 trillion infections. ("Every now and again, Dr. Meltzer loses control of his computer," one colleague remarked.)

Something about our geometric progression story is wrong.

Let's go back to the magic number R_0, which measures how many new infections each infected person generates. R_0 is not a constant of nature. It depends on the biological features of the particular infection (which may itself vary between different strains), and it depends on how many people each infected person encounters during their contagious period, which depends on how long that contagious period is (can we shorten it with appropriate treatment?) and it depends on what happens during those encounters. Were people standing close to each other or six feet apart, as current guidelines recommend? Wearing masks or not? Outdoors or in a poorly ventilated building?

But even if *nothing* about the disease or our behavior changes, R_0 changes with time.* The virus simply starts running out of new people to infect. Let's say we reach the point where 10% of the population has already been infected. The sufferer blithely and asymptomatically gamboling through his usual routine may still cough on the same number of people as he did before, but now one in every ten of those people either is already sick or has recovered, and is thus immune from reinfection.† So, on average, instead of infecting two people over the course of his contagiousness, he infects only 90% of that number, about 1.8. When 30% of the population is infected, R_0 goes down to $(0.7) \times 2 = 1.4$. And when it's 60%, R_0 becomes $(0.4) \times 2 = 0.8$, and we've crossed the critical line. Instead of R_0 being a little bigger than 1, it's a little smaller, and just like that we're riding the good kind of geometric progression instead of the bad one.

In fact, the proportion of infected people may not even make it to 40%. Because whatever that proportion is—let's call it P—our new R_0 is

$$(1 - P) \times 2$$

and when that number hits 1, the epidemic flips to exponential decay. That happens when $1 - P$ is 1/2, which means P is also 1/2; so an epidemic with a "natural R_0" of 2 will start to fade out once half of the population is infected. They call this "herd immunity." Once enough

* Strictly speaking, the name "R_0" refers to the average number of new cases per case in a population where nobody's had the virus yet, and what we call this number that's changing with time is "R" or sometimes "R_t," but lots of people talk about R_0 changing as the epidemic develops and unless you're going to start writing papers in mathematical epidemiology based only on what you learn from this book, it's probably okay not to make the distinction. Also, don't write papers in mathematical epidemiology based only on what you learn from this book.

† Or so we hope. We don't really know yet whether getting COVID-19 and recovering confers long-term immunity; without this assumption, you get a different long-term scenario, which is, as you might imagine, not as nice.

people are impervious to a disease, an epidemic can't sustain itself. But how much "enough" means depends on the original R_0. If it's 14, like measles, you need (1 – P) = 1/14, which means 93% of the population needs to be immune; that's why even a small number of kids skipping their measles shot leaves the general population vulnerable to outbreaks. For a disease with a more modest R_0 of 1.5, the turnaround comes at 33% infection. And if we're right that COVID-19 has an R_0 between 2 and 3, then the current pandemic will start burning itself out on its own once half to two-thirds of the world has come down with it.*

But that's a lot of people, a lot of sickness, and a lot of death. So the world's epidemiologists, while they differ in many material particulars, are basically unanimous in saying no, we should not just let this thing run its natural course, no, no, *no*.

THE GAME OF CONWAY

It's easy, especially if math is your thing, to think of a pandemic as *really being* a curve drawn on graph paper or a screen, the numbers just abstract quantities varying in time. But they represent individual people, people who have gotten sick from the disease or died of it. You have to stop every so often and think about those people. One of them was John Horton Conway, who died of COVID-19 on April 11, 2020. He was a geometer—well, he was a lot of things, but almost all the mathematics he did somehow involved drawing pictures.

I knew Conway when I was a postdoctoral fellow at Princeton. I asked him questions about mathematics all the time. He always had a long, informative, and illuminating answer. It was never the answer to the question I had asked. But I learned a lot all the same! He wasn't being willfully difficult; it was just the way his mind worked, more associative than deductive. You asked him something and he told you what your question reminded him of. If there was a particular piece of information you needed, a reference or a statement of a theorem, you were in for a long,

* That threshold might be lower, though probably not *radically* lower, for reasons having to to with "heterogeneity"; not everyone infects the same number of people. More on this in chapter 12.

circuitous trip, destination unknown. His office was overrun with funny puzzles, games, and toys, which were in a way recreational but which were part of his math, too. He seemed never to not be thinking about mathematics. He was once struck dead still in the middle of the street with the idea for a theorem in group theory, and was knocked over by a truck. Forever afterward, he called that theorem "the murder weapon."

All working mathematicians experience mathematics as a kind of play, but Conway was singular in his insistence that play could be a kind of mathematics. He was a compulsive inventor of games, which he liked to give funny names to: Col, Snort, ono, loony, dud, sesqui-up, Philosopher's Football. But fun was never just for fun. He made theory out of fun. We've met his mathematical gaming already in this book: it was Conway who developed the notion that a game like Nim is a kind of number, an idea his colleague Donald Knuth wrote about in a 1974 book with the extremely 1974 title *Surreal Numbers: How Two Ex-Students Turned On to Pure Mathematics and Found Total Happiness*. The book is styled as a dialogue between two students who come upon a sacred text outlining Conway's theory: "In the beginning, everything was void, and J. H. W. H. Conway began to create numbers . . ."

And it was Conway, too, who in in the late 1960s was the first to write down a list of all the knots that can be drawn on a sheet of paper with eleven or fewer crossings between strands; he accomplished this by inventing his own notation (he invented a lot of his own notation) for little pieces of the knot where two strands intertwined, which he called "tangles"—here are some of them:

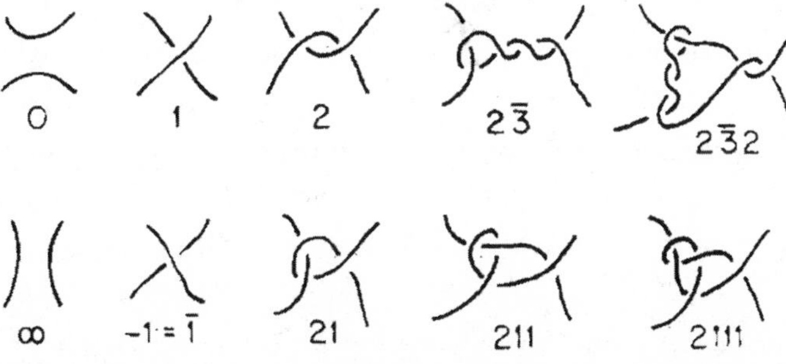

FIG. 1

One of the knots in his census is the knot that later acquired his name, the one the neural net warned was hard to understand, and the one Lisa Piccirillo proved a theorem about nonetheless.

Conway is probably most famous in the world outside theoretical math for the Game of Life, a simple algorithm that produces fabulously complex ever-changing patterns, which almost seem to be developing organically—whence the name.* But he hated being known for the game, which he saw (correctly) as much less deep than much of his other mathematics. So instead of ending there, I'll tell you one of my favorite theorems of his, a really geometric one he proved with Cameron Gordon in 1983. Take any six points in space. There are ten different ways to break up the six points into two groups of three. (Check!) For each such partition, you can join up each of the two triples of points to form two triangles. What Conway and Gordon proved is that there's at least one way to do this so that the triangles are linked together, like loops in a chain.

Maybe even more charming to me than the fact itself is the method of proof. What Conway and Gordon really prove is that, of the ten ways to partition the six points, the number that yield linked triangles is odd. But zero is even! So there must be at least one partition where the triangles are linked. It seems very weird to prove a thing exists by proving that the number of such things there are is odd, but it actually turns out to be pretty common. If you come into a room with a toggle light switch

* One fan of the game was Brian Eno, who saw a demonstration at a science museum in San Francisco in 1978 and became "completely addicted" to it, watching the patterns flow and move for hours at a time. Two years later he would cowrite "Once in a Lifetime." You may ask yourself. . . .

and the light isn't how you left it, you know someone has flipped the switch; but the *reason* you know that is that the state of the light tells you it's been flipped an odd number of times.

WHITE PEOPLE ARE OLD

Not everyone faces identical risks from COVID-19. The risk of serious symptoms, hospitalization, and death is much higher among older people, much lower among the young and middle-aged. In the United States, there are racial and ethnic differences, too. As of July 2020, confirmed cases of COVID-19 in the U.S. broke down along racial lines like this:

34.6% Hispanic
35.3% non-Hispanic white
20.8% Black

The distribution of *deaths* from COVID-19 looked different.

17.7% Hispanic
49.5% non-Hispanic white
22.9% Black

These numbers are startling on their face if you know anything about health disparities in the U.S., which in almost all cases involve differentially bad health outcomes for people of color. But white people, who made up only 35% of all confirmed COVID-19 cases, made up 49.5% of all COVID-19 deaths. So among the white subpopulation, a COVID-19 case was substantially *more* likely to be fatal than a COVID-19 case in the general population. Why?

The answer, as I learned from mathematician and writer Dana Mackenzie, is age. White people with COVID-19 are more likely to die of COVID-19 because old people with COVID-19 are more likely to die of COVID-19, and white people, in the aggregate, are old. If you break cases down by age groups, things look really different. Among Americans between eighteen and twenty-nine, the "Spring Break COVID Party" set, white people made up 30% of COVID-19 cases but just 19% of the deaths. Among people eighty-five and up, 70% of COVID-19 cases and 68% of deaths were white people. In fact, within *every single* age band of adults recorded by the CDC, a COVID-19 case in a white person was less likely to be fatal than it would be for the typical American that age. Yet, when you combine the groups together, the disease appears to be falling harder on white people. This phenomenon is called *Simpson's paradox*, and you have to watch for it with steel eyes whenever the phenomenon you're studying affects a heterogeneous population. "Paradox" isn't the right name for it, really, because there's no contradiction involved, just two different ways to think about the same data, neither of which is wrong. Is it incorrect, for instance, to say that COVID-19 has hit Pakistan less hard than it has the United States, because Pakistan has a younger and thus less vulnerable population? Or is the right comparison the likelihood that an elderly Pakistani will fall to COVID-19 relative to that person's American coeval? The lesson of Simpson's paradox isn't really to tell us which viewpoint to take, but to insist that we keep both the parts and the whole in mind at once.

WHICH COIN HAS SYPHILIS?

One thing people agreed on from the beginning: there's no way to avoid the direst possible pandemic futures without testing, a lot of testing, much more testing than we were for a long time able to do. The more tests we have, the better we know what kind of progression COVID-19 is following and where we stand on it.

Here's another math chestnut: You have sixteen gold coins. Fifteen of them are honest coins with an ounce of gold each, but one is a shaved-down counterfeit that weighs just 0.99 of an ounce. You have a very accurate scale, but it costs a dollar every time you use it. How can you most inexpensively find the impostor?

Spending $16 to weigh every coin would certainly do the trick, but that's expensive. In fact, it's *unnecessarily* expensive; if you had the bad luck to weigh fifteen coins and find them all legit, you would know, without having to shell out another dollar, that the sixteenth coin was fake. So there's no need to spend more than $15.

You can do better, though. Split the coins into two groups of eight and weigh just the first group. The total weight is either 7.99 ounces or the full 8; and whichever it is, you know which group has the counterfeit coin. So now you've narrowed down to eight coins. Split those coins into two groups of four and weigh one group, and you've narrowed it down to four with only $2 outlay. And by doing two more splits, bringing your total outlay to $4, you can pin down the fake coin for sure.

Like a lot of word problems, this one relies on some artifice to make the story work; in real life, weighing things on a scale isn't expensive!

But biological assays are, which brings us back to infectious disease. Suppose instead of sixteen coins you had sixteen army recruits. And suppose instead of one of them being slightly lighter than the rest, one of them had syphilis. In the World War II era, this was a serious issue: a 1941 *New York Times* article blamed "a great band of panzer prostitutes operating in mechanized units among the roadhouses and juke joints from Chicago to the Dakotas" for the thousands of soldiers infected with syphilis or gonorrhea: "at large, untreated, infectious, and menaces to their fellow citizens."

You can find the menaces by testing the men's blood, one by one, with a Wasserman test. That's fine for sixteen recruits, but not so good for sixteen thousand. "The inspection of the individual members of a large population is an expensive and tedious process," is how Robert Dorfman put it. Dorfman was a well-known Harvard economics professor who in the 1950s and '60s pioneered the application of mathematical models to problems of commerce. But back in 1942, he was a U.S. government statistician, six years out of college, where he had decided to focus on mathematics after concluding he had no future in his first-choice avocation, poetry. Quoted above is the first sentence of his classic paper, "The Detection of Defective Members of Large Populations," which introduced the idea of the coin puzzle to epidemiology. You can't use *exactly* the same strategy that worked for coins; half of sixteen thousand soldiers is still a lot of soldiers! But suppose, Dorfman suggests, you break the recruits up into groups of five. Then you blend blood from each group into a little serum cocktail and test it for syphilitic antigen. No antigen means you can tell all five members of the group they're clean; but if the test comes back positive, you call those five recruits back and test them one by one.

Whether this is a good idea depends on how common syphilis is in the population. If half the troops are poxy, almost all those grouped samples will come back positive, and you're going to test everybody twice; detecting the defective members is even more expensive and tedious than before. But what if only 2% of recruits have syphilis? The chance that any given sample comes back negative is the chance that all five people tested are syphilis-free, which is

$$98\% \times 98\% \times 98\% \times 98\% \times 98\% = 0.90.$$

If there are sixteen thousand soldiers, you have 3,200 groups; of those, 2,880 get cleared, leaving 320 groups consisting of 1,600 soldiers you have to go back and test one by one. So in all you ran the test 3,200 + 1,600 = 4,800 times, a big saving over testing 16,000 soldiers one by one! And you can do even better; Dorfman works out that with a 2% prevalence rate, the optimum group size is eight, which brings you down to about 4,400 tests.

The relevance to coronavirus is clear: if we don't have enough tests to test everyone one by one, maybe we can swab seven or eight people's nostrils, put the specimens all in a single container, and test them all at once.

Caveat: the Dorfman protocol for detecting syphilis was never actually used. Dorfman wasn't even working for the army; he was at the Office of Price Control when he and his colleague David Rosenblatt hatched the idea of group testing for syphilis, the day after Rosenblatt reported for his own induction and had his Wasserman test. But it turns out the idea didn't work in practice; diluting the sample made it too hard to detect the trace of antibody that remained.

The coronavirus is a different story. The polymerase chain reaction test that detects it multiplies even a tiny trace of viral RNA by a huge factor. That makes group testing feasible—and, in situations where prevalence of the disease is low and testers and equipment are in short supply, very appealing. There was group testing in Haifa and in German hospitals, and a state lab in Nebraska tested 1,300 specimens a week in pools of five, reportedly cutting the number of tests they needed in half. Wuhan, the city in central China where the pandemic began, used pooled samples to test almost 10 million people in a matter of days.

The people who really know group testing are the veterinarians, who have to identify small outbreaks in large, densely packed groups of livestock swiftly and accurately. They sometimes assess hundreds of samples with a single test. A veterinary microbiologist I know told me there was no reason their protocols couldn't be used to rapidly test people for coronavirus, though some of the implementation would have to be modified. "You can't put a thousand people on a conveyor belt and rectally sample each one as they come by," he told me—a little ruefully, I thought.

BLOOP-BLOOP

We're ready now to really get our hands into Ross and Hudson's theory of happenings, as applied to the spread of a pandemic. We need to start by making up some numbers. (A real epidemiologist would estimate these numbers as best they can, a process which gets less and less similar to "making up some numbers" as the pandemic progresses and we learn more about the dynamics of the disease.) On day one of our attempt to chart the virus's course, let us say 10,000 out of our state's population of 1 million are infected, while the remaining 99% of the population is still susceptible to infection. So:

susceptible (day 1) = 990,000
infected (day 1) = 10,000

If I keep typing "susceptible" and "infected" again and again the words are going to swim and lose their meaning, so I'm going to switch to S and I for short: S(day 1) = 990,000 and I(day 1) = 10,000.

Each day, some new people get infected. Let's say each infected person, on average, coughs on someone once every five days, or 0.2 people per day. And the chance that the coughed-on person is susceptible to infection is the fraction of the population that's susceptible, which is S/1,000,000. So the number of new infections you expect is (0.2) times I times S/1,000,000.

Every infection decreases the number of susceptible people:

$$S(\text{tomorrow}) = S(\text{today}) - (0.2) \times I(\text{today}) \times S(\text{today}) / 1{,}000{,}000$$

and increases the number of infected people:

$$I(\text{tomorrow}) = I(\text{today}) + (0.2) \times I(\text{today}) \times S(\text{today})/1{,}000{,}000$$

except we're not done yet, because—fortunately!—people who get sick get better. Time to make up one more number. Let's say the period of infectiousness lasts ten days, so that on any given day, one in ten of the currently infected people recover. (This means that each infected person, over their ten days of contagiousness, should infect about two people; so R_0 is 2.) Then we actually have

$$I(\text{tomorrow}) = I(\text{today}) + (0.2) \times I(\text{today}) \times S(\text{today})/1{,}000{,}000 - (0.1) \times I(\text{today})$$

This kind of rule is called a *difference equation,* because what it tells us is exactly the difference between the situation tomorrow and the situation today. If we can compute that every day, we can project the pandemic as far forward as we like. You should think of that lump of algebra as a machine, ideally one with a lot of lights and bloopy sounds. You put today's situation into the machine and it goes bloop-bloop and you get tomorrow's situation. Then you pick that up and stuff it back in the hopper and out comes the day after tomorrow, and so on.

On day two, the number of new infections is

$$(0.2) \times I(\text{day } 1) \times S(\text{day } 1)/1{,}000{,}000 = (0.2) \times (10{,}000) \times (990{,}000/1{,}000{,}000) = 1{,}980$$

so

$$S(\text{day } 2) = S(\text{day } 1) - (0.2) \times I(\text{day } 1) \times S(\text{day } 1)/1{,}000{,}000 = 990{,}000 - 1980 = 988{,}020$$

There are 1,980 new infected people on day two, but also one-tenth of the currently infected people, or 1,000, get better.

$$I(\text{day } 2) = I(\text{day } 1) + (0.2) \times I(\text{day } 1) \times S(\text{day } 1)/1{,}000{,}000 - (0.1) \times I(\text{day } 1) = 10{,}000 + 1{,}980 - 1000 = 10{,}980$$

Now we know the story at day two; put that in the machine and out comes the projection for day three.

$$S(\text{day } 3) = S(\text{day } 2) - (0.2) \times I(\text{day } 2) \times S(\text{day } 2)/1{,}000{,}000 = 988{,}020 - 2{,}169.69192 = 985{,}850.30808$$

$$I(\text{day } 3) = I(\text{day } 2) + (0.2) \times I(\text{day } 2) \times S(\text{day } 2)/1{,}000{,}000 - (0.1) \times I(\text{day } 2) = 10{,}980 + 2{,}169.69192 - 1{,}098 = 12{,}051.69192$$

That 69.192% of a person is a good reminder that we're just doing probabilistic projection here, making a best guess; we should not expect correctness down to the last decimal point!

You can keep this up as long as you're willing to turn the crank on the machine. The number of infected people day by day (rounding off, because who has time for that many decimal places) is

10,000, 10,980, 12,052, 13,223, 14,501, . . .

which you can check is very close to a geometric progression, increasing by about 10% each day. But it's not *quite* a geometric progression; that rate of increase is very slightly slowing. The number 10,980 is 9.8%

more than 10,000, but 14,501 is only 9.7% more than 13,223. That's not a rounding error; that's the effect of the susceptible population shrinking, providing the virus fewer opportunities to make more of itself.

You probably don't want to see page after page of S(day this) and I(day that), any more than I want to type them. Carrying out cumbersome but purely repetitive calculations like this is what computers are for. You can run this machine with just a few lines of code, and get a projection as many days into the future as you like. I did that and I got this picture:

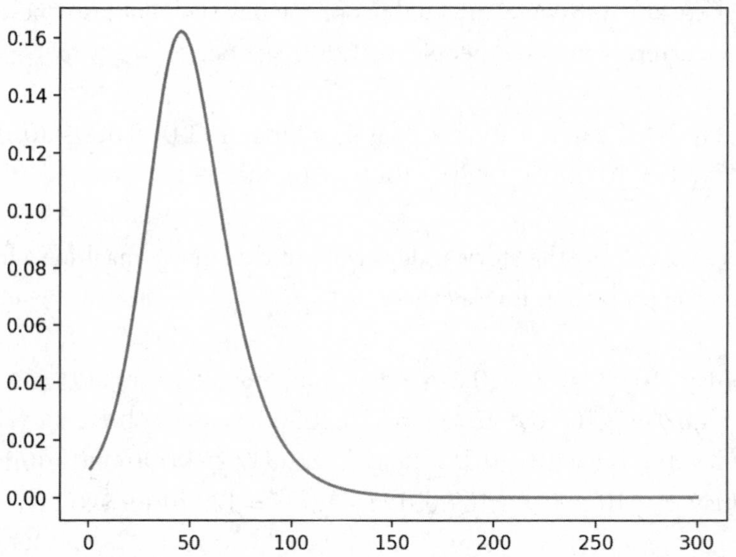

The infection peaks on day forty-five, when just over 16% of the population is infected. At that point, about 34% of the population has already recovered from the disease,* and just about exactly half is still susceptible. So R_0, which started at 2, is cut in half, and is now 1; exactly the threshold at which new infections start to drop. Though it's not quite visible in this picture, the drop-off in a model like this is typically not as steep as the ramp-up; it took forty-five days to get from 1% infected to the peak, but another sixty days to get down to 1% again.

* A more involved version of this model would accept the grim reality that some infected people are going to die instead of recover. For COVID-19, the proportion of deaths is thankfully small enough that it's okay to run the model without this for starters.

Nowadays, scientists usually ascribe this model of disease not to Ross and Hudson, but to Kermack and McKendrick. Anderson McKendrick, the recipient of Ross's letters about unlocking the door to a new science of epidemics, was, like Ross, a mathematically minded Scottish doctor; he had served with Ross in Sierra Leone. William Ogilvy Kermack was yet a third mathematical Scottish doctor, blinded as a young man in a laboratory accident by caustic alkali, but, like Hudson, possessed of a tremendous geometric intuition. He went nowhere without his heavy wooden stick, whose tapping was a familiar sound at the Royal College of Physicians Laboratory in Edinburgh, although, when it suited him, "he also had the habit of hooking his stick over his arm and of appearing silently and unexpectedly, sometimes indeed inconveniently, at the elbow of one of his assistants." Kermack and McKendrick, in their 1927 paper on the subject, acknowledge Ross and Hudson's prior work; but their paper, besides adding important new ideas, is simpler, written in less obscure notation, and altogether more usable. We call it the *SIR model*—the *S* and the *I* are the numbers we've been discussing, and the *R* stands for "recovered"—that part of the population which is, for the moment, immune. More complicated models have more compartments to put people in, and correspondingly more letters in their names.

Just as Ross had hoped, the mathematical understructure he had helped set up to study disease spread has been useful for understanding all kinds of human happenings. Nowadays we use SIR models for other contagious things, too, like tweets. In March 2011, the Tōhoku earthquake and the following tsunami destroyed the Fukushima nuclear power plant and drowned thousands of people in northeastern Japan. Panicked people shared information on Twitter, not all of it sound. There were rumors that rain would be dangerous to the touch. A widely shared tweet: "For prevention of side effects from radioactivity, it is good to drink mouthwash including iodine and to eat as much seaweed as you can." These rumors, even when they started from users with few followers, spread rapidly, as did corrections from scientific authorities. A rumor is a lot like coronavirus. You can't share it unless you've been exposed to it, and there's a degree of immunity—after you've caught it once, further encounters with the infectious agent aren't likely to touch off a new round of sharing. So it makes sense that researchers in Tokyo found the SIR model did a pretty decent job of modeling the spread of earthquake rumor tweets. You could call the average number of times a rumor is shared per person who sees it the "R_0" of the rumor. A moderately interesting rumor has a low R_0, like flu; a really juicy one is more like measles. We call the latter kind of rumor "viral," but the truth is, all rumors are viral! It's just that some viruses are more infectious than others.

LAGHU LAGHU LAGHU LAGHU

Difference equations aren't just for modeling disease. They underlie a multifarious zoo of sequences of mathematical interest. You like arithmetic progressions? You can get one by taking the difference to be a fixed number:

S(tomorrow) – S(today) = 5

which, if you start at 1, gives

1, 6, 11, 16, 21 . . .

If you want a geometric progression, you take the difference to be proportional to the current value, say:

S(tomorrow) – S(today) = 2 × S(today)

which gives

1, 3, 9, 27, 81, . . .

a sequence in which each term is triple the previous one. You can make any difference equation you want! Maybe for some reason you want the difference to be the *square* of the current value:

$$S(\text{tomorrow}) - S(\text{today}) = S(\text{today})^2$$

which gives a very swiftly growing sequence indeed:

1, 2, 6, 42, 1,806 . . .

This is not any kind of progression known to Plato, but many more progressions than that are known to the On-Line Encyclopedia of Integer Sequences, which is both a critical research tool and a fabulously successful procrastination device for every mathematician I know. The combinatorist Neal Sloane* started the project as a graduate student in 1965 and has been developing it ever since, first on punched cards, then as a book on paper, and now in online form. You give the machine a list of whole numbers, and it tells you everything the math world has ever figured out about it. The sequence above, for example, is sequence A007018 in the OEIS, from which entry I learn that the nth term of the sequence is the "number of ordered trees having nodes of outdegree 0,1,2 and such that all leaves are at level n." (Trees again!)

If you want to dress this up a little more (and in disease modeling with any pretense to realism, you probably do), you can make the difference between today and tomorrow depend not only on what happened today, but what happened yesterday. Try

$$S(\text{tomorrow}) - S(\text{today}) = S(\text{yesterday})$$

In order to get started, we need *two* days' worth of data. If today is 1 and yesterday was 1, tomorrow is going to be 1 more than 1, or 2. One day later, S(today) is 2 and S(yesterday) is 1, so S(tomorrow) is 3. The sequence continues

1, 1, 2, 3, 5, 8, 13, 21, . . .

* Also a collaborator of John Conway's, on the geometric problem of packing very high-dimensional oranges as tightly as possible into a very high-dimensional box.

with each term the sum of the two before it. This is the Fibonacci sequence, aka A000045, so renowned that it literally has an entire mathematical journal devoted to it.

It might not be clear what kind of real-world process would produce a difference equation like Fibonacci's. Fibonacci himself produced it, in his book *Liber Abaci* of 1202, from a thoroughly unconvincing biological model of multiplying rabbits. But there's a better, older way! I learned it from Manjul Bhargava, who is not only a famous number theorist but a serious student of classical Indian music and literature. He plays a mean tabla, and he knows his Sanskrit poetry. As in English, the metric structure of Sanskrit poetry is controlled by different types of syllables. In English poetry, we typically keep track of the patterns of stressed and unstressed syllables, which are called *feet*; a foot could be something like an *iamb*, where an unstressed is followed by a stressed (ba-DUM, or if you like, "To BE, or NOT to BE") or a *dactyl*, a stress followed by two unstressed (JUGG-a-lo, or "THIS is the FOR-est pri-ME-val"). In Sanskrit poetry, the key distinction is between a *laghu* (light) and a *guru* (long) syllable, and a guru is twice as long. A meter, or *mātrā-vrrta*,* is a sequence of laghus and gurus adding up to some fixed length. If that length is two, for example, there are only two possibilities: a pair of laghus, or a single guru.

In English, there are four ways you can put together two syllables: "ba-DUM," which is an iamb; "BUM-bum," which is a trochee; "DUN-DUN," a *spondee*; or the totally unstressed "bum-bum," which I have just learned is called a *pyrrhus*.† If you have three syllables to work with, each of those four possibilities spawns two more: a *trochee*, for instance, can be followed by an unstressed syllable, to form a dactyl, or by a stressed syllable, which gives you a rarely used English metrical foot called the *cretic*. ("Why ask why? Try Bud Dry" is probably the most

* A good reminder that Sanskrit is an Indo-European language, which shares a common ancestor with English and the Romance languages. "Mātrā" means measure, an English word it sounds a lot like (not to mention "meter"); "vrrta" comes from the proto–Indo-European root *wert, which means "turn," and which also gives us the English "verse." And laghu and guru are cousins to the English words "light" and "grave."

† You've never heard of poetry written in "pyrrhic pentameter" because nobody does it. Edgar Allan Poe, who knew his way around tightly metered verse, said, "The pyrrhic is rightfully dismissed. . . . The insisting on so perplexing a nonentity as a foot of two short syllables, affords, perhaps, the best evidence of the gross irrationality and subservience to authority which characterise our Prosody."

well-known example in contemporary American verse.) So there are eight possibilities for a three-syllable meter, sixteen for four syllables, thirty-two for five syllables, and so on.

Sanskrit is more complicated. There are three different meters of length 3:

laghu laghu laghu
laghu guru
guru laghu

and five of length 4:

laghu laghu laghu laghu
laghu guru laghu
guru laghu laghu
laghu laghu guru
guru guru

The same problem, in musical terms: How many ways could you put quarter notes and half notes together, with no rests, to fill up a measure in 4/4 time?

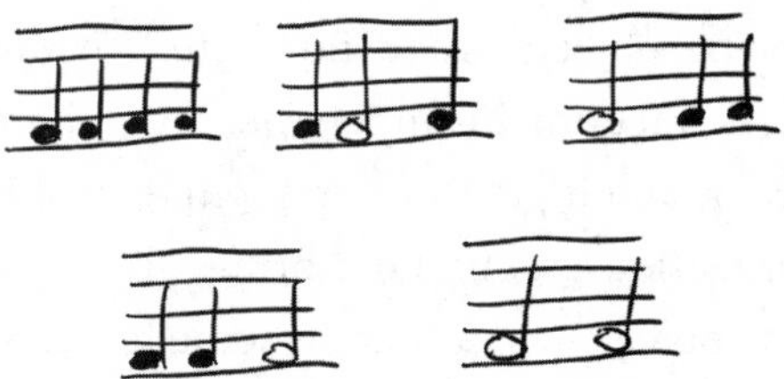

How many variations are there when the mātrā-vrrta has length 5? The order in which I wrote the possibilities above should give us a clue. The meter might end in a laghu, which means that a length-four meter comes before it; there are five of those. Or the meter could end in a guru, which uses up two units of length, so what comes before it has length three; there are three possibilities. The total number of variations is 5 + 3 = 8, the sum of the two previous terms. And now we are back to the Fibonacci sequence, or, as Bhargava likes to call it, the Virahanka sequence, after the great literary and religious scholar who first computed these numbers, five centuries before Fibonacci ever contemplated rabbits going at it.

LAWS OF HAPPENINGS

With the SIR model we've departed from the strict geometric progression, but not from the philosophy that what happens today will happen tomorrow. We just have to interpret it a bit more broadly. In an arithmetic progression, the increase each day is the same. In a geometric progression, the increase each day is different, but is the same when considered as a *proportion* of the number today. The rule for figuring out each day's increase is the same tomorrow as it was today. And under our slightly more dressed-up model, what happens tomorrow is whatever the bloop-bloop machine makes out of what happens today. The rate of growth may differ from day to day, but *it's always the same machine.*

Taking this outlook makes us heirs to Isaac Newton. His first law asserts that an object in motion will keep moving at the same speed and in the same direction, unless some force is applied to it. Tomorrow's motion is the same as today's.

But most moving objects we're interested in don't groove along through frictionless vacuum in an eternal line. Throw a tennis ball straight up in the air, and it goes up for a while, peaks, and goes down, kind of like the infection graph. That brings in the *second* law, which tells us how objects behave when there *is* some force, like gravity, acting on them.

From a pre-Newtonian perspective, the tennis ball's behavior is constantly changing. But the nature of the change never changes! If we know the ball's upward speed now, its upward speed a second later will be 16 meters per second less. For downward speed it's the opposite; the downward speed a second from now is 16 m/s *more* than it is right now.

If you want a more uniform way of saying this, you can (and should!) think of downward motion at 20 meters per second as upward motion at *negative* 20 meters per second. One second later, the velocity has dropped 16 m/s, so it's now –36 m/s. This phenomenon really confuses people when they first learn negative numbers; when you make a negative number less, it somehow feels like it's getting larger! To keep this straight, I like to use two different words: a number is *higher* if it's more positive, *lower* if it's more negative; *bigger* if it's farther from zero, *smaller* if it's closer to zero. Positive numbers get smaller as they go lower, but lowering a negative number makes it bigger.

The difference between the velocity now and the velocity a second from now is always the same 16 meters per second, because the force acting on the tennis ball is always the same: the Earth's gravity. That's another difference equation! The ball's velocity isn't constant from second to second, but the difference equation projecting its future course never changes. Throw the ball on Venus and you get a different difference equation;* but you still get one. What happens now happens again a second from now.

Unless, that is, you hit the ball! A model like this, by its very nature, predicts how a system behaves under conditions already established. A shock to the system, or even a mild nudge, changes those conditions, and takes you off the model's projection. And real systems are subject to all kinds of shocks. When there's a pandemic, we *don't* let it burn through the population—we take steps! That doesn't make models useless. If we want to know what happens to the tennis ball after we hit it, we'd better have a very solid understanding of how the ball moves under gravity alone. Models of disease can't predict the future because they can't predict what we'll do. But they can definitely help us decide *what* to do, and when we need to do it.

A TIPPING IN EVERY POINT

Data about COVID-19 comes to us day by day, not hour by hour or minute by minute. But the location of a thrown ball can be measured on

* The common difference would be 8.87 meters per second, the gravity at the surface of Venus being slightly weaker than ours.

much smaller time scales than a second. We could ask how the ball's velocity is changing every half second, or every tenth of a second, or every picosecond; most ambitiously of all, we might want to describe something like the instantaneous rate at which the velocity is changing, the speed of the change of the speed. Newton had that handled as well. The whole point of his theory of fluxions, which we now call differential calculus, is to make sense of questions like that. We won't go into it here except to say that a difference equation, shaved down to infinitesimal time increments to adequately describe continuous change, takes on a new name: it's called a *differential equation*. Any physical system whose evolution in time can be described in terms of its current state is governed by a differential equation. Tennis balls on Venus, water flowing through a tube, heat diffusing through a metal bar, satellites orbiting planets orbiting suns: each has a differential equation of its own. Some are easy to solve in explicit terms, others are hard, most are impossible.

The language of differential equations was what Ross, Hudson, Kermack, and McKendrick used in their models. Ross had left St. Louis by the time Henri Poincaré delivered his lecture on the final day of the 1904 exposition, but had he seen it, he might have gotten a ten-year head start on his work on epidemics. Poincaré told his audience that day:

> What did the ancients understand by a law? It was to them an internal harmony, statical as it were, and unchangeable; or else a model which nature tried to imitate. To us a law is no longer that at all; it is a constant relation between the phenomenon of to-day and that of to-morrow; in a word, it is a differential equation.

The differential equations that Ross and Hudson applied to pandemics have "tipping point" behavior; there's a threshold level of immunity, the herd immunity point, which divides two very different kinds of behavior. A disease introduced into a population with immunity below that level will explode exponentially, at least at first. But if the population has immunity above that point, the disease dies out. The dynamics of two bodies in space obey a simple dichotomy, too; they orbit each other stably in an ellipse, or they fly apart from each other on a hyperbolic path. But the switch from two bodies to three turns out to generate a fantastic range of new dynamical possibilities. These were the differential equations Poincaré had wrestled with in the work on the three-body problem that made his name. The complex behavior Poincaré described was to be the beginning of a new field, chaotic dynamics. When there's chaos, the tiniest perturbation to the present condition of a system can give it a drastically different future. *Every* point is a tipping point.

Poincaré already knew what Ross was to learn: that differential equations were the natural language for any attempt to create something like a Newtonian physics of disease—or, given Ross-like ambitions, a physics of all happenings. Tomorrow's happening depends on today's.

Chapter 11

The Terrible Law of Increase

On May 5, 2020, the White House Council of Economic Advisers posted a chart showing deaths from COVID-19 in the U.S. through early May 2020, together with several potential "curves" that roughly fit the data so far.

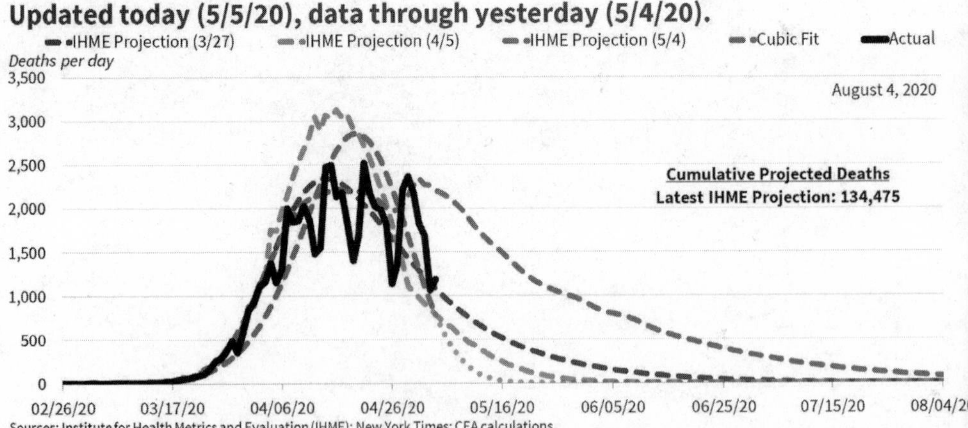

One of those curves, marked "cubic fit" in the chart, represented a stance of extreme optimism; it shows deaths from COVID-19 dropping to essentially zero in just two weeks' time. That curve was roundly mocked, especially once it came to be known that the "cubic fit" was the work of White House adviser Kevin Hassett. Hassett's biggest previous brush with fame was his coauthorship of the book *Dow 36,000*, published in October 1999, which argued that based on past trends the stock market was due for a tremendous near-term rise in prices. We know now what happened to the people who rushed to invest their life savings in Pets.com. The bull market stalled shortly after Hassett's book came out, then started to drop; it would take the Dow five years just to return to its 1999 high point.

The "cubic fit" curve was a similar overpromise. Deaths from COVID-19 in the United States decreased through May and June, but the disease was far from gone.

What's mathematically interesting about this story isn't that Hassett was wrong—it's *how* he was wrong. Understanding that is the only way we can learn strategies for avoiding this genre of wrongness in the future, beyond the limited-in-application "don't trust Kevin Hassett." To understand what went awry with the cubic fit we have to go back to the great British rinderpest outbreak of 1865–66.

Rinderpest is a disease of cattle, or was, until it was finally eradicated from the Earth in 2011, the culmination of a fifty-year program.* Water buffalo and giraffes can get it, too. It originated in central Asia, probably before recorded history, and was carried around the world by Huns and Mongols. Some hold it to be the fifth biblical plague suffered by the stubborn Egyptians. Sometime in the middle of the Middle Ages, a variant of the disease jumped the species barrier to humans; that spin-off virus is what we now call measles. Like measles, rinderpest is really contagious, which means it can tear through a population extremely fast. On May 19, 1865, a shipment of infected cattle arrived at the port of Hull in east Yorkshire. By the end of October, almost twenty thousand cows had

* Which makes it a great trivia question: "The only viruses ever eradicated from nature are smallpox and what?" Only works if no veterinary epidemiologists come to your trivia night.

fallen ill. Robert Lowe, a Liberal MP and later chancellor of the exchequer and home secretary, warned the House of Commons on February 15, 1866, in words that would have sounded uncomfortably familiar in 2020: "If we do not get the disease under by the middle of April, prepare yourself for a calamity beyond all calculation. You have seen the thing in its infancy. Wait, and you will see the averages, which have been thousands, grow to tens of thousands, for there is no reason why the same terrible law of increase which has prevailed hitherto should not prevail henceforth." (Lowe had an undergraduate degree in math and knew his way around a geometric progression.)

William Farr disagreed. Farr was a leading British physician of the mid-nineteenth century, the architect of the country's vital statistics office and an advocate for health reforms in the nation's crowded cities. If you've heard his name, it's probably in connection with the great success story of early epidemiology, John Snow's discovery of the source of the 1854 London cholera outbreak at the Broad Street water pump. Farr represented the British medical consensus on the wrong side of that argument, committed to the belief that cholera was spread not by a living organism but by a fermented miasma emanating from the filthy water of the Thames.

In 1866, it was Farr who was rowing against conventional wisdom. He wrote a letter to the London *Daily News*, insisting that the rinderpest, far from threatening to burn through the whole bovine population, was about to start guttering out of its own accord. "No one can express a proposition more clearly than Mr. Lowe," Farr wrote, "but the clearness of a proposition is no evidence of its truth. . . . It admits of mathematical demonstration that the law of increase which has hitherto prevailed, instead of implying 'that the averages which have been thousands will grow to tens of thousands,' implies the reverse; and leads us to expect that the subsidence will begin in the month of March." Farr went on to make specific numerical predictions, down to the very cow, for the cases of rinderpest for the five months to come. By April, he said, that number would be down to 5,226, and by June a mere 16.

Parliament ignored Farr's claim, and the medical establishment rejected it. The *British Medical Journal* ran a short, dismissive response:

"We will venture to say, that Dr. Farr will not find a single historical fact to back his conclusion that in nine or ten months the disease may quietly die out—may run through its natural curve."

They ventured wrong! This time, Farr had the best of the argument. Just as he'd predicted, cases declined through the spring and summer, and the outbreak was done by the end of the year.

Farr relegated his "mathematical demonstration" to a terse footnote, correctly guessing that the readers of the *Daily News* would rather naked formulas be shielded from their view. We need not be so discreet. But to see what Farr was doing, we have to go further back, to the beginning of Farr's career. In the summer of 1840, he submitted a report to the Registrar-General summarizing the causes and distribution of the 342,529 human deaths known to his office to have occurred in England and Wales in 1838: "a more extensive series," he convincingly boasts, "than has ever before been published in this or any other country." He records deaths from, among other things, cancer, typhus, delirium tremens, childbirth, starvation, old age, suicide, apoplexy, gout, dropsy, and something terrifyingly named "the worm-fever of Dr. Musgrave." He specially notes that the rate of tuberculosis (then called consumption) is higher among women than men, which he blames on the practice of corset-wearing. Here the recitation of statistics gives way to a passionate plea for reform: "*Thirty-one thousand and ninety* English women died in one year of this incurable malady! Will not this impressive fact induce persons of rank and influence to set their countrywomen right in the article of dress, and lead them to abandon a practice which disfigures the body, strangles the chest, produces nervous or other disorders, and has an unquestionable tendency to implant an incurable hectic malady in the frame? Girls have no more need of artificial bones and bandages than boys." (Farr doesn't reveal here—not directly, at any rate—that his wife had died of tuberculosis three years earlier.)

The report's final section, concerning the smallpox epidemic of 1838, is the one for which it is chiefly known today; it is here that Farr first addresses the progress of epidemics, which, in Farr's words, "suddenly rise, like a mist from the earth, and shed desolation on nations—to disappear as rapidly or insensibly as they came." It is Farr the statistician's goal to apply some numerical sense to the insensible, even if the ultimate origins of the epidemic can't be known. (He does mention, in a footnote, the theory that epidemics are caused by "minute insects transmitted from one individual to another, through the medium of the atmosphere," but sets aside that hypothesis on the grounds that the best microscopists of the day had failed to observe any such "animalcules.")

Farr recorded the deaths from smallpox, from one grim month to the next, through the epidemic's slow decline. The numbers looked like this:

4365, 4087, 3767, 3416, 2743, 2019, 1632

Farr guessed that, like many natural processes, the decline would follow the law of geometric progression, in which the ratio between each pair of consecutive terms is the same. The first ratio is 4,365/4,087 = 1.068. But the second is different: 4,087/3,767 is 1.085. The sequence of ratios looks like this:

1.068, 1.085, 1.103, 1.245, 1.359, 1.237

Those ratios clearly aren't the same each time, or even close; it looks they're growing (at least until the very last term), which violates the geometric law. But Farr wasn't ready to give up hunting for a geometric progression. What if the ratios themselves, stubbornly nonconstant, are actually themselves growing geometrically? This gets somewhat meta, because we are now asking whether the ratios *of the ratios* are always the same. Are they? You start with 1.085/1.068 = 1.016 and the sequence goes on in the following way:

1.016, 1.017, 1.129, 1.092, 0.910

I have to be honest with you here; that sequence doesn't look constant to me. But at the same time, it's not obviously increasing or decreasing; and for Farr, that was enough. By modifying the sequence a little bit, he was able to find a sequence of numbers

4,364, 4,147, 3,767, 3,272, 2,716, 2,156, 1,635

that matched the actual smallpox deaths pretty well, and where the ratios of ratios really did all share a common value, 1.046. (Does tweaking the numbers sound fishy? It's really not. Real-life data is messy and seldom—when it involves people, I'd say never—follows a precise mathematical curve to the nth decimal. Your goal is to find a law that's close enough.) And this rule of 1.046, Farr argued, matched the real data closely enough to be called the law of the epidemic.

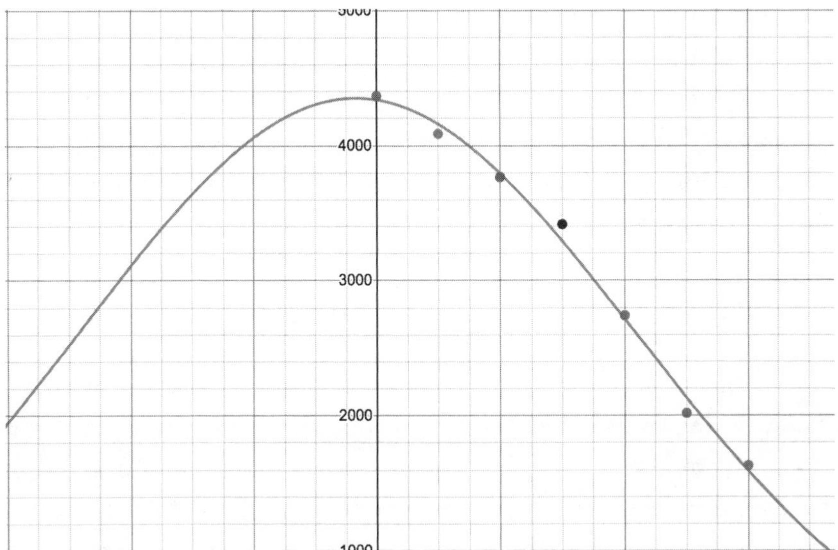

The curve here shows Farr's model for the course of smallpox, and the dots are the actual numbers of deaths each month; his nice smooth curve matches the real data reasonably closely.

You might be able to guess now what Farr must have done with the rinderpest data. But you're probably going to guess wrong! Farr had the sequence of rinderpest case counts for each of the first four months of the outbreak:

October 1865: 9,597
November 1865: 18,817

December 1865: 33,835
January 1866: 47,191

and found that the ratios of cases from month to month were 1.961, 1.798, and 1.395. If this were a geometric progression, the "terrible law of increase" Lowe had warned Parliament about, those numbers would all be the same. In fact, they were decreasing, signaling to Farr that some kind of diminution was already in effect. So he took the ratios of the ratios:

1.961/1.798 = 1.091
1.798/1.395 = 1.289

But Farr didn't stop here. Those ratios of ratios don't look constant; the second is notably bigger than the first. So he worked out the ratio of the ratios of the ratios!

1.289/1.091 = 1.182

Now the single ratio 1.182 is certainly a constant sequence, by virtue of only having one number in it. And Farr, confident as always, declared this number to be the law that governed all, the ratio of ratios of ratios that would guide the whole progress of rinderpest. Since the last ratio of ratios on the list was 1.289, the next would be 1.289 × 1.182, around 1.524. Which meant that what followed 1.961, 1.798, and 1.395 in the decreasing sequence of ratios should be 1.395/1.524 or 0.915. In other words, the disease was already due for a decrease! In February, Farr reasoned, you ought to see 0.915 × 47,191 or about 43,000 new cases.

You have my permission to feel a little queasy about Farr's argument here. Why does he get to assume the ratio of ratio of ratios will stay fixed at 1.185 for all future time? I'm not going to say that kind of assumption is justified, but it does have a history. Let me start by explaining how I won the neighborhood talent show.

THE GREAT SQUARE ROOTIO

Every January, in the heart of the cold dead Wisconsin winter, my neighborhood holds a talent show. Kids play violin, parents do goofy sketches. I computed square roots in my head, under the name the Great Square Rootio. And I won! Mental computation of square roots was a party trick I learned in college. Its social utility in that context was not as great as I'd anticipated. But I'm going to teach it to you anyway.

Say you're asked to compute the square root of 29. For the trick to work, you have to know your squares pretty well, because you need to be able to roll off your tongue that 5 squared is 25, and 6 squared is 36. Now consider a sequence of numbers:

$$\sqrt{25}, \sqrt{26}, \sqrt{27}, \sqrt{28}, \sqrt{29}, \sqrt{30}, \sqrt{31}, \sqrt{32}, \sqrt{33}, \sqrt{34}, \sqrt{35}, \sqrt{36}$$

We know only the first and last of the twelve terms; they are 5 and 6. It is the fifth term we're trying to figure out.

Now suppose the sequence is an arithmetic progression. It's not, but suppose it is. The Great Square Rootio grants you permission. The eleven jumps from the first term to the last get you from 5 to 6, so if the difference between any two consecutive terms is the same, each of those differences had better be 1/11. So the square root of 29, four jumps up from 5, would be 5 4/11. Oh, did I mention you have to be able to divide in your head a little? You might know that 1/11 is about 0.09, so 4/11 is about 0.36, or you might just reckon that 5 4/11 is a little less than 5 4/10, which is 5.4, and either way you say, "It's 5.3 something, probably almost 5.4." (The actual answer is about 5.385.)

You can see, I hope, the conceptual similarity with Farr's argument, though here we're using differences instead of ratios. We baselessly decide, just as Farr did, that in fact all the differences are the same, and then, using the meager facts we have at hand, we figure out what that common difference must be. That seems unjustified. And yet it kind of works!

It is fair to ask: Why, besides my inner need to best the neighbor kid who taught himself to play "Free Fallin'," would I do this? Can't I just hit the square root button on my calculator? Yes, I can. But William Farr couldn't. And neither could astronomers of the seventh century. That's how far back this idea goes. In order to track the motions of celestial bodies, you need the values of trigonometric functions; those values were kept in voluminous tables, compiled at great expense of effort and time. To compile those tables, you need better accuracy than my square root party trick can provide. Around the year 600, a new idea emerged, in the kingdom of Gurjaradesa in India, from the astronomer and mathematician Brahmagupta, and in China from the Sui dynasty astronomer and calendar maker Liu Zhuo.

We don't want to get into the guts of the imperial calendar, so I'll stick with the square root example to explain their method. This is going to be the arithmetically grittiest part of the whole discussion; you are *not* supposed to be able to do this in your head at a college party while chugging a Keystone Light.

To execute the Brahmagupta-Liu approach, you need to take into account *three* square roots you know instead of two: say, $\sqrt{16} = 4, \sqrt{25} = 5, \sqrt{36} = 6$. From $\sqrt{16}$ to $\sqrt{36}$ there are twenty steps, and you need to get from 4 to 6; so you might take the Great Square Rootio's counsel, and presume these square roots lie in an arithmetic progression, so that the difference between each and the next is 2/20. I told you this wasn't exactly true, and here's the proof: if the sequence were an arithmetic progression, the square root of 25, nine steps on from $\sqrt{16}$, would be 4.9, which is not, as it should be, 5.

Here's the fix. We've seen that we can't insist that the square roots form an arithmetic progression if we want to match the three values we already know; that is, we can't make all twenty-one differences equal. The next best thing, then, is for the differences *themselves* to form an arithmetic progression; that is, we want the differences between the differences to be all the same! This is just the same as Farr's idea about ratios of ratios.

$\sqrt{16}$	$\sqrt{17}$	$\sqrt{18}$	$\sqrt{19}$	$\sqrt{20}$	$\sqrt{21}$	$\sqrt{22}$	$\sqrt{23}$	$\sqrt{24}$	$\sqrt{25}$	$\sqrt{26}$	$\sqrt{27}$	$\sqrt{28}$	$\sqrt{29}$	$\sqrt{30}$	$\sqrt{31}$	$\sqrt{32}$	$\sqrt{33}$	$\sqrt{34}$	$\sqrt{35}$	$\sqrt{36}$
	?	?	?	?	?	?	?	?	?	?	?	?	?	?	?	?	?	?	?	?

For this to work, we need the second row to be an arithmetic progression of twenty numbers whose sum is 2; but you also need the sum of the first nine numbers in the progression, which gets you from $\sqrt{16} = 4$ to $\sqrt{25} = 5$, to be 1. It turns out there's just one arithmetic progression that fills the bill. Here's a slick way to figure it out. Since the first nine terms sum to 1, their average is 1/9. But the average of an arithmetic progression has to be the middle term, which in this case is the fifth term; so that term is 1/9.

On the other hand, the last eleven terms also sum to 1, so their average is 1/11, and that must be the middle term of the final eleven, which is the fifteenth term of the sequence overall.

$\sqrt{16}$	$\sqrt{17}$	$\sqrt{18}$	$\sqrt{19}$	$\sqrt{20}$	$\sqrt{21}$	$\sqrt{22}$	$\sqrt{23}$	$\sqrt{24}$	$\sqrt{25}$	$\sqrt{26}$	$\sqrt{27}$	$\sqrt{28}$	$\sqrt{29}$	$\sqrt{30}$	$\sqrt{31}$	$\sqrt{32}$	$\sqrt{33}$	$\sqrt{34}$	$\sqrt{35}$	$\sqrt{36}$
?	?	?	?	$\frac{1}{9}$	?	?	?	?	?	?	?	?	?	$\frac{1}{11}$	?	?	?	?	?	

That's enough to determine the whole progression! From the fifth term to the fifteenth is ten steps, and the distance we need to traverse is the decrease from 1/9 to 1/11, or 2/99, so each step must be 2/990. That means the first difference, which is four steps up from 1/9, is 1/9 + 8/990 = 118/990, and the last, five downward steps on from 1/11, is 1/11 – 10/990 = 80/990.*

$\sqrt{16}$	$\sqrt{17}$	$\sqrt{18}$	$\sqrt{19}$	$\sqrt{20}$	$\sqrt{21}$	$\sqrt{22}$	$\sqrt{23}$	$\sqrt{24}$	$\sqrt{25}$	$\sqrt{26}$	$\sqrt{27}$	$\sqrt{28}$	$\sqrt{29}$	$\sqrt{30}$	$\sqrt{31}$	$\sqrt{32}$	$\sqrt{33}$	$\sqrt{34}$	$\sqrt{35}$	$\sqrt{36}$
$\frac{118}{990}$	$\frac{116}{990}$	$\frac{114}{990}$	$\frac{112}{990}$	$\frac{110}{990}$	$\frac{108}{990}$	$\frac{106}{990}$	$\frac{104}{990}$	$\frac{102}{990}$	$\frac{100}{990}$	$\frac{98}{990}$	$\frac{96}{990}$	$\frac{94}{990}$	$\frac{92}{990}$	$\frac{90}{990}$	$\frac{88}{990}$	$\frac{86}{990}$	$\frac{84}{990}$	$\frac{82}{990}$	$\frac{80}{990}$	
				$\frac{1}{9}$										$\frac{1}{11}$						

So what's the square root of 29, according to state-of-the-art seventh-century astronomy? To get from $\sqrt{16}$ to $\sqrt{29}$ you need to add up the first thirteen differences:

* Yes, these fractions are not in lowest terms, so your tenth-grade teacher might have marked this answer incorrect. It's not incorrect! The expressions 80/990 and 8/99 are different names for the same ratio, and if it's nine-hundred-ninetieths we're talking about there's not a thing wrong with using the former name.

118/990 + 116/990 + 114/990 + . . . + 94/990

and add that to 4. You end up with 4 and 1,378/990, which is about 5.392. That's roughly three times closer than our first estimate of 5 4/11.

The method of successive differences was transmitted from India to the Arab world, then rediscovered again multiple times in England, most notably by Henry Briggs. In 1624, Briggs used the method to produce the *Arithmetica Logarithmica,* which compiled the logarithms of thirty thousand numbers to fourteen decimal places each. (Briggs was the inaugural holder of the Gresham Chair in Geometry, the same position Karl Pearson would later occupy when he introduced the public to the geometry of statistics.) Like much else in seventeenth-century European math, the method was formalized and improved by Newton, to the point that nowadays we usually call it "Newton interpolation." There's no evidence in Farr's writing that he knew any of this history. Good ideas in mathematics often bubble up naturally when the problems of the world create the need for them.

The need for logarithms didn't end with Briggs. A table is a finite thing, and you might always find yourself in need of a logarithm of some number that fell in between the ones treated in the *Arithmetica.* The genius of the method of differences is that it allows you to estimate the values of quite complicated functions like cosines and logarithms using only the basic operations of addition, subtraction, multiplication, and division; so you can fill in the entries between the ones in the printed book, as needed. But as the example of the square roots shows, you have to do a *lot* of addition, subtraction, multiplication, and division—and that's if you're only studying differences of differences! To get better approximations, you might want differences of differences of differences, or differences of *those* triple differences, and so on until your head spins.

You wouldn't want to do this by hand. You might even want some kind of mechanical engine to compute these differences in your stead. That brings us to Charles Babbage. Babbage had been fascinated by automata ever since his childhood, when a "a man who called himself

Merlin"* allowed Babbage entry to his private workshop and showed the boy his most ingenious mechanical creation: "an admirable *danseuse,* with a bird on the fore finger of her right hand, which wagged its tail, flapped its wings, and opened its beak. This lady attitudinized in a most fascinating manner. Her eyes were full of imagination, and irresistible."

In 1813, Babbage was twenty-one years old and a student of mathematics at Cambridge. He and his friend John Herschel (who excelled Babbage in his studies and would later invent the blueprint) had founded a mathematical society as a kind of parody of the many student societies hotly contesting the proper interpretation of Scripture; *their* society's mission would be to exalt the mathematical notation of Leibniz's calculus over the competing system the hometown hero Newton had developed. The Analytical Society quickly outgrew its satirical origin and became an actual intellectual salon, aimed at bringing the new ideas of France and Germany into a country that had become a bit of a math backwater since Newton's time.

"One evening," Babbage recalls in his memoirs, "I was sitting in the rooms of the Analytical Society, at Cambridge, my head leaning forward on the Table in a kind of dreamy mood, with a Table of logarithms lying open before me. Another member, coming into the room, and seeing me half asleep, called out, 'Well, Babbage, what are you dreaming about?' to which I replied, 'I am thinking that all these Tables (pointing to the logarithms) might be calculated by machinery.'"

It didn't take long for Babbage, like his inspiration Merlin, to turn the dream into copper and wood. The machine, thought of now as the first mechanical computer, computed logarithms by means of the method of differences; that's why he called it "The Difference Engine."

There's one big difference between the work of the Great Square Rootio and that of Farr. When we estimated square roots we were finding values *between* square roots we already knew, a process called *interpolation.* Farr, using the method on counts of sick cows, was trying to

* In fact this was John Joseph Merlin, a prolific Belgian tinkerer who also invented the roller skate. The grown-up Babbage, decades later, bought the automaton at auction from a defunct museum and installed it in his drawing room.

extrapolate—to estimate the value of a function into the future, beyond the bounds where its value was known. Extrapolation is hard and is undertaken at one's peril.* Just think of what would happen if we'd used our party trick to guess the square root of 49, a number larger than the two whose square roots we took as input. Our heuristic, remember, was that the square root of a number grows by 1/11 every time the number grows by 1. Since 49 is 24 more than 25, its square root should be 24/11 more than 5, or 7.18. The true value is 7. What about 100? That's 75 more than 25, so the square root of 100 should be 5 + 75/11 = 11.82. The real square root is 10. The trick stinks now!

That's the danger of extrapolation. It tends to get less reliable the further out you get from the known data your differences are anchored in. And the deeper you go into the differences of differences of differences, the more jerky and weird the extrapolations get.

And that's what happened to Kevin Hassett. Although he was not a student of nineteenth-century epidemiology, the "cubic fit" he used was founded on exactly the same heuristic argument William Farr had adopted to model rinderpest. His model guessed that the ratio between the ratios between the ratios of successive data points would remain constant through the life of the epidemic. (You don't need to read antique papers from the history of medicine to carry out this strategy, because you can do it nowadays with a few keystrokes in Excel.) Hassett's curve was roughly right about the past—deaths from COVID-19 in the U.S. had indeed peaked, at least in the short term—but when it came to the extrapolation, he was wildly wrong about the epidemic's staying power.

Naive extrapolation can lead you far from the truth in a pessimistic direction, too. Justin Wolfers, an economist at the University of Michigan, who called Hassett's model "UTTER MADNESS" (caps lock his), had just a month earlier written: "Project the U.S. line forward just 7 days, and we'll be at 10,000 deaths in total. Project it forward a week after that, and we'll be at 10,000 per day." Wolfers was extrapolating in an even simpler fashion than Hassett, projecting deaths along an un-

* We already saw in chapter 7 the way even a very powerful neural net can fall flat on its lack of a face when asked to extrapolate beyond the bounds of its training data.

abated geometric progression. And the results show how quickly extrapolation can go out of whack. The U.S. *really did* hit 10,000 total COVID-19 deaths one week after Wolfers's projection. But a week after that, the death rate had hit its spring peak, at 2,000 deaths a day, a fifth of the figure Wolfers had gotten to by blind extrapolation.

BUT SOME ARE USEFUL

When I explained Farr's reasoning to my teenaged son, he said, but Dad, why didn't Farr wait until the end of February, which was already half over, and get one more data point? Then he would have had two ratios of ratios of ratios to work with instead of just one, giving him a more solid basis to assert his figure of 1.185 as the true "law of increase."

Good question, son! My best guess: Farr's choice was a victory of feeling over reason. Farr, a proud man, as we've seen, thought the peak of the infection would be visible in the next month's figures, and he wanted to predict the peak before it happened, not after.

Farr's prediction was premature, it turned out; the number of new cases in February was just over 57,000, still quite a bit more than the previous month's total of 47,191. Had he waited for more data to come in, he'd have found that the last ratio was 57,000/47,191 = 1.208, and the last ratio of ratios 1.395/1.208 = 1.155, and the new ratio of ratio of ratios 1.155/1.289 = 0.896. Finding that non-constancy, would he have gone so far as to compute the ratio of the two ratios of ratios of ratios? We can't know.

What's for sure is that Farr, while his timing was off, got the big question right; the epidemic was approaching its peak and would soon start to tumble. There were only 28,000 new cases of rinderpest in March, and it continued dropping from there, though not as quickly as Farr had projected; his curve, which I graph on the following page, has the disease just about gone by the end of June, when in fact that took the rest of the year.

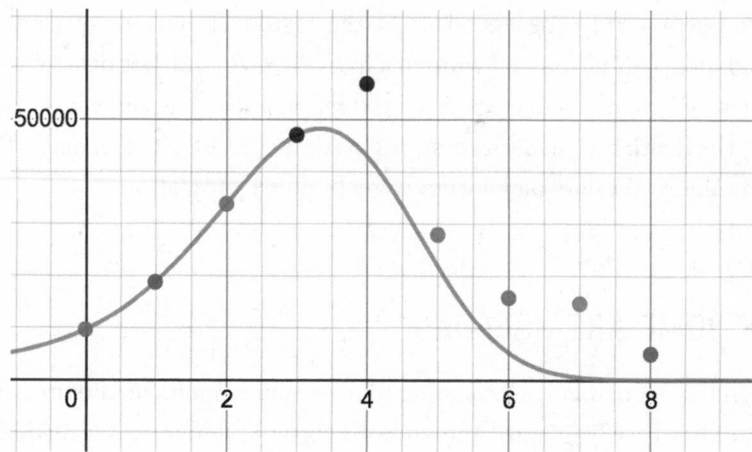

You can see here the danger of extrapolation. Farr's calculation did well on the near-term question (is this thing going to turn around soon?), worse on the long-term (when's this going to be over?).

Why *did* rinderpest die down? Farr, still not fully sold on the germ theory of disease, said it was because whatever poisonous substance was passing from cow to cow gave up some of its noxiousness with each animal it passed through. That is not, we now know, how viruses work. When the *British Medical Journal* scoffed at Farr's letter, they weren't disputing his conclusion about the outcome, but his reasoning. "He quite forgets to take into account," the snippy anonymous reviewer writes, "the fact that at the present time every one is satisfied as to the virulently contagious nature of the disease, and consequently takes measures to prevent it." Farr predicted rinderpest would "subside spontaneously"; all we can say for sure is that it subsided.

Farr's method was mostly forgotten for decades, until it was brought back into epidemiology by John Brownlee at the beginning of the twentieth century. Brownlee noticed something Farr had not: that if you model an epidemic by taking the ratio of ratios to be constant, as Farr had done with smallpox, you get a beautifully symmetric curve, which falls off exactly as quickly as it mounted. In fact it is none other than the normal distribution, or bell curve, which plays a central role in probability theory. The bell curve is treated, by people who know a little bit of math, with a kind of fetishistic reverence. And it *does* describe an astonishing array of natural phenomena. But the rise and fall of epidemics isn't one of them. Farr knew this: even in 1866, Farr argued on the third ratio instead of the second, and predicted the rinderpest wave would be asymmetric, decreasing more quickly than it had risen. Brownlee, too, recognized that strict adherence to the normal curve was rare in real-life epidemics. But somehow "Farr's Law" has come to mean the belief that epidemics follow the bell curve, all the way up and all the way down, a view Farr himself was far too savvy to adopt. I tend to call it Farr's "Law" to emphasize that it's not actually a law—but it might be better to call it "Farr's" "Law."

Such rigidity creates the danger of bad extrapolation. In 1990, Dennis Bregman and Alexander Langmuir, the latter a legendary epidemiologist who valued "shoe leather" on-the-ground research over pure lab work, published a paper called "Farr's Law Applied to AIDS Projections." Invoking Farr's success with rinderpest, they carried out a similar analysis on the statistics of AIDS in the United States. But they had adopted the too-narrow view that an epidemic had to be symmetric, and that AIDS would decline as rapidly as it had swept through the population. Their conclusion was that AIDS had already peaked, and that in 1995 there would be only about nine hundred cases in the United States.

In fact, there were 69,000.

Which brings us back to COVID-19 and 2020. Many forecasts* chose to model the progression of deaths from COVID-19 in each state as a perfectly symmetric bell curve. Not because they were bell curve fetishists, but because they found that curve best fit the sparse available data from the initial weeks of the outbreak. In some epidemics, that might have worked fine. But the COVID-19 curve has been steadfastly asymmetric, rocketing in each region to its high point and then declining with a painful slowness, dealing sickness and fear all the way down.

* Most notably, the very widely shared model from the Institute for Health Metrics and Evaluation at the University of Washington–Seattle; later in the pandemic's course, IHME wisely dropped the symmetry assumption. To be fair, the rise and fall of the epidemic was roughly symmetrical in the initial outbreak in Wuhan; the rapid decrease is now thought to be a result of China's extraordinarily severe suppression measures.

This epidemic comes up the elevator and goes down the stairs. If your projection insists otherwise, it's going to keep on missing the truth as you lunge along the bell curve into the future, and it's going to keep changing its tune as it encounters new data that conflicts with the predictions it's committed to.

We're encountering here a deep problem common to every attempt to mathematically project the present into the future. To make a projection is to make a guess about the law that governs the variable you're tracking. Sometimes that law is simple, as with the motion of the tennis ball. It enjoys a lovely symmetry; the time between the toss and the apex is the same amount of time it takes to return to your hand. What's more, if you carefully measured its height above the ground at each second, and wrote down those numbers in a sequence, you would find, as Farr approximately did, that the differences between the differences were always the same, through the ball's whole parabolic arc. In fact, that property is exactly what *makes* the arc parabolic, not a semicircle, or a catenary like the Gateway Arch in St. Louis. If you're lucky, you can discover regularities like this even if you don't understand the underlying mechanism; Galileo discovered the parabolic law of projectile motion by careful observation, decades before Newton worked out the general theory of force and acceleration.

But sometimes laws aren't simple! If the range of laws we're willing to consider is too narrow—for instance, if we insist that the pandemic follows a symmetrical course whose ratios of ratios are exactly constant—we are going to jitter and glitch as we attempt to match our too-rigid rule to reality. It's a problem called *underfitting.* It's the same thing that happens to a machine-learning algorithm that doesn't have enough knobs, or has the wrong ones.

It makes me think of Robert Plot, who in 1677 was the first person to publish a drawing of a dinosaur bone. Plot's repertoire of possible explanations for the bone's origin wasn't broad enough to include the truth, which was that he was looking at the partial femur of a gigantic reptile now called *Megalosaurus*. He considered the possibility that an ancient Roman elephant had gone astray and expired in Cornwall, but his comparison with a real elephant femur ruled that out. It had to be a person, he thought, and so the only question was which *kind* of person. His answer: an extremely tall person.

To be fair to Plot, he was dealing with a truly new phenomenon, and it's hard to blame him for not grasping what he was looking at. True underfitting would be a more serious error, as if Plot had simply taken as his model that bones in the ground belonged to people, despite numerous examples of nonhuman bones in the ground. Digging up the skeleton of a garter snake, the underfitting paleontologist would exclaim: my goodness, how extraordinarily sinuous this tiny person must have been!

The point of a model isn't to tell us whether the total number of COVID-19 deaths in the United States would be 93,526 (as a widely watched model from the Institute for Health Metrics and Evaluation projected on April 1) or 60,307 (April 16) or 137,184 (May 8) or 394,693 (October 15) or to give us the precise day and hour when the number of hospital beds in use will peak. For that kind of work, you need a soothsayer, not a mathematician. But that doesn't make models useless. Zeynep Tufekci, not a mathematician or a modeler but a sociologist, summed it up in an article with the shareworthy and utterly correct title "Coronavirus Models Aren't Supposed to Be Right." A better goal for models is to make much broader and qualitative assessments: Is the pandemic spiraling out of control right now, growing but flattening, or dying out? That's the goal Kevin Hassett and his cubic model failed at.

We're a lot like AlphaGo. The program learns an approximate law that assigns a score to each position of the board. The score does not tell us, on the nose, whether a position is a W, an L, or a D; that's beyond the capacity of any machine to compute, whether it's implemented on a cluster or inside our skull. But the job of the program isn't to get that answer exactly right; it's to give us good advice about which of the many paths before us is most likely to have victory at the end.

Modeling a pandemic is harder than AlphaGo in at least one way; in Go, the rules stay the same the whole game. In an epidemic, you make a model based on certain facts about who's transmitting to whom and when, and those facts can suddenly change, by mass human action or by government decree. You can use physics to model the flight of a tennis ball, and tennis players, if they are any good at all, are rapidly and unconsciously computing in that physics model to figure out where a certain shot is going to place the ball. But you can't use physics to predict who's going to win a long tennis match; that depends on how the players *react* to the physics. Real modeling is always a dance between predictable dynamics and our unpredictable responses.

I saw on the news a photo of a protester in Minnesota, angry about the stay-at-home order the governor had issued to limit transmission of the virus there, and presumably skeptical that COVID-19 represented a serious threat. He was carrying a big sign that said STOP THE SHUTDOWN and another one that said THE MODELS ARE *ALL* WRONG. He was paraphrasing, I imagine not on purpose, a famous slogan of the statistician George Box, one that's deeply appropriate to the COVID-19 moment. It goes like this: All models are wrong. But some are useful.

CURVE FITTERS AND REVERSE ENGINEERS

There are two ways to predict the future. You can try to figure out the way the world works, and derive from that understanding good guesses about what comes next. Or you can . . . not do that.

Ronald Ross lays out this distinction very clearly, as a way of separating himself from predecessors like Farr whom he means to supplant. Ross plants a flag for the first approach, which you might call "reverse engineering": his project is to start with facts he knows about the spread of disease, and from there reason his way to the differential equations that the curve of the epidemic must necessarily satisfy. William Farr was in the opposite camp. He was not a Reverse Engineer, but a Curve Fitter. The "curve fitting" mode of prediction is to look for regularities in the past and guess that those patterns will persist in the future, without troubling yourself too much as to why. What happened yesterday will happen today. In this fashion, you can make predictions without gaining or even seeking any insight into what's happening in the guts of the system. And your predictions might even be right!

Most scientists feel natural sympathy with Ross and the Reverse Engineers. Scientists like understanding things. So it's cold water to a lot of faces that the Curve Fitters are enjoying a resurgence, driven by progress in machine learning.

You might have noticed that Google can now translate documents from language to language pretty well. Not perfectly, not the way a human speaker would do it, but with an acuity that would have been science-fictional a few decades ago. Predictive text is getting better, too; you type, and the machine leaps ahead of you and offers you a chance, with a single keystroke, to insert the word or phrase the machine predicts you intend to type next. Quite often it's right. (I, out of pride or spite, change my phrasing when the machine correctly guesses what I was going to say, or, if I have no choice but to concede the model's correctness, at least type the word out myself letter by letter as God intended.)

If you asked Ronald Ross how this worked, he might say something like this: We know a lot about the internal structure of an English sentence—those of us of a certain age can even diagram one—and about the meanings of words, which are recorded in dictionaries. Given all this information, it ought to be possible for a native speaker to understand the mechanism of a sentence well enough to guess that when I type, "I was hoping we could get together next week and have," the next word is likely a noun serving as the object of the verb "have," and that among the things one can have, it is one of the things that people tend to get *together* to have: so "lunch" or "coffee" but not "possessions" or "turnips" or "COVID." (Okay, I guess it could be "COVID.")

But that's not how Google's language machine works at all. It's more like Farr. Google has seen billions of sentences, enough to observe statistical regularities concerning which combinations of words are likely to be meaningful sentences and which are not. And among the meaningful ones, it can assess which sentences appear most likely. Farr looked at previous epidemics; Google looks at previous emails. People who came before you in the vast history of English utterance have said, "I was hoping we could get together next week and have" and most of them followed up with "lunch" or "coffee." Nobody tells the machine what a noun or a verb is, or what a turnip is, or what lunch is. And it doesn't *know* what those things are, in any meaningful way. It works anyway. Not yet as well as a human writer or translator, maybe not ever. But pretty well!

The machine works even if you're typing something completely original, as we all like to think we are. There was an intellectual dustup in 2012 between Noam Chomsky, who more or less invented the modern academic discipline of linguistics, and Peter Norvig of Google, who's leading a massive engineering effort to avoid that discipline. Chomsky, in the 1950s, introduced a famous example illustrating the rule-governed nature of human language: "Colorless green ideas sleep furiously." This is a sentence an English speaker has never seen before (or at least before Chomsky made it famous it was), and there's no way to attach to it a meaningful interpretation as a statement about the physical world. And yet our minds clearly recognize it as a grammatical sentence, and even "understand" it—we could correctly answer questions based on it, like "Are the colorless green ideas resting calmly?" and recognize (because *we* know what nouns and verbs and adjectives are) that "Furiously sleep ideas green colorless" needs to be rearranged in order to make any kind of sense. But contra Chomsky, a modern machine can come to the same conclusions without learning structural rules about language. A language program develops a way to rate a string of words as "sentency" or "non-sentency," based on its resemblance to other sentences actually produced by humans. Like the machine trained to distinguish a cat from a non-cat, it uses some form of gradient descent to find its way step by step to a strategy that does a good job identifying the sentences it's already seen as more sentency than other strings of words. And not just that; the strategy that it finds (for reasons that remain not totally clear even to practitioners) tends to do a good job rating the sentenciness of strings of words that weren't part of the training. "Colorless green ideas sleep furiously" gets a much higher sentenciness score than "Furiously sleep ideas green colorless," even without a formal model of grammar, even though neither of those two sentences, if you train on data gathered pre-Chomsky, are ever encountered in the observed data. Even the component pieces, like "Colorless green," are seen rarely if at all.

Norvig observes that, when it comes to real-world machine translation or autocompletion, statistical methods like this decisively outperform all attempts to reverse-engineer the underlying mechanisms of human language production.* Chomsky retorts that, be that as it may, methods like Google's provide no insight into what language *really is*; they're like Galileo observing the parabolic arc of a projectile before Newton stepped in to lay down the laws.

They're both right, about language, and about pandemics, too. You can't do without a certain amount of both curve fitting and reverse engineering. One of the most successful modelers of the pandemic in 2020, a recent MIT graduate named Youyang Gu, ably combined both approaches, using a Ross-style model of differential equations designed to mimic the known mechanics of COVID-19 transmission, but using machine learning techniques to tune the many unknown parameters in that model to match the observed pandemic-so-far as well as possible. We need to catalog as much as we can about what happened yesterday if we want to predict what's happening tomorrow, but we are never going to have billions of past pandemics to look at, and if we want to be prepared for the next viral novelty, we had better look for laws.

* And of course both are still crushed by human beings, who learn language more accurately than any AI, with a billionth of the training input.

Chapter 12

The Smoke in the Leaf

In 1977, a group of Dutch math team members at the International Mathematical Olympiad in Belgrade posed the following puzzle to their British counterparts: What's the next number in the following sequence?

1, 11, 21, 1211, 111221, 312211 . . .

Does it make it easier if I tell you that the next few terms are

13112221, 1113213211, 31131211131221, 13211311123113112211 . . . ?

Most people don't get this. I sure didn't, when I first saw it. But the solution, when you hear it, is silly and charming. This is the "Look-and-Say" sequence. The first term is 1: "one one." So the next term is 11, or "two ones." That makes the next term 21, or "one two, one one," which you spell out as 1211, "one one, one two, two ones," and so on.

This is just an amusement, or so it seemed to the Dutch math team. But sometime in 1983 the Look-and-Say sequence made its way to John Conway, for whom making amusements into mathematics (and mathematics into amusement) was a way of life. Conway showed that the

Look-and-Say sequence never contains a number greater than 3, and that the long-term behavior of the sequence is controlled by the behavior of exactly ninety-two special digit strings, which Conway called "atoms" and named after chemical elements (1113213211 is "hafnium" followed by "tin"). What's more, the number of digits in the terms of the sequence itself behaves in a predictable way. The terms of Look-and-Say we've written down so far have length

1, 2, 2, 4, 6, 6, 8, 10, 14, 20 . . .

It would be very nice if this were a geometric progression. It isn't. The ratio of each term to the one before it is

2, 1, 2, 1.5, 1, 1.33, 1.25, 1.4, 1.42857 . . .

But as we go further, a regularity starts to set in. The forty-seventh, forty-eighth, and forty-ninth Look-and-Say numbers have

403,966, 526,646, 686,646

digits, respectively. The second number is 1.3037 times the first. The third is 1.3038 times the second. The ratios seem to be settling down. What Conway proved, by ingenious manipulations of his ninety-two atoms as they underwent what he called "audioactive decay" by the Look-and-Say process, is that those ratios do indeed approach a fixed constant, which Conway computed exactly.* The lengths of the Look-and-Say numbers don't form a geometric progression, but they do form a progression that gets more and more geometric as time goes on.

Geometric progressions are elegant and pristine. But in the real world they're rare. Kinda sorta geometric progressions like the length of the Look-and-Say numbers are a lot more common. They bring us into contact with a mathematical notion of fundamental importance called

* For algebra fans: it is the largest root of a certain polynomial of degree 71!

an *eigenvalue*. We can't avoid them if, for instance, we want to make the Ross-Hudson models of disease spread even slightly realistic.

DAKOTA AND DAKOTA

Ross and Hudson's theory of happenings, as applied to disease, relies on keeping track of the proportion of the population currently infected. This already creates some ambiguity. What is that population? Your neighborhood? Your city? Your country? The world?

You can see this really matters by a simple exercise in addition. Let's say a novel disease, the Wall Drug Flu, is raging through the Upper Plains. Suppose the number of cases in North Dakota is tripling every week, but in neighboring South Dakota, for whatever reason, cases are only doubling every week. The cases in North Dakota might look like this:

10, 30, 90, 270

and in South Dakota:

30, 60, 120, 240

Then the total number of cases, if Dakota were just one state, would be

40, 90, 210, 510

which isn't a geometric progression at all: the ratios between successive terms are 2.25, 2.33, 2.43. If you saw these Dakota numbers as a unit, you might think some sinister force was causing the virus to get more contagious with every passing week. You might begin to freak out. Will the growth rate ever stop growing?

Don't freak out. The growth of cases isn't a geometric progression, but it's a kinda sorta, like the Look-and-Say lengths. In the four weeks we've covered, the total cases are roughly evenly split between the two Dakotas. But that won't last. The next four weeks in North Dakota yield caseloads of

810, 2,430, 7,290, 21,870

and in South Dakota, just

480, 960, 1,920, 3,840

The total cases in both Dakotas in the eighth week was 25,710, which is 2.79 times the 9,210 cases the week before. That ratio is pretty close to 3, and it's only going to get closer. The faster growth in North Dakota absolutely swamps that in South Dakota. Ten weeks into the epidemic, almost 95% of cases will be there. At some point, you might as well just ignore South Dakota entirely; the disease is more or less entirely in North Dakota, tripling every week.

The two Dakotas are a reminder that to get pandemics right we need to think about space as well as time. In the basic SIR story, any two people in the population are equally likely to encounter each other and mingle their exhalations. We know that's not really true. South Dakotans mostly meet South Dakotans, and North Dakotans North Dakotans. That's exactly what makes it possible for the rate of spread to be different in different states, or in different localities within the same state. Uniform mixing of the population would cause the dynamics of the disease to equalize, much as mixing hot water with cold results quickly in lukewarm water, not a hot-cold swirl.

Here's a more complicated Dakota scenario. Suppose South Dakotans are incredibly compliant with all guidelines of social distancing, so much so that no infection events at all take place between two South Dakotans. In North Dakota, though, they're all over each other, breathing each other's air and just generally flouting the rules. Each North Dakotan who has the virus transmits it to one other person in the state. What's more, the North Dakotans like to wander across the border and get in the face of whomever they meet, and via this practice each North Dakotan who's infected gives it to one South Dakotan, and each South Dakotan gives it to one North Dakotan.

Got all that? If not (or even if so) let's see how this works, starting from one North Dakotan with WDF, and South Dakota flu-free. During the following week, that North Dakotan infects one fellow Peace Garden Stater and one South Dakotan, while South Dakota, having no infected people, provides no new infections. To make this simpler, let's assume that flu sufferers recover after their contagious week, so at week's end the only people infected are the *newly* infected ones; one in North Dakota and one in South.

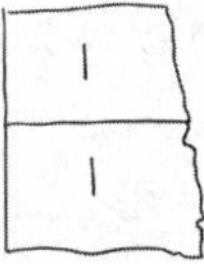

The next week, that North Dakotan infects two more people, one in North Dakota and one in South Dakota, while the sick South Dakotan infects one North Dakotan who came too close; so we have

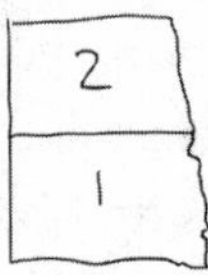

Time marches on and the infection spreads more broadly. The next few weeks bring:

Do you hear Sanskrit poetry in the air? The number of Wall Drug Flu–afflicted North Dakotans each week:

1, 1, 2, 3, 5, 8, 13 . . .

are the Virahanka-Fibonacci numbers. And so are the South Dakota case counts, shifted backward by a week. The rules of transmission we've chosen enforce this: each week, the number of South Dakotans with Wall Drug Flu is the number of North Dakotans who had it last week, and the number of North Dakotans with Wall Drug Flu is the number of cases among both North Dakotans and South Dakotans last week, which is the number of North Dakota cases last week combined with the number of North Dakota cases the week before that.

Now the Fibonacci sequence is *not* a geometric sequence. The ratios between successive terms bob up and down:

1, 2, 1.5, 1.66 . . .

Except it sort of actually is a geometric sequence! Especially if we let it go on for a while. The twelfth Fibonacci number is 144, the thirteenth is 233, and the fourteenth is their sum, 377. The ratio 233/144 is about 1.61806. The next ratio, 377/233, is about 1.61803. Those are pretty close. And if you followed the infection's growth week after week, you'd find that the ratio from one week to the next settles down almost exactly to a common ratio, very close to 1.618034. Once again, we encounter this phenomenon of not-quite-but-almost exponential growth.

What is that mysterious number hiding in the Fibonacci sequence? It is not just any number. It's a number with a fancy name: the *golden ratio*, or the golden section, or the divine proportion, or φ.* (The more famous a number is, the more names it tends to have.) If you want an exact description, there's one in terms of the square root of five; the golden ratio is $\left(1+\sqrt{5}\right)/2$.

People have been making a fuss over this number for centuries. In Euclid, the proportion goes by the more mundane name of "division into the extreme and mean." He needed it to construct a regular pentagon, since the golden ratio is the proportion between the diagonal of such a pentagon and its side. Johannes Kepler rated the Pythagorean Theorem and Euclid's ratio the top two achievements of classical geometry: "the

* A Greek letter which is pronounced "fee" if you're a number theorist and "fie" if you are in a fraternity.

first we may compare to a mass of gold, the second we may call a precious jewel."

Somewhere along the way, the ratio stopped being a jewel and started being golden; a text from 1717 says "the ancients called this section the golden one." (There's no evidence any ancient *actually* called it that, but assigning your coinage an unspecified weight of tradition gives it a little cultural oomph.) A golden rectangle is one whose length is φ times its width; it has the agreeable feature that if you cut it crosswise so that one of the two pieces is a square, the other one is a smaller golden rectangle. And if you like you can cut a square out of *that* golden rectangle, yielding a yet smaller one, and go on that way, forming a kind of spiral made out of squares:

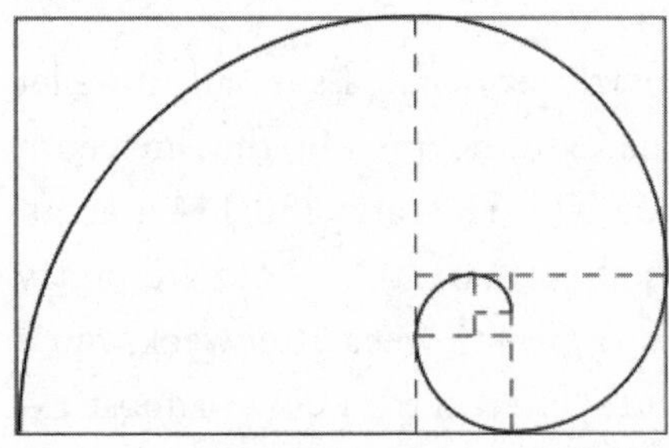

Kepler appreciated the golden ratio for both its geometric and arithmetic properties; he discovered the Virahanka-Fibonacci sequence independently, and found that the ratios between consecutive terms got closer and closer to the golden ratio. The relationship between the geometry and the arithmetic of the sequence is made visible if you write down an almost-golden rectangle whose length and width are two consecutive Fibonaccis, like this 8 x 13 specimen:

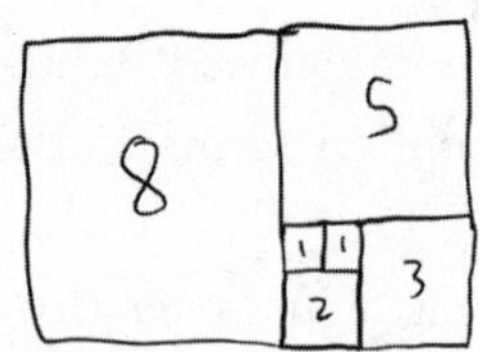

which you might call "golden until it's not"—cut off a square and you get a 5 x 8 rectangle, cut out a smaller one and you're down to 3 x 5, moving your way backward along the Fibonacci sequence with every cut. Eventually you get back to zero and your square spiral ends, instead of going on forever.

My very favorite feature of the golden ratio is one that gets comparatively less attention, so now's my chance to publicize it! The reason I have to keep on typing 1.618 . . . with that annoying ellipsis is that the golden ratio is an irrational number; you can't express it as one whole number divided by another, which also means you can't write the golden ratio as a decimal of finite length, or even one that repeats, like 1/7 = 0.142857142857142857 . . .

But that doesn't mean there aren't rational numbers that are pretty close to it. Of course there are! A decimal expansion, after all, *is* a way of writing down fractions that are close to a number:

16/10 = 1.6 (pretty close)
161/100 = 1.61 (closer)
1,618/1,000 = 1.618 (closer still)

The decimal expansion gives you a fraction with denominator 1,000, which is within 1/1,000 of the golden ratio;* if we let the denominator be 10,000 we can get within 1/10,000, and so on.

We can do better than decimals! Remember, the ratios between Fibonacci numbers are also fractions that get closer and closer to the golden ratio

8/5 = 1.6
13/8 = 1.625
21/13 = about 1.615

Go farther in the sequence and you get to

233/144 = 1.6180555555 . . .

which is only about 2 in 100,000 away from the golden ratio, a substantially better approximation than 1,618/1,000, with a way smaller

* Actually, we can get within 1/2,000, by rounding the final digit up or down instead of just truncating the decimal expansion; details left to the interested reader! At any rate we are not going to be so careful that we need to sweat factors of two.

denominator. In fact, the difference is less than a hundredth of the fraction 1/144.

Some celebrity irrationals can be approximated even more closely. Zu Chongzhi, a fifth-century astronomer in Nanjing, observed that the simple fraction 335/113 is incredibly close to π, only about 2 in *10 million* away. He called it *milü* ("very close ratio"). Zu's book on mathematical methods is lost, so we don't know how he came up with the milü. But it was no simple find; it would be a thousand years before the approximation was rediscovered in India, another hundred before it was known in Europe, and another century after that before it was conclusively proved that π was actually irrational.

How closely should we expect an irrational number to be approached by rationals? It's an arithmetic problem but it's best to think about it geometrically. There's an amazing trick for this, invented by Peter Gustav Lejeune Dirichlet in the early nineteenth century. We found a fraction, 233/144, whose distance from φ was less than a hundredth of 1/144. Can we find a fraction p/q whose difference from the golden ratio is less than a *thousandth* of 1/q? We can, and Dirichlet's proof that we can is so simple I can't not present it to you.* Draw the segment of the number line that covers the numbers between 0 and 1, and then cut it into a thousand equal pieces. (I can't draw a thousand equal pieces, so just imagine them):

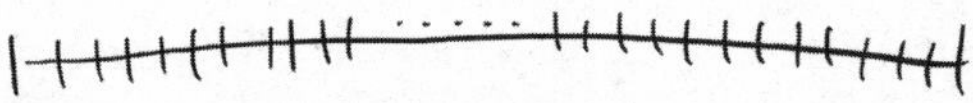

Now start writing down multiples of φ:

$$\varphi = 1.618\ldots,\ 2\varphi = 3.236\ldots,\ 3\varphi = 4.854\ldots,\ 4\varphi = 6.472\ldots$$

and plot the *fractional part* of each of these numbers—the part that comes after the decimal point—on the number line. If I draw the fractional parts first three hundred multiples of φ on the number line,

* This section gestures at the very beginning of the subject of Diophantine approximation; if it appeals to you, some things to read about are Dirichlet's approximation theorem (the one we prove here), continued fractions, and Liouville's Theorem.

marking each one with a vertical bar to make it a little more visible, I get a kind of "bar code":

Each one of those bars lands in one of the thousand boxes. The golden ratio itself lies in the 619th box. (Not the 618th box, for the same reason we're now in the twenty-first century despite the year starting with 20; the first box corresponds to numbers between .000 and .001, the second box to numbers between .001 and .002, and on you go.) The next multiple, 2φ, lands in box number 237, 3φ in box 855. Keep on putting these numbers in the boxes. If any of those multiples of φ winds up in the *first* box, we win; because to say a multiple qφ has fractional part between 0 and .001 is to say that the difference between qφ and some integer p is at most .001, which, after you divide both those numbers by q, is to say the difference between φ and the fraction p/q is at most one-thousandth of 1/q.

But why should any of our multiples land in that first box? Maybe, like the Scottie dog marching around a Monopoly board longing to land on Boardwalk, the multiples traverse the territory again and again without ever touching the crucial square!

This is where Dirichlet's remarkable insight intervenes. He called it the *Schubfachprinzip* ("chest-of-drawers principle"), but Anglophone mathematicians nowadays call it the pigeonhole principle. It says this: if you have a bunch of pigeons, and a bunch of holes, and you put all the pigeons into holes, and there are more pigeons than there are holes, then some hole has to have two pigeons in it.

That statement is so obvious it's hard to believe it's good for anything. Sometimes the very deepest math is like that.

For us, the pigeons are the multiples of φ and the thousand boxes are the holes. And what we learn by thinking about Dirichlet's drawers is that, if we look at a thousand and one multiples, at least two of them have to share a hole. Let's say 238φ and 576φ are the crowded pigeons. They're not (those two perch in box 93 and 988 respectively) but let's say they were. Then the difference between those two numbers must be within 1/1000 of some integer, which we call p; but that difference is 338φ, which must land in the first box—or, to be fair, in the very last box of numbers ending in 0.999. . . . Either way, p/338 is our close-enough approximation.

It doesn't matter *which* two multiples of φ land in the same box; whatever pair there is gives you a fraction that is really close to φ. In reality, it turns out the first pigeon collision is between φ and 611φ = 988.6187 . . . , which shares the 619th box with φ. Their difference, 610φ, is about 987.00007, and so 987/610 is a really good approximation to φ. You won't be surprised to learn that 610 and 987 appear as consecutive terms of the Fibonacci sequence, just past the point where we left off computing it.

There was nothing special about the number 1,000. If you want a rational number p/q whose difference from φ is less than a *millionth* of 1/q, you can do that too, though you might have to let q be as big as 1 million.

The difference between Zu Chongzhi's "close ratio" 355/113 and π is only about one thirty-thousandth of 1/113. As far as Peter Gustav Lejeune Dirichlet is concerned, you might have to look at fractions with denominator as big as 30,000 to find an approximation that good. But you don't! The milü isn't just a good approximation to π, it's a *shockingly* good approximation.

Let's see how this plays out on the number line. If I look at the first three hundred multiples of 1/7, and plot their fractional parts the way I did the multiples of φ, marking each one with a bar, I get a picture that looks like, well, seven bars; because no matter what I multiply 1/7 by, I get some number of sevenths, whose fractional part is either 0, 1/7, 2/7, 3/7, 4/7, 5/7, or 6/7.

| | | | | | |

It's the same for any rational number; we may consider more and more multiples, but the bars will be constrained to a finite collection, evenly spaced from 0 to 1.

What about π? Here are its first three hundred multiples:

That's a lot of bars. But not three hundred. In fact, if you were to count the bars visible here, you would see that there are exactly 113 of them. What you're seeing here is the signature of the milü. Because π is so very close to 355/113, its first three hundred multiples are also very close to some number of one-hundred-thirteenths, which means those bars are going to stay very close to the numbers 0, 1/113, 2/113, (pretend I wrote down all 113 possibilities), and 112/113. Since π isn't exactly *equal* to the milü, its multiples don't hit those fractions on the nose; the bars in the picture that look a little fatter are those that are actually several bars clustered so close together they look like one on the page.

Which brings us back to the golden ratio. The bar code formed by the first three hundred multiples of φ, which I already drew above, is nicely spread out, not clustered like the π bars. Draw a thousand multiples, and it's the same story, just with more bars:

And no matter how many multiples of φ I take, a thousand, a billion, or more, those bars are never going to line up along a small set of evenly spaced positions, the way the bar code of a rational number does, or even cluster *near* those positions, the way the bar code for π does. There is no golden milü.

Here's a beautiful fact, a bit too hard to prove here: you won't find any better rational approximations to φ than the ones the Fibonacci sequence provides, and these approximations are never much better than the ones Dirichlet's theorem guarantees. In fact, in a way that can be made quite precise, but not here, φ, out of all real numbers, is the one that's least well approximated by fractions; it is the *most irrational* irrational number. That, to me, is a jewel.

LOOKING FOR A CERTAIN RATIO

One day in the nineties I had dinner with a friend of a friend at the Galaxy Diner in New York. The friend of a friend said he was making a movie about math and wanted to talk to a practitioner about what the mathematical life was really like. We ate patty melts, I told him some stories, I forgot about it, years went by. The friend of a friend was named Darren Aronofsky, and his movie *Pi* came out in 1998. The main character of the movie is a number theorist named Max Cohen who thinks extremely intensely and twirls his fingers in his hair a lot. He meets a Hasidic man who gets him interested in Jewish numerology, the practice called *gematria** where a word is turned into a number by adding up the numerical value of the Hebrew letters it contains. The Hebrew word for "east" adds up to 144, the Hasid explains, and "The Tree of Life" comes to 233. Now Max is interested, because those are Fibonacci numbers. He doodles some more Fibonaccis across the stock market pages of the newspaper. "I never saw that before," says the impressed Hasid. Max feverishly programs his computer, which is named Euclid, and draws golden rectangle spirals, and stares for a good long while into the similar spirals of milk in his coffee. He computes a 216-digit number which seems to be the key to forecasting stock prices and is also possibly God's secret name. He plays a lot of Go with his thesis advisor. ("Stop thinking, Max. Just *feel*. Use your intuition.") He gets a bad headache and twirls his hair even more intensely. The beautiful woman in the apartment next door is intrigued. I forgot to mention it but this movie is in black and white. Somebody tries to kidnap him. Finally he drills a hole in his own skull to let some of the math pressure out and the movie arrives at what appears to be a happy ending.

I don't remember what I told Aronofsky about math, but it wasn't that.

(Full disclosure: there were occasions in my twenties, after *Pi* came

* You'd think this would obviously be a Hebraization of the Greek word "geometry," but apparently this account of the word's origin is controversial.

out, when I'd sit in a coffee shop, my well-worn copy of Robin Hartshorne's *Algebraic Geometry* tactically placed in a visible spot on the table, thinking very intensely, and twirling my fingers in my hair. No one was ever intrigued.)

Aronofsky learned about the Fibonacci numbers in a high school course on "Math and Mysticism," and felt an instant kinship with the sequence, because his home zip code was 11235. That kind of attention to coincidence and pattern, meaningful or not, is characteristic of golden numberism. Somewhere along the way, the understandable attraction to the mathematical properties of 1.618 . . . bled out into more grandiose claims. The number theorist George Ballard Mathews was already complaining about Aronofsky's movie back in 1904:

> The "divine proportion" or "golden section" impressed the ignorant, nay even learned men like Kepler, with a sense of mystery, and set them a-dreaming all kinds of fantastic symbolism. Even to the Greeks it was *the* section; and their philosophers, doubtless infected by the East, speculated about atoms and regular solids in a way that seems to us childish but was serious enough to them. At any rate, the man who first found out an exact construction for a regular pentagon had reason to feel proud of his exploit; and the superstitions which have gathered around the *pentagramma mirificum* are grotesque echoes of his fame.

Figures with lengths in golden proportion to one another are sometimes said to be inherently the most beautiful. The nineteenth-century German psychologist G. T. Fechner presented subjects with piles of rectangles to see if they'd find the golden ones most pleasing. They did! It's a handsome rectangle. But the claims that the Great Pyramid of Giza, the Parthenon, and the *Mona Lisa* were all designed on this principle aren't well substantiated. (Da Vinci did illustrate a book by Pacioli about the number, which the Italians knew as "the divine proportion," but there's no evidence he paid special attention to the ratio in his own artwork.) The name φ for the golden ratio is a twentieth-century coinage honoring the Greek sculptor Phidias, who was said to, but probably actually didn't, use the golden proportion to fashion classically perfect bodies of stone. An influential 1978 paper in *The Journal of Prosthetic Dentistry* suggests that a set of false teeth, for maximum smile appeal, should have the central incisor 1.618 times the width of the lateral incisor, which should in turn be 1.618 times as wide as the canine. Why settle for a gold tooth when you can get a golden one?

Golden numberism really took off in 2003 when Dan Brown published his megahit novel *The Da Vinci Code,* the story of a "religious symbologist" and Harvard professor who uses the Fibonacci sequence and the golden ratio to unwind a conspiracy involving the Knights Templar and modern-day descendants of Jesus. After that, "put a φ on it" was just good marketing. You could buy jeans whose golden proportions were flattering to your rear (they go with your false teeth!). There was a "Diet Code," which argued that Leonardo would have wanted you to lose weight by eating proteins and carbs in golden-ratio proportions. And there was perhaps the greatest work of mystical geometric hoo-hah ever produced: "BREATHTAKING," the Arnell Group marketing firm's twenty-seven-page explanation of the new Pepsi "globe" logo they designed in 2008. The pitch explains that Pepsi and the golden ratio are natural partners, because, as you no doubt knew, "the vocabulary of truth and simplicity is a reoccurring phenomena [*sic*] in the brand's history." A timeline situates the unveiling of the new Pepsi logo as the culmination of five thousand years of science and design including Pythagoras, Euclid, da Vinci, and somehow the Möbius strip. We're just lucky Arnell didn't know about Virahanka, because one trembles to think what pseudo-subcontinental philosophy he would have tossed into the mix if he had.

The new Pepsi logo would be built out of arcs from circles whose radii were in golden ratio to each other, a ratio that the pitch declares would now, in a truly impressive rebranding bid, be known as "The Pepsi Ratio." And that's when things get really weird. On subsequent pages we find "Pepsi Energy Fields" and their relation to the Earth's magnetosphere, and this diagram illustrating the relevance of Einstein's take on gravitation to the brand's attractiveness on the grocery aisle:

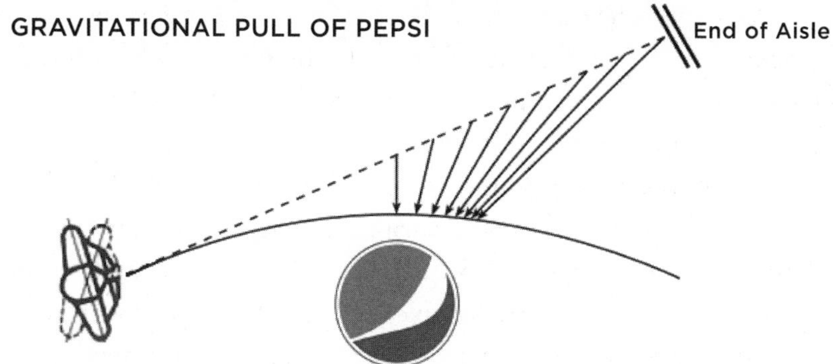

As absurd as all this is, Arnell's Pepsi globe is still on the cans a decade later. So maybe the golden ratio really is the true natural arbiter of what's beautiful and good! Or maybe people just like Pepsi.

Ralph Nelson Elliott was an accountant from Kansas who, for the first three decades of the twentieth century, bounced back and forth between the United States and Central America, working for railroad companies in Mexico and financial reorganization in U.S.-occupied Nicaragua. In 1926 he fell ill from a nasty infestation of parasitic amoebas, and he had to move back to the United States. A few years afterward, the stock market went haywire and threw the world into depression, so Elliott had a lot of free time, and a lot of motivation to restore some order to a financial world that no longer fit into neat double-entry bookkeeping. Elliott surely didn't know about Louis Bachelier's work on stock prices as a random walk, but if he had, he wouldn't have given it a minute. He didn't want to believe stock prices were randomly jittering like dust suspended in fluid. He wanted something more like the comforting physical laws that kept the planets safely in their orbits. Elliott compared himself to Edmond Halley, who figured out in the seventeenth century that the apparently random comings and goings of comets actually obeyed a rigid timetable. "Man is no less a natural object than the sun and moon," Elliott wrote, "and his actions too, in their metrical occurrence, are subject to analysis."

Elliott pored through seventy-five years' worth of the stock ticker, down to its minute-by-minute movements, trying to make the ups and downs resolve into a story that made sense. The result was the Elliott

Wave Theory, which posited that the stock market was governed by an interlocking collection of cycles, from the subminuette, which shuttles back and forth every few minutes, to the "grand supercycle," the first of which started in 1857 and is still ongoing.

For any of this to help an investor make money, one needs to know when the market's about to take a turn for the better or the worse. The wave theory has an answer. Elliott believed the movement of the stock market to be controlled by predictable patterns of up-and-down trends, which in-the-know wave theorists can anticipate by the principle that the ratio between the length of the present trend and the length of the last one tends to be golden: 1.618 times as long. Elliott, in this respect, was a kind of precursor to Max Cohen in *Pi*, scribbling Fibonacci numbers across the stock pages.

The rule of 1.618 wasn't absolute; the next wave might be 61.8% as long, because that makes the last trend 1.618 times as long as this one. Or 38.2% as long, because that's 61.8% of 61.8%. There's a lot of wiggle room, and the more wiggle room your theory has, the easier it is to describe what's already happened with a confident *Just as I thought!* To be honest, it's hard for an outsider to penetrate exactly what Elliott does and does not predict. The Wave Theory, like all theories developed by people who spend a lot of time alone, is dense with idiosyncratic terminology: ("**Third of a Third**—Powerful middle section within an impulse wave. **Thrust**—Impulsive wave following completion of a triangle.") Elliott, not content to have solved the stock market, spent the last ten years of his life writing his real life's work, *Nature's Law—The Secret of the Universe.* (Spoiler: it's waves.)

This might be just another story about a weird old theory from the ash heap of financial history, like those of Roger Babson, who believed Newton's laws of motion governed the stock market, predicted the great crash of 1929, then predicted the imminent* end of the Depression in 1930, founded Babson College in Massachusetts and Utopia College in Eureka, Kansas (at the geographic center of the continental United States, where he thought it would be safe from the atom bomb), ran for

* Not actually imminent.

president as the Prohibition Party candidate in 1940, and spent much of the money he made selling stock tips on an attempt to develop an antigravity metal.

The difference is that the Elliott Wave Theory is still a going concern. A guide to "technical analysis" from Merrill Lynch includes an entire chapter about it, "The Fibonacci Concept," which trots out the usual golden ratio sizzle reel:

> As with all other methods of analysis, the Fibonacci relationship is not 100% reliable. It is uncanny, though, how often it predicts significant turning points. Speculation abounds as to why the Fibonacci Ratio and its derivatives keep showing up. The fact is, this mysterious ratio is found repeatedly in nature. It was pervasive in the paintings of the Renaissance period, defining proportions and perspective. It was also found in the architecture of Ancient Greek temples—much before the time of Fibonacci.

Your Bloomberg terminal, should you be flush enough to have one, will draw little "Fibonacci lines" on the stock charts for you, so you know what level a price can rise to before it's compelled to repeat a φ-sized copy of its most recent trend, in what wavists call a "Fibonacci retracement." In April 2020, *The Wall Street Journal* warned its readers there was "likely more pain ahead" for the coronavirus-battered S&P 500 stock index; prices were up 23% since the market bottomed out in late March, but the Fibonacci retracement foretold further losses. Two months later, the S&P was up another 10%.*

I have a friend, a pretty rich friend if you must know, who uses Fibonacci methods in her investments. Her argument is that it doesn't matter if it "really" works—it only matters that enough people *think* it works to make markets correlate ever so slightly with what the Elliot waves predict. The waves, like Tinker Bell, are wished into life by their true believers. Maybe my friend is right, but the evidence for her view is

* And what about two months after that? Maybe Fibonacci will turn out to be right! Well, look. If what you mean by "my prediction was right" is "At some unspecified time in the future the stock market is going to dip," your prediction is both correct and unimpressive.

pretty slim. If your investment manager is an adherent of Fibonacci retracement, I would say—forgive me—they're not earning their φ.

DAKOTA AND DAKOTA REVISITED

What if we adjust our model to make North Dakotans slightly filthier? Say each infected person in North Dakota infects *two* more fellow North Dakotans, instead of one. If we start, as we did before, with one infected person in North Dakota and no infections in the South

(1 ND, 0 SD)

the next generation sees two new North Dakotans infected and one new South Dakotan:

(2 ND, 1 SD)

and then the two North Dakotans combine to infect four more North Dakotans and two more South Dakotans, while the already infected South Dakotan infects one new North Dakotan:

(5 ND, 2 SD)

The infected North Dakotans per week is a sequence

1, 2, 5, 12, 29 . . .

where each number is *twice* the number before it in the sequence added to the number two ticks back. This sequence has a name, too; it's the Pell sequence. It's not a geometric progression, but like the Fibonacci sequence, it gets close to one. The ratios between consecutive terms are

2/1 = 2
5/2 = 2.5

12/5 = 2.4
29/12 = 2.416666 . . .

Go out a ways in this sequence and you'll find 33,461 followed by 80,782; the ratio is 2.4142 . . . which is just about exactly $1+\sqrt{2}$. And the farther you go in the sequence, the closer the ratios get to that governing constant.

It would be the same if each North Dakotan infected three other people in the state; the magic ratio would be $(1/2)(1+\sqrt{13})$, a number that's a little more than 3.3. Or if we expanded our original model to include Nebraska, and declared that each Nebraskan infects one South Dakotan and each South Dakotan one Nebraskan, but the Nebraskans don't infect each other; the intricate interaction between all three states gives you a sequence of infected people in North Dakota that goes

1, 1, 2, 3, 6, 10, 19, 33 . . .

This sequence has no name,* but it has a similar property to the ones above; the successive ratios get closer and closer to a quantity 1.7548 . . . , which, if you really insist on an exact expression, is

$$\frac{1}{3}\left(2+\sqrt[3]{\frac{25}{2}-\frac{3\sqrt{69}}{2}}+\sqrt[3]{\frac{25}{2}+\frac{3\sqrt{69}}{2}}\right)$$

This kind of regularity, not the golden ratio in particular, is an underlying principle that appears everywhere in nature. It doesn't matter how many states you include, exactly how many Utahns the average Wyomingite infects, etc.—the number of infections in each state will

* Though it does appear in the On-Line Encyclopedia of Integer Sequences as entry A028495, where I learn that the nth entry is the number of ways to achieve checkmate in n + 1 moves starting from a particular carefully chosen position. Weird!

always tend toward a geometric progression.* Plato was right all along—geometric progressions really are the kind nature favors.

That strangely complicated number that governs the rate of geometric growth is called an *eigenvalue*. The golden ratio is just one possibility; its agreeable features arise from the fact that it's an eigenvalue of a particularly simple system. Different systems have different eigenvalues; in fact, most systems have more than one. In the very first Dakota scenario we considered, the epidemic was really the combination of two separate outbreaks, each growing geometrically, one tripling each week and the other doubling. As time went on, the faster-growing outbreak dominated the numbers so thoroughly that the overall count was more or less a geometric progression with common ratio 3. That's a situation where you have two eigenvalues, 3 and 2; the *biggest* one is the one that matters.

In systems where multiple parts interact, it's not so easy to see how you can tease the process apart into separate, perfectly geometric progressions like that. But you can! For instance, here's a geometric progression that starts with a number that's approximately .07236 and where each term is the golden ratio times the previous one.

0.7236 . . . , 1.1708 . . . , 1.8944 . . . , 3.0652 . . . , 4.9596 . . .

And here's another one, which starts with 0.2764 and has a negative common ratio, −0.618. (This ratio is actually just what you get when you subtract the golden ratio from 1.) In this latter sequence, you have exponential decay toward zero instead of exponential growth, like an epidemic with small R_0. (Well, maybe not *that* much like it, since every other number is going to be negative.)

0.2764 . . . , −0.1708 . . . , 0.1056 . . . , −0.0652 . . . , 0.0403 . . .

Add those two geometric series and something quite wonderful happens; the mess after the decimal points completely cancels out and you get, on the nose, the Fibonacci sequence

* At least at first. The geometric progression models the early part of an epidemic, where the virus hasn't started running out of susceptible hosts.

1, 1, 2, 3, 5 . . .

In other words, the Fibonacci sequence isn't a geometric progression; it's *two* geometric progressions, one governed by the golden ratio and the other by –0.618. Those two numbers are the two eigenvalues. In the long run, only the biggest one really matters.

But where do these two numbers come from? There's not a North Eigenvalue and a South Eigenvalue; each eigenvalue, 1.618 and –0.618, captures something deep and global about the system's behavior. They're not properties of any individual part of the system, but emerge from interaction among its parts. The algebraist James Joseph Sylvester (about whom more very soon) called these numbers the *latent roots*—"latent," as he vividly explained, "in a somewhat similar sense as vapour may be said to be latent in water or smoke in a tobacco-leaf." Unfortunately, English-speaking mathematicians have preferred to half translate David Hilbert's word *Eigenwert*, which is German for "inherent value."*

We don't have to divide the pandemic by geography; we can use whatever categories we like. Instead of dividing Dakotans into North and South, we might divide them into two age groups, or five age groups, or ten, keeping track in each case of how much interaction there is within each age group and between the different ones. With ten age groups this is quite a bit of information; to organize it all, you might want to make a 10 × 10 box of numbers, where in the third row and seventh column you place the number of close personal contacts between members of the third age group and members of the seventh. (So this would be *slightly* redundant, in that the number you put in the seventh row and the third column is the same as the one in the third row and the seventh column; or maybe you think young people are more likely to transmit to their elders than vice versa, in which case you'd want to put two different numbers in those positions after all.) A box of numbers like this is what Sylvester called a *matrix*; in this case, his coinage has stuck. Computing an eigenvalue of a matrix—the latent number that governs the growth of the many-part system described by

* And *Eigenwert* is, in turn, probably a Germanized version of Poincaré's earlier term *nombre characteristique*.

the box of numbers—has come to be seen as one of *the* fundamental calculations. Most mathematicians compute eigenvalues daily.

The eigenstory can give you a much more refined picture of a pandemic's progress and its expected future than the basic models we discussed earlier. In particular, if some population subgroups are much more likely to contract and transmit the virus than others, an initially high R_0 might not necessarily indicate a pandemic that eventually spreads through most of the population. Maybe, instead, the early numbers are being driven by that highly susceptible subgroup, and once the virus permeates that small sliver of the population, which comes out the other side at least temporarily immune, transmission among the remaining people isn't swift enough to keep the pandemic growing. You can make models like that, where the pandemic stops after infecting a much smaller fraction of people, just 10 or 20%, even with a high R_0. To really work out these numbers involves eigenvalue wrangling among the different subgroups, but you can see the main idea just by imagining this simple case: just 10% of the population was susceptible to the virus, but each of them exposes twenty other people when they get infected, while the other 90% are immune. You'd see initial growth with an R_0 of 2, because each infected person would encounter twenty other people but only infect the two who were susceptible. But after infecting just about everybody in that 10% of the population, the virus would run out of susceptible people.

Geometric progressions aren't the whole story, as we've seen. The R_0 of the epidemic may change with time, as governments and individuals modify their strategies of infection control. Beyond that, there's the rise and fall predicted by the Ross-Hudson-Kermack-McKendrick model as the virus saturates the population, arrives at herd immunity, and slowly, painfully subsides. You can do all this analysis with populations split into spatial or demographic subgroups, at which point you are studying not so much one epidemic as an ensemble of epidemics, each one feeding into all the others. The result, when you put all this together, is something that actually looks vaguely realistic: outbreaks and lulls in different populations at different times.

And all this modeling, if you really want to get it right, is carried out *stochastically,* which means, for instance, that you're not just assigning each individual their own precise R_0—as if you, partying twenty-five-year-old, are certain to infect precisely six of your fellow youngsters and one senior citizen this week—but a random variable. If the random variable isn't varying *too* wildly, this might not matter; maybe half of sufferers infect one new person and half infect two, and you don't lose much by thinking of next week's infections as 1.5 times this week's and modeling with an R_0 of 1.5. But what if 90% of infected people don't infect anyone, 9% infect ten new people each, and the remaining 1% infect sixty people each? That still averages to 1.5 new infections per person, but the dynamics of the epidemic are different. Maybe that small fraction of people is ultra-infectious for some biological reason, maybe they're the ones who choose to attend big indoor weddings; doesn't matter, the math can handle it. Superspreading events are big but rare. In any given region, you might go a while without having such an event, and the disease might burble along for a while, with a case here and there brought in from outside, without exploding. But a few superspreading events in close succession and all of a sudden there's a local surge. This leaves you with a thoroughgoing uncertainty about causes. When two places are hit very differently by the disease, that might be because one place had a better suite of policies in place. But it might just be the stochasticity. The greater the extent to which infection is dominated by superspread, the more dumb luck there is in which populations suffer and which are spared.

That doesn't mean a local health department should just throw up its hands, burn some offerings, and pray random fate is merciful. Knowing that an epidemic is driven by superspread can actually be useful. If superspread is where disease transmission is coming from, you can suppress transmission by suppressing superspread. No big indoor weddings, no bars, no yodeling contests, and maybe you can get away with milder restrictions on other forms of human contact.

HOW GOOGLE WORKS, OR: THE LAW OF LONG WALKS

There was the internet before Google and the internet after; for people who first logged on after the mid-1990s it's almost impossible to convey the drastic, immediate difference it made. Suddenly, instead of having to know what sequence of links to follow or what HTML address to manually type in to reach a certain piece of information, you could just . . . ask. It seemed like a miracle. It was actually eigenvalues.

The best way to see how this works is to go back to the pandemic. Suppose you had a more refined model, where you don't just break the population into two Dakotas or ten age bands; you go further, divvying up the population into slimmer and slimmer categories until in the end each person is their *own* box. When you get to this point you have what's called an *agent-based model,* which is great, if you can somehow keep track of—or meaningfully approximate—the monumental data of each individual's interactions with everyone else. Such a model is in many ways like the random walks Ronald Ross studied. But instead of an infected mosquito moving around, it's the virus itself that's randomly walking, jumping with some probability from an infected individual to a susceptible person they come in contact with. And the same kind of eigenvalue analysis applies, though now the size of your box of numbers is gigantic, having a row and column for each person in the population!

You might imagine that a person's likelihood of getting infected, in a model like this, would depend on how many contact events they had with others. And that's true, to a point. But it matters *who* those interactions are with. A person and their spouse probably have an infection-risking interaction with each other just about every day. But if they rarely interact with people outside the relationship, those contacts don't matter to overall spread. If you cut your social contacts to a minimum, just seeing your best friend, that might seem pretty safe; but if your best friend is regularly going to unmasked warehouse parties, you're at high infection risk despite your low number of contacts.

Agent-based models haven't actually dominated COVID-19 modeling, because in fact we don't have (and shouldn't have!) anything like the fine-grained data about individual contacts we'd need to make it work.

But we aren't really talking about COVID-19 anymore. We're talking about searching the internet. The network of links between web pages is much easier to measure than the network of contacts between people. But the structure is similar. There are many, many individual pages, and each pair of pages is either linked or not.

If you're searching for "pandemic," you don't want a web page randomly selected from every single one on the internet mentioning that word. You want the best one! You might naturally think the best page on a subject is the one with the most links to it. But that can be illusory. The purveyor of a "pandemics are really just side effects of municipal water fluoridation" pamphlet is perfectly capable of putting up a hundred different websites on that theme and linking them all to each other. If you give the "Dentifrice or DEATH?!?!" page a high score based on that, you're making a big mistake.

It matters where the links come from. The fluoridation pages, multiply linked to each other but not from the outside world, are like the sequestered couple whose contacts are all in-house. Having a friend who parties a lot is the analogue of having a link to your page from CNN; a link should count for a lot if it comes from a page that has a lot of links *to* it. You can model importance on the internet by a random walk, much like the agent-based models for pandemic spread. If you do a random walk on the internet, following a randomly chosen link from each page, which pages do you tend to visit a lot, and which pages do you almost never encounter?*

It's a very charming feature of random walks that this question has an answer. This goes all the way back to Andrei Andreyevich Markov and the Law of Long Walks: If a mosquito has a finite set of bogs it can land in, and if each bog has a fixed set of bogs it is connected to, and if the mosquito, at each moment, chooses a destination at random from the bogs it can reach from its current bog, then each bog has a *limiting probability*. That is, each bog has a percentage attached to it, and the

* What happens if you arrive at a page with no links? This is like the problem of getting stuck at a local optimum in gradient descent, and the same fix applies; you start over at a random spot and keep going.

mosquito, wandering for a long time, is likely to spend almost exactly that percentage of the time in that bog.

It's a little easier to grasp what this says in the context of Monopoly. That's a random walk; your little wheelbarrow moves among the forty positions according to the dictates of the dice. Robert Ash and Richard Bishop computed the limiting probabilities of this walk in 1972. The most likely place for the wheelbarrow to be is in jail; on average it spends about 11% of its time there.* But if you want to know where you should put your houses and hotels, you want to know which of the *property* spots is most frequently arrived at: that's Illinois Avenue, where the wheelbarrow spends about 3.55% of its time, substantially more than the 2.5% you'd expect if the forty locations occurred equally often. In any given game, of course, you might miss those spots entirely (at least this is what always seems to happen for my lucky kids when I, obeying probability, pile houses on Illinois Ave.). But overall, if you were to keep track of where *all* the players land in *all* your games over a long stretch of time, the Law of Long Walks says those are the proportions you'll approach.

There's a limiting probability for each of the forty squares, which means you have a list of forty numbers; this is the kind of gadget we called a *vector* in an earlier chapter, and this vector is not just any vector: it's called an *eigenvector.* Like an eigenvalue, it captures something inherent to the long-term behavior of a system that's not apparent just by looking at it, something latent like the smoke in a leaf.

What Ash and Bishop did for Monopoly, the builders of Google did for the entire internet. I should say *do* for the entire internet, since the web, unlike Monopoly, is constantly spawning new locations and shedding old ones. The limiting probability for a website gives you a score, which they called PageRank, and which captures the true geometry of the internet as no one had before.

It's really beautiful how this plays out. The probability of being at any given place on the internet is a complicated sum of geometric progressions, just like the total number of infected people in the two Dako-

* Ash and Bishop assume you stay in jail for three turns or until you roll doubles, rather than paying the $50 to waltz out immediately.

tas, but if there were billions of Dakotas instead of two. That sounds impossible to analyze. But remember: a geometric progression can be exponentially exploding or exponentially decaying, or, at the precise boundary between those two behaviors, it can stay exactly constant. In the case of a random walk like this one, it turns out, one of the geometric progressions is constant, and *all the others* are exponentially decaying away. Their contribution gets smaller and smaller as time progresses and the walker keeps walking. We can see this even in a simple random walk like the mosquito from chapter 4 which flits between two bogs. Markov's analysis showed that, in the long run, the mosquito would spend one third of its existence in the first bog. But we can be more precise: If the mosquito starts out in bog 1, the chance that it's in bog 1 after a single day is 0.8, after two days 0.66, after three days 0.562;* we can put these together into a series

1, 0.8, 0.66, 0.562, 0.493 . . .

which will, in time, converge to 1/3, the long-term chance that the mosquito is there. This sequence isn't a geometric progression. But it is (no surprise by now, I'm guessing) a combination of *two* geometric progressions; that is, the result of adding the two progressions together term by term. One of the progressions is constant:

1/3, 1/3, 1/3, 1/3, . . .

and the other is not, each term being 70% of the previous one.

2/3, 14/30, 98/300, . . .

In time, that second geometric progression inexorably dwindles to near-nothing, leaving behind the constant refrain of 1/3.

As for two bogs, so for a billion websites. The operation of the random

* This is not supposed to be obvious! But you can work it out step by step, by hand, or, if you like matrices, by raising a "transition matrix" to a power.

walk melts away all the inessential complications of the network. What's left at the end is the constant geometric progression, a single number that sits there, unchanging, while everything else dies away, like the pure tone that remains when you hold down a piano key until the harmonics subside. The number that remains is the PageRank.

THE NOTES IN THE CHORD

This intricate superimposition of hundreds or thousands of interlinked models, geometric progressions or something hairier than that, can seem a little baroque at first, like the pre-Newtonian theory of epicycles in which planetary motion was retrofitted to a complex combination of smaller circular motions layered atop larger ones, wheels rolling on wheels. Or, for that matter, like the Elliott Wave Theory with its small and medium waves wriggling atop the ultra-two-ply-megacycles. But the eigenvalue story is real math, and is everywhere. It's at the heart of quantum mechanics—*there's* a geometry story I wish there were room to tell here. But maybe I'll tell one small part of it, because it gives me a chance, here at the end of the chapter, to make an actual mathematical definition. Enough vagueness, let's compute!

Consider a sequence of numbers which is infinite, and not just that, infinite in both directions, like

...	1/8	1/4	1/2	1	2	4	8	...

Any such sequence can be *shifted* one spot to the left:

...	1/4	1/2	1	2	4	8	16	...

In this case, something very handsome happens: shifting the sequence one spot to the left is the same thing as doubling every term. That's because the sequence is a geometric progression! If I'd used a geometric progression with a ratio of 3 between successive terms, the shift would multiply every term of the sequence by 3. But if I'd used a sequence that wasn't a geometric progression, like

...	-2	-1	0	1	2	...

the shifted version

...	-1	0	1	2	3	...

wouldn't be a multiple of the original sequence. The sequences with the special property that shifting multiplies them by something—the geometric progressions, that is—are the *eigensequences* for the shift, and the number the eigensequence gets multiplied by is its eigenvalue.

Shifting isn't the only thing we can do to a sequence. We can, for instance, multiply each term of its sequence by its position; the zeroth term by 0, the first term by 1, the second term by 2, the negative first term by –1, and so on. Let's call that operation the pitch. If we pitch our geometric progression, using the convention that 1 is the zeroth term of the sequence, we turn

1/8	1/4	1/2	1	2	4	8

into

-3/8	-2/4	-1/2	0	2	8	24

This isn't a multiple of the original sequence, so our geometric progression isn't an eigensequence for the pitch. What *is* an eigensequence for the pitch is something like

...	0	0	0	0	0	1	0	...

with a 1 in position 2 and a 0 everywhere else. Pitch this sequence and you get

...	0	0	0	0	0	2	0	...

which is twice the original sequence. So this is an eigensequence of the pitch with eigenvalue 2. In fact, one can show (can you show it?) that sequences with only one nonzero term are the *only* eigensequences for the pitch. (What about the sequence of all zeroes? That does indeed get sent to a multiple of itself by both the pitch and the shift, but the zero sequence doesn't count; there's no well-defined way to say which multiple of itself it is, for one thing.)

You may have heard that, at the very bottom of physics, a particle typically doesn't have a well-defined position or momentum, but rather exists in a sort of cloud of uncertainty as to one or both. One can think of "position" as an operation we can do on a particle, just as shifting is an operation on a sequence. More precisely, the particle has a "state," which records everything about its current physical situation, and the operation called "position" changes the particle's state in some way. For the present discussion, it doesn't matter what kind of entity a state is,* but it does matter that a state is the kind of thing which, like a sequence, you can multiply by a number. And just as an eigensequence for shift is a sequence that gets multiplied by a number when you shift it, an eigenstate for position is a state that gets multiplied by a number—the eigenvalue—when you position it. It turns out that a particle acts like it has a precise location in space exactly when its state is an eigenstate. (And what *is* that location? You can figure it out from the eigenvalue.) But most states aren't eigenstates, just like most sequences aren't geometric progressions. As we've seen above, though, more general sequences like Virahanka-Fibonacci can often be broken up as combinations of geometric progressions; in just the same way, a state that's not an eigenstate can be decomposed as a combination of eigenstates, each one with its own eigenvalue. Some of the eigenstates appear with greater intensity, others with less, and this variation is what governs the probability of the particle being found at any particular spot.

The momentum of a particle is a similar story; *momentum* is another

* If you must know, it's a point in a certain kind of space called a *Hilbert space*—the same Hilbert we last saw messing around with the foundations of geometry at the fin de siècle.

operation on states, which you can think of as being analogous to the pitch. And a particle with a well-defined value of momentum instead of a vague probability cloud would be an eigenstate for that operator, the analogue of an eigensequence for the pitch.

So what kind of particle would have *both* its position and its momentum well defined? That would be like a sequence of numbers that's an eigensequence for both the shift and the pitch.

But there's no such thing! An eigensequence for the shift is a geometric progression. An eigensequence for the pitch is a sequence with exactly one nonzero element.

No nonzero sequence can be both at once.

There's another way to prove this fact, which brings us even closer to quantum physics. (The rest of this chapter would be a great time to have paper and pencil out, or you can just skim, I won't judge.) We start by asking: What happens if we both shift and pitch our sequence? Say we start with any old sequence, like

...	4	2	1	-3	2	...

then shift it

4	2	1	-3	2	...	...

and pitch it (remembering that –3, which was in the first position before, is in the zeroth position now, 1 is in the negative-first position, and so on . . .)

-12	-4	-1	0	2	...	....

One might call this combined operation shift-then-pitch, or shiftpitch for short.* But why did we do it in that order? What if instead we carried out a pitch-then-shift? Our original sequence, pitched, becomes

...	-8	-2	0	-3	4	...

and when you follow up with the shift, you get

* Exercise: Can you come up with an eigensequence for the shiftpitch?

-8	-2	0	-3	4	. . .	. . . .

The pitchshift and the shiftpitch are not the same! We're looking here at a phenomenon called *noncommutativity*, which is the fancy math word for the fact that doing one thing and then doing a different thing doesn't always have the same result as doing the latter thing followed by the former. The math we learn in school is mostly commutative; multiplying by two and then multiplying by three is the same thing as multiplying by three and then multiplying by two. Some operations in the physical world commute, too, like putting on your left and your right glove. Do it in whichever order you like and you end up in the same both-gloves-on situation. But try to put on your shoes before your socks and you'll encounter noncommutativity.

What does any of this have to do with eigenvalues, though? It comes down to the difference between the pitchshift and the shiftpitch. Subtract the shiftpitch from the pitchshift

-8	-2	0	-3	4	. . .	. . .

-12	-4	-1	0	2	. . .	. . .

and the resulting sequence is

4	2	1	-3	2	. . .	. . . .

Exactly the sequence we started with! (Or, to be more careful about it, its shift.) In fact, no matter what sequence you start with, the difference between the pitchshift and the shiftpitch is the shift of the original sequence. Now suppose you had somehow managed to find a mysterious sequence S that was an eigensequence for both pitch and shift; perhaps the shift of S is three times S and the pitch of S is twice S. In that case, the pitch of the shift of S is the pitch of three times S, which must be six times S,* but the same reasoning shows that the shiftpitch S is also six times S, the same thing as the pitchshift. So the difference between the pitchshift of S and the shiftpitch of S is the sequence of all

* We're using here that the pitch is *linear*, which means that multiplying by three and then pitching is the same as pitching and then multiplying by three; another commutativity issue!

zeroes. But that difference is also (the shift of) S itself! So S must have been zero, which, as we stipulated earlier, doesn't count.

The idea of an eigensequence is to capture those circumstances where operations like shifting and pitching act like multiplication. But multiplications commute with each other, while shifting and pitching do not. This creates a tension! The operations are alike, yet not alike. It was the same tension William Rowan Hamilton had to face down in order to formulate his beloved quaternions. He wanted to treat a rotation as a kind of number, but rotations don't commute; rotating 20 degrees around one axis and then rotating 30 degrees around another just isn't the same as doing the two rotations in the other order. In order to get "numbers" that modeled rotations, he had to discard an axiom, the axiom of commutativity. (Of course, two rotations might commute; they do so, for instance, if they are rotations around the same axis. And it's worth noting that, in this case, any point on that common axis is left unchanged by both rotations; it is an eigenthing for both rotations at once, with eigenvalue 1 in each case.)

The situation in quantum physics is much the same. The operators representing momentum and position do not commute. And the difference between the positionmomentum and momentumposition of a particle's state is just—well, not quite the state itself, but the state multiplied by a number called Planck's constant, and denoted $\hbar$. In particular, that means the difference can't be zero,* which in turn implies, just as with the sequences, that a particle's state can never be an eigenstate for both the position and momentum operators. In other words: a particle can't have both a well-defined position and a well-defined momentum. In quantum mechanics we call this the Heisenberg uncertainty principle, and it walks around in a cloak of mystery and intrigue. But it's just eigenvalues.

We have left a lot out, obviously.† We keep on talking about how lots of interesting sequences can be decomposed as combinations of

* Although, relative to the scale of our sensory apparatus, Planck's constant is *pretty close* to zero, which is why objects do seem, to our direct perception, to both be in particular place and moving in a particular way.

† And if you want to learn about what I've left out, Sean Carroll's book *Something Deeply Hidden* is a great nontechnical primer in the mathematics underlying quantum physics.

geometric progressions, and states of particles can be decomposed as combinations of actual eigenstates. But how, in practice, do we actually *carry out* this decomposition? Here's an example from a more classical part of physics. A sound wave can be decomposed into pure tones, which are eigenwaves for some operation; their eigenvalue is determined by their frequency, the note they play. If you hear a C major chord, that's a combination of three eigenwaves, one with eigenvalue C, one with eigenvalue E, and one with eigenvalue G. There's a mathematical mechanism, called the Fourier transform, for separating a wave into its component eigenwaves. It's a rich story, intertwining calculus, geometry, and linear algebra, and it wasn't developed until the nineteenth century.

But you can *hear* the individual notes in the chord, even if you don't know calculus! And that's because this deeply geometric computation, which took mathematicians hundreds of years to develop, is also carried out by a curled-up piece of meat in your ear called the cochlea. The geometry was there in our bodies before we knew how to codify it on the page.

Chapter 13

A Rumple in Space

One of the first examples considered by early adopters of Markov's theory of the random walk, like the Hungarian George Pólya and his student Florian Eggenberger, was the spread of phenomena through two-dimensional space. Ignoring the furious Russian's disdain for real-world applications, they used Markov processes to model smallpox, scarlet fever, train derailments, and steam boiler explosions. Eggenberger called his thesis "The Contagion of Probability." (Since his thesis was in German, that was just one word: *Die Wahrscheinlichkeitsansteckung.*)

Here's how to think of disease spread as a random walk in space. Suppose we start with a dot on a rectilinear grid, like the street map of Manhattan. The dot is a person infected with a virus. A person's personal contacts are the four people adjacent to them in the network. To make this as simple as possible, suppose each person, each day, infects all those people unlucky enough to be their neighbor.

Each person has four neighbors, so you might think we'd see an exponentially growing pandemic with $R_0 = 4$. But it's not like that at all. After one day, five people have been infected:

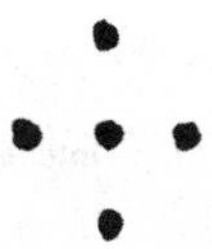

after two days, thirteen:

and after three days, twenty-five:

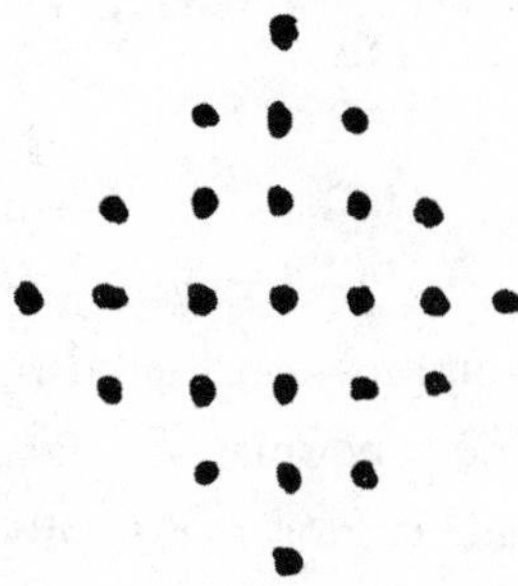

The sequence goes on: 1, 5, 13, 25, 41, 61, 85, 113 . . . The growth is faster than an arithmetic progression (the difference between consecutive terms increases each time)* but much slower than any geometric progression. At first, each term is more than twice the last, but the ratios decrease as you go: 113/85 is just 1.33.

When we first set up our disease model, we saw infections grow exponentially in a geometric progression. This model is different, because we're not just thinking about how many people are infected, but *where* they are, and how distant they are from one another. We're taking geometry into account. The geometry of *this* kind of epidemic is a diagonally oriented square,† centered at patient zero, methodically expanding day by day at a constant pace. It's totally different from what we've seen with COVID-19, which seemed to kindle all over the globe in a span of a few weeks.

* You might have fun checking, à la William Farr, that the differences *between* those differences are always the same, namely 4.

† Though in the Manhattan geometry, as we saw in chapter 8, this square is a circle!

Why is the growth so slow? Because the four people you encounter aren't four people chosen at random from the widespread denizens of North Dakota; they're the people near you. If you're this person:

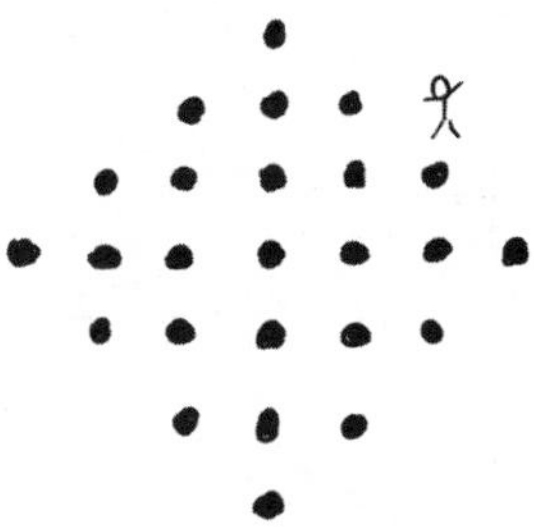

two of the four people you come in contact with tomorrow are already infected. And the uninfected person to your north is also getting a helping of virus from their western neighbor at the same time they're getting it from you. The virus is spreading *redundantly*, encountering the same people again and again.

And this should remind you of our old friend the dithering mosquito, who keeps visiting and revisiting the same city block and only very slowly ventures far from its spawning ground. The total number of places the mosquito can visit if it flies for n days is no more than the number of filled-in squares in a diamond of radius n, which is just not that many. It's hard to explore a geometric network rapidly, whether you're a mosquito or a virus.

Pandemics used to work this way. The Black Death arrived in Europe in Marseilles and Sicily in 1347, then rolled northward across Europe in a steady wave, taking about a year to get to northern France and cover Italy, another year to get make its way across Germany, and another to reach Russia.

Things were already different by 1872, the year of the great North American horse flu epizootic. It's an epizootic, not an epidemic, because "demos" is Greek for "people" and people don't get horse flu.* The term isn't much used anymore, but the 1872 horse flu left such a mark

* So is a plant plague an "epiphytic"? Should be, you'd think, but the word seems to be rarely used in that sense.

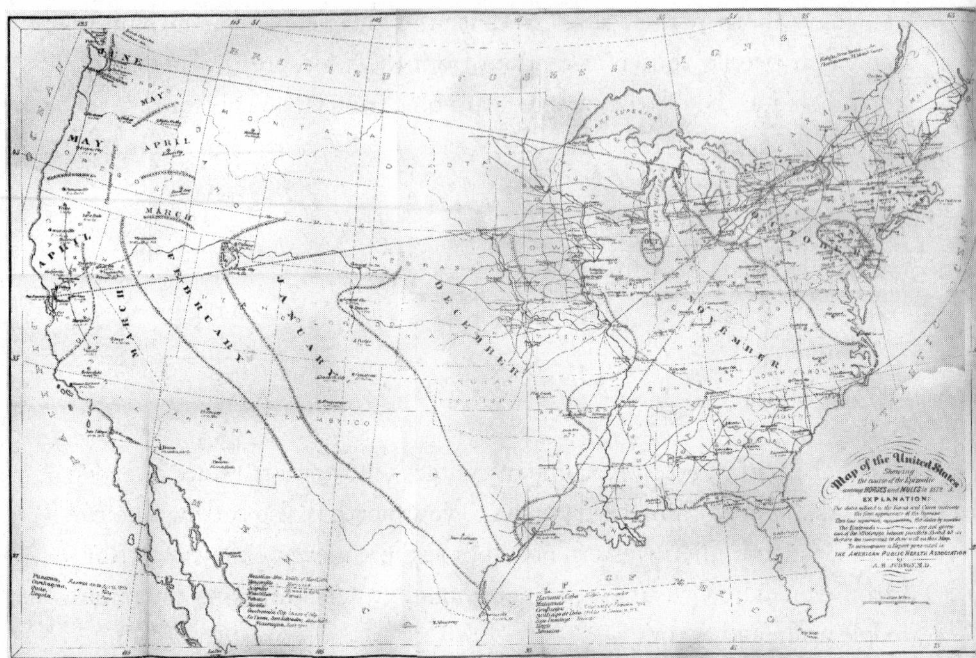

on American life that the word, sometimes pronounced "episoozick" or "epizootiack," was slang for an unclassifiable illness, animal or human, well into the twentieth century. A correspondent in Boston reported, "At least seven eighths of the entire number of animals in this city were suffering from the disease," and Toronto, where the epizootic started in the fall of 1872, was called a "vast hospital for diseased horses." Imagine if all the cars and trucks got the flu and you can start to imagine the impact.

The epizootic expanded outward from Toronto to cover most of the continent, but not in a smooth expanding wave, as the Black Death had done. The disease hopped the border and landed in Buffalo on October 13, 1872; it was in Boston and New York City by October 21, and a week after that was seen in Baltimore and Philadelphia. But it didn't get to inland places like Scranton and Williamsport, which are closer to Toronto, until early November. By that time horses as far south as Charleston were already sick. The westward progress was just as uneven; the flu got to Salt Lake City by the second week in January and arrived in San Francisco in mid-April, but didn't make it to Seattle, at roughly the same distance from Toronto as the crow flies, until June.

That's because flu doesn't travel as the crow flies. It travels as the train runs. The transcontinental railroad, then only three years old, carried horses and disease from the center of the country directly to San Francisco, and rail lines joining Toronto with the big coastal cities and Chicago seeded early outbreaks in those cities, too. Unmechanized travel to places away from the rail lines was slower, and so the epizootic got there later.

A WRINKLE IN PIZZA

"Geometry" means "measuring the Earth" in Greek, and that's exactly what we're doing. To assign a patch of land, or a set of people, or a set of horses a *geometry* is, at bottom, to assign a number to any two points, which we are to interpret as the distance between them. And a fundamental insight of modern geometry is that there are different ways to do this, and a different choice means a different geometry. We have seen this already, when we charted the distance between cousins on the family tree. Even when we're taking about dots on the map, we have multiple geometries to choose from. There's the crow-fly geometry, where the distance between two cities in the United States is the length of a straight line* joining one to the other. And there's the geometry where the distance between two cities is "how long it would take you to get from the former to the latter in 1872," which is the relevant one for the epizootic.† In that metric (for a *metric* is the term of art we use, in geometry, instead of the cumbersome "assignment of a distance to each pair of points"), Scranton is farther from Toronto than New York is, while being much closer as the crow flies. You can do whatever you like—this is math, not school! Maybe your metric is "the distance

* "But what about the curvature of the Earth's surf—" If that's an issue for you, wait a few pages, we'll get there!

† And the relevant ones for drawing isochrone maps like the one we saw in chapter 8.

between the two places on a list of all U.S. cities in alphabetical order"—now Scranton is once again closer to Toronto than it is to New York.

The idea that geometry is not fixed, but can be altered by our will, is familiar to several generations of bookish American children, from the following picture:

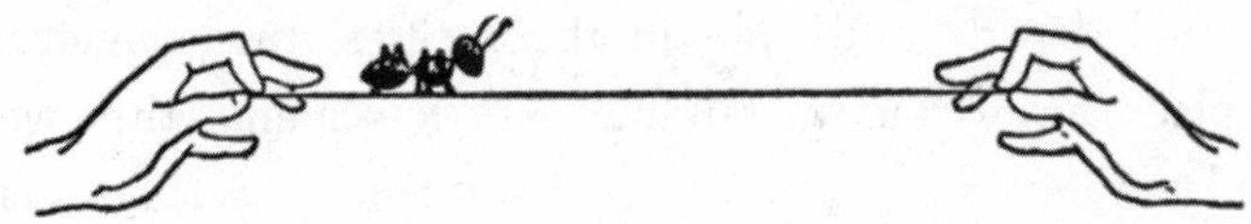

This is a geometrical demonstration by Mrs Whatsit, one of the three interplanetary witch/angels who helps three children defeat a cosmic evil in *A Wrinkle in Time.** How do they traverse the universe faster than the speed of light? "We have learned to take shortcuts whenever possible," she says. "Like in math."

The ant, she explains, is close to one end of the string, but very far from the other. But move the string in space and that distance collapses to almost nothing, allowing the ant to hop straight from one hand to the other. "Now, you see," Mrs Whatsit explains, "he would be there, without that long trip. That is how we travel." The bending of the string is the wrinkle that gives the book its title. The witches call it a "tesseract." In the context of 1872 it's called a "railroad." The rails joining Chicago to San Francisco are an alteration of the continent's geometry, a change in the metric that makes two points closer to each other than we might naively have thought. Or points might get farther apart! The epizootic of 1872 got as far south as Nicaragua, but never crossed into South America. That's because the Isthmus of Panama presented to would-be crossers "an almost impassable swamp intersected by rough and difficult mountain ranges." Colombia and Nicaragua are quite nearby on the surface of the Earth, but in the horse-travel metric they were effectively an infinite distance apart.

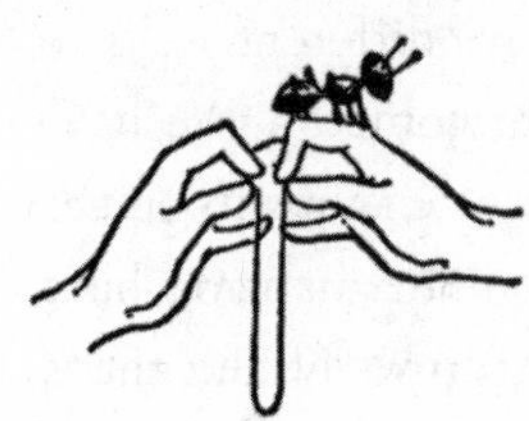

* Mrs Whatsit is also the answer to a great trivia question: What character has been portrayed on-screen by both Alfre Woodard and Reese Witherspoon?

The contemporary world is wrinkled as heck. Before we even knew there was a pandemic, COVID-19 was on planes between China and Italy, between Italy and New York, between New York and Tel Aviv. If we somehow didn't know there was such a thing as an airplane, we could infer it from the nature of the pandemic spread. And yet the standard geometry of the Earth's surface still plays a role. The hardest-hit parts of the United States in the spring of 2020 weren't the cities with the international airports and the jet-setting residents; they were the places you could drive from New York. Pandemics travel both fast and slow, in whatever vehicle they can hitch a ride on.

"To put it into Euclid, or old-fashioned plane geometry," Mrs Whatsit explains a little later in the book, "a straight line is *not* the shortest distance between two points."* But in *new*-fashioned geometry we can stand up for Euclid. What is the shortest distance between two points on the surface of the Earth, like Chicago and Barcelona? It can't be a straight line in the usual sense unless you're a really good digger, because the Earth's surface, unlike Euclid's plane, has some curvature to it. There *are* no straight lines on the surface of the sphere.

But there has to be a shortest path. And it might not be the one you think. Chicago and Barcelona are at nearly the same latitude, 41 degrees north. If you connected those two cities by a straight line on a map, you'd travel due east about 4,650 miles along the forty-first parallel. But that's going the long way! The actual shortest path appears on the map to arc northward, departing North America near the small cod-processing town of Conche, Newfoundland, and reaching its northernmost point in the Atlantic at a latitude of roughly 51 degrees. That shaves more than two hundred miles off your trip.

The idea that moving east or west along a latitude line is "straight" is one of those superficially appealing formulations that collapses when you think about what it would really imply.† Suppose you start walking due west when you're two meters from the South Pole. In a few seconds

* Why do we always say this, when a line is not a "distance"? Apparently the strange phrasing originates in a bad nineteenth-century translation of Legendre's geometry textbook, which describes the line more accurately as "le plus court chemin" (the shortest path). The translator who blew it here? The historian and essayist Thomas Carlyle, in his pre-fame gig as a high school math teacher in Kirkcaldy, Scotland.

† Insert comparison with your least favorite political ideology here.

you will have described a very small, very cold circle. You will not feel like you're walking in a straight line. Trust this feeling.

The best idea of what a straight line should mean in the spherical setting was there in Euclid all along; we simply *define* a straight line to be a shortest path. (Actually, that's a little more like a line segment, which unlike a line has a definite beginning and end.) All shortest paths on the sphere, it turns out, are pieces of "great circles," so-called because they're the biggest circles you can draw on a sphere, the ones that pass through two directly opposite points. And a great circle is what we mean by a line on the sphere. The equator is a great circle, but other latitude lines are not so great. A line of longitude is a great circle, once you pair it with its antimeridian, the longitude line directly across the Earth from it. So due north or due south travel really *is* straight-line motion. If the asymmetry between north-south and east-west bothers you, just remember that it's built into the way we reckon longitude and latitude. Meridians meet; parallels don't. There is no West Pole.

Although we would be free to make one up! We can put up a pole wherever we like. Nothing stops us, for instance, from declaring one pole to be in the middle of the Kyzylkum Desert in Uzbekistan, and the other directly across the Earth in the South Pacific. Harold Cooper, a software engineer in New York, made a map like that. Why? Because then about a dozen of your meridians—or, as Cooper calls them, "avenues," go straight up and down along the length of Manhattan, and so the perpendicular lines of latitude are crosstown streets. In this way you can extend the New York street grid to the remainder of the globe.* The University of Wisconsin math department is near the corner of 5086th Avenue and West Negative 3442nd Street, which perhaps accounts for our extremely downtown vibe.

That we draw lines of latitude as straight lines on our world maps is an inheritance from the person who devised those maps, Gerardus Mercator. Born Gerhard Cremer, he did the fashionable thing among scientists of his time and adopted a Latinized version of his name: "Mercator" is Latin for

* You can Manhattanize your own location at extendny.com.

"merchant," which is what "Cremer" means in low German. (If I did the same I'd be Jordanus Cubitus,* which does have a certain ring to it.) Mercator studied mathematics and cartography with the Flemish master Gemma Frisius, wrote a popular manual for cursive handwriting, was imprisoned by religious zealots for most of 1544 on suspicion of Protestantism, developed and taught a geometry course to high school students in Duisburg, and made lots and lots of maps. He is known today for the one he made in 1569, which he called *Nova et Aucta Orbis Terrae Descriptio ad Usum Navigantium Emendata* ("New and Expanded World Map Corrected for Sailors") and which we now call the Mercator projection.

Mercator's map was good for sailors because what mattered to them wasn't taking the absolutely shortest route; what mattered to them was not getting lost. At sea, you can use a compass to keep yourself at a fixed angle to the north (or at least the magnetic north, which is hopefully not too far off). In a Mercator projection, north-south lines of longitude are vertical lines, latitudes are horizontal, and all angles on the map are the same as they are in real life. So if you set a course due west, or 47 degrees from north, or whatever, and stick to it, the path you travel—called a *loxodrome* or a *rhumb line*—is a straight line on the Mercator map. If you have the map and a protractor, it's easy to see where on shore the rhumb line will land you.

Mercator's map gets some things wrong, though. It has to, because meridians are depicted on the map as parallel lines, which of course cannot meet. But meridians do meet—twice, in fact, at the North and South Poles. So something has to go wrong for Mercator's map as you go far to the north and south. Indeed, Mercator cut his off at parallels well short of the poles to avoid the painfully apparent Arctic and Antarctic distortions. Latitude lines near the poles get farther and farther apart on the map, while in real life they're separated by the same distance. That makes things in the polar region look bigger than they really are. In the Mercator projection, there's as much Greenland as there is Africa. In reality, Africa is fourteen times bigger.

* "Ellenberg" is German dialect for "elbow," or at least so family lore has it.

Couldn't there be a better projection? You might want great circles to show up as straight lines (a *gnomonic* projection) and you might want the relative areas of geographic objects to match real life (an *authalic* projection) and you might want the projection to get angles right (a *conformal* projection, of which Mercator's is one). But you can't have all those things. The reason comes down to Carl Friedrich Gauss's proof of the pizza theorem. Gauss didn't call it the pizza theorem, though surely he would have, had New York–style slices been available to him in nineteenth-century Göttingen. Instead he called it the *Theorem Egregium*, which in contemporary English is roughly "The Awesome Theorem." I will not trouble you with the precise statement, but instead draw a picture:

A smooth curved surface, if I zoom in on it enough, is going to look like one of these four pictures. On the left we have a section of a sphere; in the middle, a flat plane and a piece of a cylinder; and on the right, we have a Pringle. Gauss devised a numerical notion of "curvature"—the flat plane has curvature 0, as does the cylinder. The spherical surface has positive curvature, and the Pringle's curvature is negative. A more complicated surface like this one

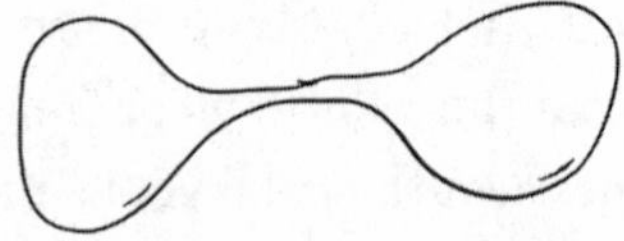

might be positively curved at once place and negatively curved at others.

It turns out that if you can map one surface to another in a way that keeps angles and areas the same, then it gets the metric right, too—in other words, the *geometry* of the two surfaces is the same. The distance between two points on one surface is the same as the distance between the corresponding points on the other.

The Awesome Theorem is this: if you can project one surface onto another in a way that keeps the geometry the same—in other words, if you are allowed to bend it or twist it but not *stretch* it—the curvature must stay the same, too. An orange peel is a piece of a sphere, and is positively curved; so you can't flatten it out to a surface of zero curvature. And a piece of pizza, cut as it is from a flat round pie, has curvature zero. It is free to bend into a curvature-0 cylindrical shape by a droop of the tip:

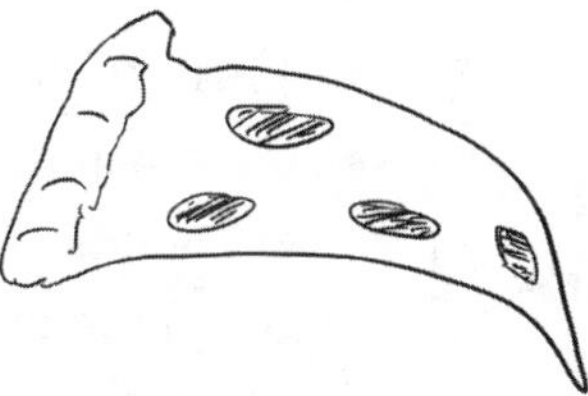

or a curl-up of both edges:

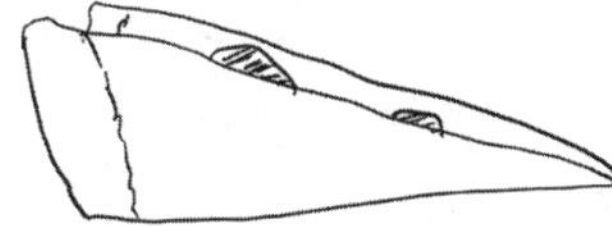

but *it cannot do both*. Because that would make it a Pringle. A pizza is not a Pringle, and cannot be made into one, because the curvature of a Pringle is negative, not zero. And that is why, when you walk down Amsterdam Avenue with your take-out slice, you curl the edges up—because the pizza's zero curvature, and Gauss's theorem, prevents the tip from dipping down and dropping hot cheese down the front of your shirt.

One doesn't quite need the full awesomeness of the Awesome Theorem to know you can't have a map of the Earth's spherical surface that satisfies all your geometric desires. The issue is captured in an old riddle: A hunter wakes up one day, crawls out of his tent, and goes looking for a bear. He walks ten miles south; no bear. Walks ten miles east; still no bear. Walks ten miles north, and finally there's a bear, right in front of his tent.

The riddle: What color is the bear?

If you don't know the riddle, here's another version. Start at Libreville,

Gabon, just about on the equator, go straight north until you get to the North Pole, turn 90 degrees right, and head back south, striking the equator again near the town of Batahan in Sumatra; finally, make another right-angle turn and head due west a quarter of the way around the Earth until you get back to Libreville.

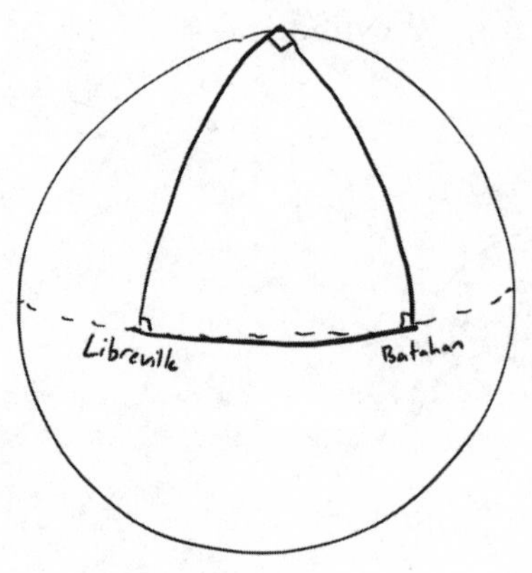

Remember, our imaginary perfect projection is supposed to render great circles as straight lines. The path we followed consists entirely of segments of great circles, so on our imaginary perfect map it has to be three straight line segments—a triangle. But each angle on the map must be what it was on the globe: namely, 90 degrees. No triangle in the plane can have three right angles; and that ends the dream of the perfect map.

Oh, and the bear was white. Because his tent had to be at the North Pole, so it was a polar bear!

(Slap knee.)

AND WHAT IS YOUR ERDŐS-BACON NUMBER?

From the geometry of the flat map to that of the sphere already involves some rich mathematics. But we have in mind yet more radical departures from Euclid's book. What about the geometry of movie stars? Not the curves and planes of their physical bodies—enough has been written about those—but the network formed by their collaborations. For actors to have a geometry, we need a metric, a notion of how far apart two stars stand in the firmament. For this we use the "costar distance." A *link* between two actors is a movie in which they both appeared, and the distance between two actors is the shortest chain of links that joins them. George Reeves was in *From Here to Eternity* with Jack Warden, who appeared in *The Replacements*—his last movie—with Keanu Reeves.

So the distance between George and Keanu is 2. Or rather, it's at *most* 2; we still have to check there's no *shorter* path between the two, which would be a single link, a movie both actors appeared in. The older Reeves died five years before Keanu was born, so 2 it is.

There's nothing special about movie stars; you can define the same kind of distance in any collaboration network. In fact, the idea is much older in the context of mathematicians, where a link joins every two mathematicians who wrote a paper together. The geometry of mathematicians has been a party game ever since Casper Goffman wrote a half-page note in the *American Mathematical Monthly*, "And What Is Your Erdős Number?," in 1969. Your Erdős number is your distance from the mathematician Paul Erdős, who's considered central to the network thanks to his immense number of collaborators—511 at last count, but even though he died in 1996 he still occasionally gains new links from authors writing up papers using ideas they learned while talking to him. Erdős was a famous eccentric, occupying no real home, unable or purportedly unable to cook or do laundry,* peripatetically crashing at this or that mathematician's house, proving theorems with his hosts, consuming healthy doses of Benzedrine. (At least once, he declined to join a group of mathematicians for a post-lunch coffee, explaining, "I have something much better than coffee.")

Your Erdős number is the length of the shortest chain of links connecting you to Erdős. If you are Erdős, your Erdős number is 0; if you are not Erdős but wrote a paper with Erdős, your Erdős number is 1, if you didn't write a paper with Erdős but you wrote one with someone whose Erdős number is 1, your Erdős number is 2, and so on. Erdős is connected to just about every mathematician who ever wrote a collaborative paper, which is to say that just about every mathematician has an Erdős number. Marion Tinsley, the master of checkers, had an Erdős number of 3. So do I; I wrote a paper in 2001 about modular forms with Chris Skinner, who as a Bell Labs intern in 1993 wrote a paper about

* A disagreeable feature of the Erdős legend: it encourages some mathematicians to see domestic work as somehow beneath our station and beyond our capabilities at once. And yet we eat the food and wear the clean shirts. Fact: thinking about mathematics while washing the dishes is good for both mathematics and, if you are as prone to reveries as most mathematicians are, the dishes.

zeta functions with Andrew Odlyzko, who wrote three papers with Erdős between 1979 and 1987. And the distance between Tinsley and me is 4. The three of us form an isosceles triangle:

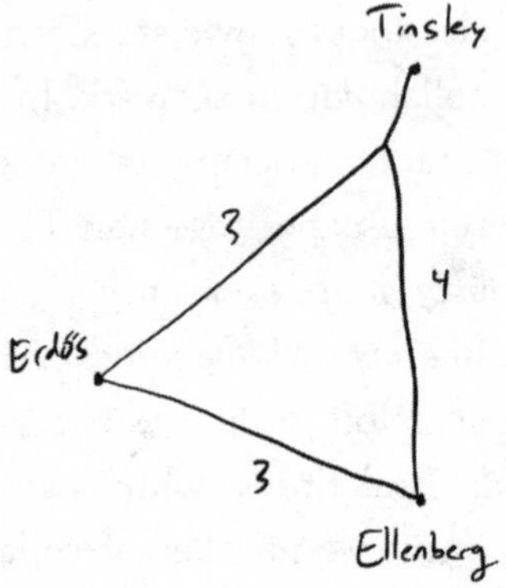

The triangle looks a little crimped, because Tinsley only wrote one joint paper in his short mathematical research career, with his student Stanley Payne, so that link forms part of the line from Tinsley to Erdős, and also part of the line from Tinsley to me.

Now zoom out, so that we can see every one of the 400,000-some mathematicians who have ever published a paper. And we connect every pair of coauthors by a link:

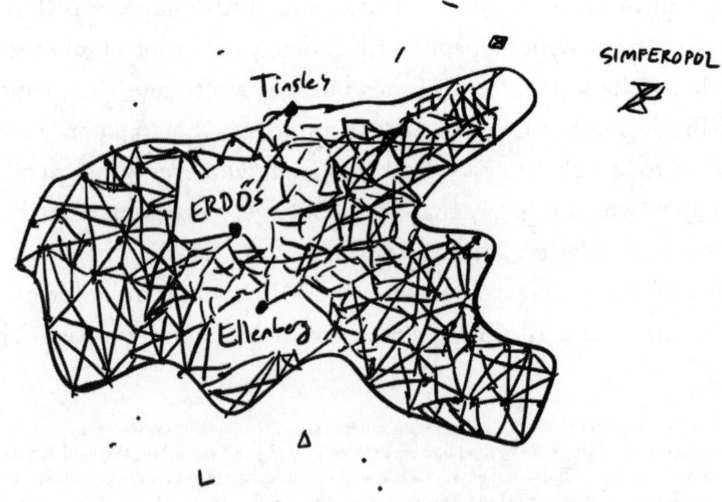

The big glob (or as the technical terminology has it, the big "connected component") is the 268,000 mathematicians who are connected by some chain of links to Erdős. What looks like dust is the collection of lonely mathematicians who have never collaborated on a paper; there are about 80,000 of those. The rest of the mathematical population is broken up into tiny clusters, the largest of which consists of thirty-two applied mathematicians mostly based at Simferopol State University in Ukraine. Every single mathematician in the big component is connected to Erdős by a chain of no greater than thirteen links; if you have an Erdős number at all, it's at most 13.

It may seem weird that there's such a big gap between the one giant glob and the almost totally disconnected collection of mathematical loners left out, instead of a bunch of globs of various sizes. But it's actually the general way of things, a fact we know thanks to Erdős himself. The notion of the Erdős number doesn't just honor Erdős's sociability; it's a shout-out to the pioneer work that Erdős, together with Alfréd Rényi,* did on the statistical properties of large networks. Here's what they showed. Suppose you have a million dots, where by "a million" I mean "some large number I don't care to specify." And suppose you have in mind some number R. To make a network out of these dots, you have to decide which pairs of dots are connected in the network and which aren't; and we do this completely by chance, saying that a pair of edges is connected with probability R in a million. Say R is 5. Each dot has a million (okay, 999,999) other dots it *can* be connected to, but has only a 5 in a million chance of being connected to each one; a million 5-in-a-million chances add up, and you'd expect each dot to be connected to about five others. R is the average number of "collaborators" each dot has.

What Erdős and Rényi discovered is a tipping point. If R is less than 1, the network almost surely falls apart into countless disconnected pieces. But if R is bigger than 1, it's just as certain that there will be one giant glob taking up a large portion of the network. Within the glob, every dot has a path to every other, the way almost every mathematician

* In case anyone who's reading this doesn't speak Hungarian, these names are pronounced, roughly, "Airdish" and "Rainy."

has a path to Erdős.* A tiny change in R, from 0.9999 to 1.0001, creates a huge change in the behavior of the network.

We've seen this before. Suppose the dots are the population of South Dakota, which is indeed about a million. And suppose two dots are connected if people come into close contact and inhale each other's breath. This is not *exactly* a good model for infection spread—it doesn't take into account that different people are infected at different times—but it's close enough for consulting work. The average number of people an infected person infects is R, which now rips off its rubber mask and reveals itself to have been R_0 all along. Less than 1? The disease stays localized in some tiny segment of the network. Greater than 1? It gets just about everywhere.

Another thing Erdős is famous for is the idea of "The Book," a volume that contains the most perfectly compact, elegant, illustrative proof of every theorem. Only God gets to see it. You didn't have to believe in God to believe in the Book; Erdős himself, though raised in a Jewish family, had no use for religion. He called God "the Supreme Fascist," and once remarked, while visiting Notre Dame, that the campus was very charming but there were too many plus signs. And yet he ended up with a view of mathematical reality not so different from that of the devout Hilda Hudson, who also believed a truly good proof to be a case of direct communication with the divine. Poincaré, neither a believer nor a mocker of belief, was more skeptical about this kind of revelation. If a transcendent being knew the true nature of things, Poincaré wrote, "he could not find words to express it. Not only can we not divine the response, but if it were given to us, we could understand nothing of it."

GRAPHS AND BOOKWORMS

The Erdős game for movie stars was invented in the 1990s by a bunch of bored college students who observed that Kevin Bacon had seemingly

* What if R is *exactly equal* to 1? There are hundreds and hundreds of papers about that; it often happens that the cases straddling a boundary between two regimes are the ones where the most baroquely rich math is hiding.

been in a movie with everyone; he was the Erdős of 1980s and '90s Hollywood. And so you can define a movie star's Bacon number to be the distance, in the costar geometry, from Kevin Bacon. Just as almost every mathematician has an Erdős number, almost every actor has a Bacon number. I happen to have both. My Bacon number is 2, thanks to being in *Gifted* with Octavia Spencer, who played "Big Customer" opposite Kevin Bacon's "Jorge" in the 2005 Queen Latifah vehicle *Beauty Shop*. So my "Erdős-Bacon" number is 3 + 2, or 5. The club of people with an Erdős-Bacon number is pretty small. Danica McKellar, who starred in *The Wonder Years* as a teen and who, by all accounts from my friends who taught her at UCLA, would have had a long career in math had she not chosen acting, has an Erdős-Bacon number of 6. Nick Metropolis* developed and gave his name to one of the most important algorithms in random walks, one that helped realize Boltzmann's dream of understanding the properties of gases, liquids, and solids via one-by-one analysis of molecules and their ceaseless billiard-ball collisions. But long after that, and more germanely to the present discussion, he played a bit part in the Woody Allen movie *Husbands and Wives*, thus beating me out with an Erdős-Bacon number of 4 (he's distance 2 from both.)†

Mathematicians don't generally call these networks networks; we call them graphs, which is incredibly confusing, since they have nothing to do with the graphs of functions you might have drawn in school. We blame the chemists for this. A *paraffin* is a molecule made up of just carbon and hydrogen atoms; a really simple one, with just one carbon atom and four hydrogens, is methane (the "cow burps cause global warming" gas). The paraffin wax the word probably reminds you of is a heftier molecule, with dozens of carbon atoms. Nineteenth-century chemists could tell how much carbon and hydrogen each compound had in it by "elemental analysis," which is a fancy word for "set it on fire and see how much of the result is carbon dioxide and how much is water." But they soon began to understand that there were molecules with the same chemical formulas

* Together with Augusta and Edward Teller and Arianna and Marshall Rosenbluth. (Arianna Rosenbluth died of COVID-19 on December 28, 2020.)

† Two coauthors of Erdős, Daniel Kleitman and Bruce Reznick, each claim Erdős-Bacon number 3 by virtue of movies in which they appeared as extras. Is this cheating? Not for me to say. (Yes, it's cheating, come on.)

that nonetheless had really different properties. The point, they came to see, is that counting isn't everything. Molecules have geometry. The same atoms can be arranged in different ways.

Butane, the stuff that burns in a Zippo lighter, is C_4H_{10}: four carbon atoms, eight hydrogens. Those carbon atoms might be linked in a chain of four

```
    H H H H
    | | | |
  H-C-C-C-C-H
    | | | |
    H H H H
```

or arranged in a Y shape, giving a molecule we call "isobutane."

```
  H   H       H
   \ /         \ / H
    C    H      C
   /     |     / \
  H   \  |  /      H
         C
         |
         C
       / | \
      H  H  H
```

The more carbon atoms there are, the more different geometries you can have. Octane, as its name suggests, has eight carbon atoms; in its standard form it has the eight carbons lined up in a line. But the C_8H_{18} that goes in your gasoline and gives you a smooth ride is this one:

```
          H       H
          |       |
        H-C-H   H-C-H
    H     |   H   |   H
    |     |   |   |   |
  H-C  -  C - C - C - C-H
    |     |   |   |   |
    H     |   H   H   H
          |
        H-C-H
          |
          H
```

whose science name is 2,2,4-trimethylpentane. I can see why they don't label gas pumps with a 2,2,4-trimethylpentane rating. But the usual nomenclature leads to the moderately strange fact that the stuff chemists call octane is extremely low-octane.

A molecule is a network; the dots are atoms, connected by bonds. In a paraffin, the carbons aren't allowed to link up into a closed loop; so the network of carbon atoms forms a tree, just like the positions in a checkers game.

It turns out each carbon atom needs to bond with four other atoms, while hydrogen atoms monogamously bond just once; given this, you can satisfy yourself that the two butanes drawn above are the *only* way four Cs and ten Hs can get together. For pentane, with five carbons, there are three ways

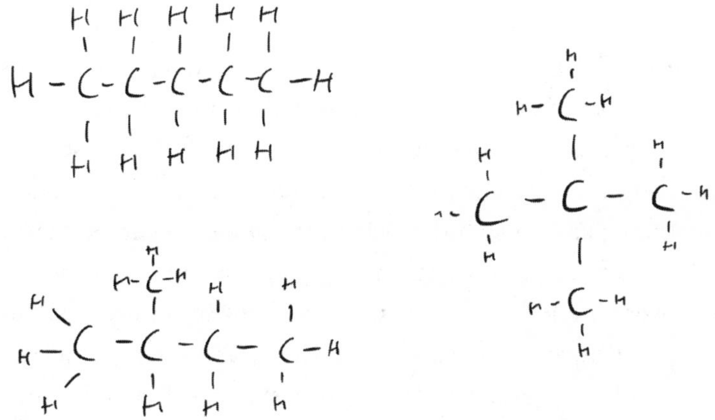

and for hexane, five ways (I'm not going to draw all the little hydrogens this time):

```
C-C-C-C-C
```

```
    C   C
    |   |
C - C - C - C
```

```
  C
  |
C-C-C-C-C
```

```
  C
  |
C-C-C-C
  |
  C
```

```
    C
    |
C-C-C-C-C
```

Virahanka-Fibonacci again! But no; there are nine seven-carbon atoms, not eight. There are just not very many small numbers, which means there's a lot of overlap between small counting problems. It's a challenge for standardized tests; I see why you might want to ask a student, "What is the next number in the sequence 1, 1, 2, 3, 5, . . . ?" but if they respond, "9, because I assumed we were counting paraffins," you will have to concede that the smart aleck deserves full credit.*

A good picture clarifies the mind marvelously. Chemists leapt forward in their understanding when they started drawing pictures like the ones on these pages, which they called the *graphic notation*. Mathematicians, too, were inspired by the new geometric questions the chemists had uncovered, and quickly transposed them into pure mathematics. How many different structures were there, and how should this wild geometric zoo be organized? The algebraist James Joseph Sylvester was one of the first to take these questions seriously. Chemistry, he wrote, has "a quickening and suggestive effect upon the algebraist." He compared its action on the mathematical mind to the inspiration poets draw from paintings:

> In poetry and algebra we have the pure idea elaborated and expressed through the vehicle of language, in painting and chemistry

* The sequence that counts the number of paraffins with more and more carbon atoms is, of course, recorded in the On-Line Encyclopedia of Integer Sequences: it is sequence A000602.

the idea enveloped in matter, depending in part on manual processes and the resources of art for its due manifestation.

Sylvester seems to have understood the phrase "graphic notation" to mean that the atomic networks the chemists were drawing were called "graphs"—he adopted that notation in his own work, and here we are, stuck with it.

Sylvester was English, but also in a sense the first American mathematician; as an established senior researcher, in his sixties, he joined the faculty of the just-launched Johns Hopkins University in 1876, a time when American mathematics barely existed, and students had to sail to Germany to learn anything serious. He looked the part of the distinguished elder scholar. One contemporary described him as "a giant gnome, beard on enormous chest, fortunately no neck, for no neck could upbear such a monstrous head, bald but for the inverted halo of hair collaring its juncture with the broad shoulders." Everybody noticed Sylvester's huge head. Francis Galton, the statistician and phrenology enthusiast, reminisced to Karl Pearson, his protégé: "It was a treat to watch the great dome." (Galton was writing in complaint about Pearson's discovery that cranial capacity was not, as the large-headed Galton had always believed, correlated with intellectual achievement.)

The American mathematical enterprise might have gotten going a lot earlier, for Sylvester had actually been hired in the United States once before, back in 1841, at the University of Virginia. This might have seemed the perfect launching spot, for Virginia was the mathophile Thomas Jefferson's university, where one of the three nonnegotiable admissions requirements was to "demonstrate a thorough knowledge of Euclid." But things went badly from the start. If you have someone in your life who likes to complain about how entitled American college students are these days, you should encourage them in the strongest possible terms to read about American college students of the early nineteenth century. At Yale, in 1830, forty-four students, including the son of Vice President John C. Calhoun, were expelled after refusing to take a final exam in geometry that had been changed from open-book to closed, an event known as the "Conic Sections Rebellion." At

Virginia, student unrest had gone past classroom insubordination to outright violence. Students gathered in masses to chant, "Down with the European professors," and it was routine for rocks to be thrown through the windows of disfavored faculty members. In 1840, student rioters shot and killed an unpopular law professor.

Sylvester was not just a European, but a Jew: a local newspaper complained that the people of Virginia were "Christians, and not heathen, nor mussulmen, nor Jews, nor Atheists, nor Infidels," and that their professors should be held to the same religious standard. Sylvester's appointment had been held up by the fact that he did not, strictly speaking, hold a degree. This, too, was a religious problem. Cambridge required graduates to swear an oath of adherence to the Thirty-Nine Articles of the Church of England, which Sylvester couldn't do. Fortunately, Trinity College Dublin, which had to accommodate not just Protestants but Catholic students, too, required no such oath, and awarded Sylvester a BA shortly before he departed for America.

Sylvester, at that time physically unimposing (despite the head) and hardly older than the students he was teaching, found his attempts to maintain classroom discipline met with insolence and scorn. His attempts to punish William H. Ballard of New Orleans for reading a book under his desk in class ballooned into a dispute the entire faculty had to adjudicate. Ballard charged Sylvester with the worst violation he could imagine, accusing his professor of addressing him the way a white man back in Louisiana would talk to a slave. To Sylvester's great frustration, many of his colleagues saw things Ballard's way. Amazingly, it got worse from there. Later the same term, Sylvester made the mistake of noting some errors in a student's oral exam, which induced the student's older brother to avenge his family honor by punching Sylvester in the face. Sylvester, surely aware of the fate of the unpopular law professor, had taken the precaution of arming himself with a sword cane, with which he launched a counterattack. The student's brother was uninjured, but that was it for Sylvester at Virginia. He meandered around the United States for some months, looking for a more suitable situation. He came close to securing a position at Columbia University, but again found himself judged a bit too Old Testament for the job. The trustees told him, in what must have seemed to them a defense, that they were in no way prejudiced against foreign professors, and would have found an American Jew just as unsuitable for hire. That failure also sank a courtship Sylvester was pursuing in New York.

"My life is now pretty well a blank," Sylvester recalled. He went back to England, lonely and unemployed, and made a living here and there—as an actuary, lawyer, and private math tutor to Florence Nightingale—while doing algebra on the side. It took him more than a decade to work his way back into university life. It didn't help that, by the time the rumors about his time at Virginia made it across the Atlantic, a lot of people thought he'd killed the kid he'd merely lunged at with a weaponized walking stick. Sylvester also had an unfortunate taste for academic squabbles, as you might guess from papers like his 1851 "An Explanation of the Coincidence of a Theorem Given by Mr. Sylvester in the December Number of This Journal, With One Stated by Professor Donkin in the June Number of the Same," which I will paraphrase: "Although I sometimes submit papers to your journal, I don't regularly read it, so I didn't notice Donkin's earlier paper, which concerns a theorem I actually proved nine years ago, but didn't tell anybody about because I thought it was too easy and must have been published somewhere else already." He concludes with a very attenuated sorry-not-sorry to Donkin—and here I have to quote, it's just too rich—"whose high and worthily earned reputation, not to speak of the disinterested love of truth for its own sake, apart from personal considerations, which animates the labours of the genuine votary of science, must make him indifferent to whatever credit might be supposed to result from the first authorship or publication of the very simple (however important) theorem in question." He applied for the Gresham lectureship in geometry that Karl Pearson would later hold, gave a sample lecture, and was turned down. He never did marry.

For all this strife, he eventually did retake his place in institutional English mathematics, and spent the middle of the nineteenth century helping to invent the subject we now call linear algebra. For Sylvester, this was barely separate from the geometry of space, a subject to which he incessantly returned. Linear algebra allows one to extend one's intuition about three-dimensional space to spaces of whatever dimension you like;* so the mind naturally turns to the question of whether some such higher-dimensional space might be where we actually live. Sylvester liked the metaphor of the "bookworm," a perfectly flat creature that lives inside a two-dimensional sheet of paper, having no idea and no way of forming the idea that there's more to the world than that. What if, Sylvester asks, we three-dimensional beings are just as limited? Does our ability to imagine enable us to be more than worms and see beyond our three-dimensional "page"? Perhaps, Sylvester suggested, our world is "undergoing in the space of four dimensions (space as inconceivable to us as our space to the suppositious bookworm) a distortion analogous to the rumpling of the page . . . ." This is, of course, the same theory laid out by Mrs Whatsit, with the bookworm standing in for the string-walking ant.

Sylvester once opened a lecture by apologizing, "An eloquent mathematician must, from the nature of things, ever remain as rare a phenomenon as a talking fish," but this is the obligatory apology of someone rather proud of his verbal skill. Indeed, like William Rowan Hamilton and Ronald Ross, Sylvester was a poet. He wrote what may have been the only sonnet ever addressed to an algebraic expression, "To a missing member of a family group of terms in an algebraical formula."† Sylvester took it even further, writing an entire book, *The Laws of Verse,* which aimed to place the technical practice of poetry on a rigorous mathematical foundation. Here Sylvester, though he gives no indication of having ever studied Sanskrit prosody, adopts the same point of view Virahanka did thirteen hundred years earlier, that a stressed syllable is just twice as long as an unstressed one. (Sylvester uses the musical terms "crotchet" and "quaver" for what Virahanka called laghu and guru.)

---

* It's linear algebra that provides us the theory of "vectors" that's so central to machine learning, and which gave Geoffrey Hinton the wherewithal to describe fourteen-dimensional space as just like a three-dimensional space to which you loudly say "fourteen!" every so often.

† Daniel Brown, in his extremely interesting book *The Poetry of Victorian Scientists,* argues that this poem can be read as addressing Sylvester's exclusion from the university system on account of his faith, casting Sylvester himself as the "missing member." This Brown is no relation to the *Da Vinci Code* Dan Brown, despite his skill at finding religious symbolism in the work of historical scientists.

I'm being careful to describe Sylvester's goal as *elevating* poetry to a mathematical subject, not *reducing* it to one, since that was certainly Sylvester's view. He was a lifelong opponent of the popular view of mathematics as a desultory trudge through deductive steps. For Sylvester, mathematics was a way to touch a transcendent reality; you propelled yourself by intuition to get there, only after the fiery moment circling back and constructing a logical scaffold to help others attain the vista. He attacks the customary pedagogy of his time, explicitly connecting it with the stultifying Anglican conventionalism that had denied him academic places:

> The early study of Euclid made me a hater of geometry, which I hope may plead my excuse if I have shocked the opinions of any in this room (and I know there are some who rank Euclid as second in sacredness to the Bible alone, and as one of the advanced outposts of the British Constitution) by the tone in which I have previously alluded to it as a school-book; and yet, in spite of this repugnance, which had become a second nature in me whenever I went far enough into any mathematical question, I found I touched, at last, a geometrical bottom.

He admired both Germany and America, where he felt the intellectual wind on his face in a way that England made impossible; he went so far as to say (to an American audience, of course—he could be impolitic but he was no fool) that despite geography, America and Germany were in one hemisphere and England another. But Sylvester returned again to England in the 1880s as the Savilian Professor of Geometry, a post first occupied by the logarithm-table maker Henry Briggs. Around that time Sylvester went to visit the young Poincaré, who more than anyone else in the late nineteenth century was freeing geometry from its Euclidean prison and insisting on its position at the foundation of all science.

> I recently paid a visit to Poincaré in his airy perch in the Rue Gay-Lussac in Paris. . . . In the presence of that mighty reservoir of pent-up intellectual force my tongue at first refused its office, my

> eyes wandered, and it was not until I had taken some time (it may be two or three minutes) to peruse and absorb as it were the idea of his external youthful lineaments that I found myself in a condition to speak.

For once in his long and eloquent life, Sylvester found himself without words.

When Sylvester died, in 1897, the Royal Society minted a medal in his honor. Poincaré was the first recipient. The younger mathematician gave a moving speech in Sylvester's honor at the society's 1901 annual dinner. It would surely have pleased Sylvester to hear the great geometer praising his mathematics as possessing "something of the poetic spirit of ancient Greece."

Also present at the dinner was Sir Ronald Ross. Imagine if he'd been seated next to Poincaré, and imagine if Poincaré, in the spirit of small talk, had told him about the interesting work of his student Bachelier on the random walk in finance, and imagine if Ross had made the connection with his still-developing ideas about wandering mosquitoes. . . .

LONG-DISTANCE MIND READING

In its issue of May 15, 1916, the magic magazine *The Sphinx* ran this ad:

> **LONG DISTANCE MIND READING.** You mail an ordinary pack of cards to any one, requesting him to shuffle and select a card. He shuffles again and returns only HALF the pack to you, not intimating whether or not it contains his card. By return mail you name the card he selected. **Price $2.50.**
>
> NOTE—On receipt of 50 cents, I will give you an actual demonstration. Then, if you want the secret, remit balance of $2.00.

The advertisement was placed by Charles Jordan, a chicken farmer in Petaluma who also built enormous radios as a hobby and maintained a lucrative sideline winning puzzle contests in newspapers. (He got so good the papers barred him from continuing to enter; he just arranged for confederates to submit his answers in exchange for a portion of the take, a scheme that almost backfired when one of his partners was called into the newspaper's offices for an in-person tiebreaker.) Jordan was also a prolific inventor of card tricks. Despite having no formal mathematical training that we know of, he was a pioneer of bringing math into magic.

I'm going to teach you how to read minds through the mail. I know, a magician never reveals the secret of a trick! But I'm not a magician. I'm a math teacher. And the secret of Jordan's trick comes down to the geometry of card shuffling.

I learned about the geometry of shuffling cards from Persi Diaconis, my undergraduate thesis advisor. A lot of academic mathematicians have a pretty predictable backstory. Not Diaconis, the son of a mandolin player and a music teacher who ran away from home at fourteen to be a magician in New York, then went to City College to study probability theory after a fellow practitioner told him it would make him better at cards. He met Martin Gardner, a fellow enthusiast of both math and magic,* who wrote him a recommendation letter saying, "I don't know a lot about mathematics but this kid invented two of the best card tricks of the past ten years. You ought to give him a chance." Some places, like Princeton, weren't impressed; but Harvard had Fred Mosteller, an amateur magician as well as a statistician, and Diaconis went there to be Mosteller's student. By the time I got to Harvard, he was a professor there.

Introductory graduate courses in math at Harvard have no set curriculum; the professors are allowed to teach whatever material they find most suitable. In my first year, the fall semester of algebra was taught by Barry Mazur, my eventual PhD advisor, and the course was devoted to his research subject, later also mine, algebraic number theory. The spring was taught by Diaconis, and it was an entire semester of card shuffling.

The geometry of card shuffling is a lot like that of movie stars and mathematicians—only much, much bigger. The points of our "space" are the ways in which the fifty-two cards can be ordered. How many

* And the premier mathematical popularizer of the twentieth century; it was through Gardner's *Scientific American* column that Conway's Game of Life became world famous, for example. Gardner is mentioned in Nabokov's *Ada*, he was declared a "suppressive person" by the Church of Scientology, and he had lunch with Salvador Dalí to talk about tesseracts. He lived on a street called Euclid Avenue and he once got a short story about topology published in *Esquire*. Fun guy.

ways is that? The first card can be any of the fifty-two cards in the deck. Having made that choice, the next card can be any of the cards that remain; no matter what card's on top, there are fifty-one unused cards left. So that's 52 × 51 = 2,652 choices just for the first two cards. And the next card can be any of the fifty cards you haven't used, giving us 52 × 51 × 50 or 132,600 choices in all. Keeping this up all the way down the line, the number of orderings is the product of all the numbers from 52 down to 1. This number is usually denoted 52! and called "52 factorial," though there was a movement in the nineteenth century to call it "52 admiration," in keeping with the excited typography. The factorial of 52 is a sixty-seven-digit number whose exact value I'm not going to trouble you with; it's a lot bigger than the number of mathematicians or movie stars, though, that's for sure.

(Of course, this geometry is in some naive sense *smaller* than the geometry of a humble line segment, which has infinitely many points!)

In order to have a geometry, we need a notion of distance. That's where shuffling comes in. A *shuffle* here is a standard riffle shuffle; you cut the deck in two pieces, then make a new pile by laying down cards one at a time, each time choosing to take from the left or the right. When the cards are all laid down, the two piles are merged into a single shuffled one. (It's not required that you strictly alternate between the two piles.) This is customarily carried out by the maneuver called a *dovetail*, where you press the two piles against each other so that the corners bend slightly up and then let them interleave themselves with a satisfying brrrrrippp sound. There are a lot of different riffle shuffles—for instance, if one of your two piles is just a single card, you can slip that card anywhere into the other pile. That counts as a riffle shuffle, though probably not one you'd be likely to do in real life. We say one ordering of the cards is linked to another if you can get from the first to the second by a riffle shuffle. And the distance between two orderings is the number of shuffles it takes to get from one to the other.

There are about four and a half quadrillion different riffle shuffles, which is a big number, but not a patch on 52 factorial. So a deck straight out of the box that you shuffle once can't be in just any order; it has to be in one of the orderings within distance 1 of the factory order. In geometry, we have a name for the set of points at distance at most 1 from a given point; we call it a "ball."*

The smallness of the ball is the key to mind reading through the mail. Here's the nature of the trick. I mail you a deck of cards. You shuffle it, then divide the shuffled deck into two piles, then choose whatever card you like from one pile, make careful note of what it was, and insert it in the other. Now pick *either one* of those two piles, throw the cards on the floor, pick them up again, put them in an envelope in whatever mixed up order, and send them back to me. I will postally reach into your mind and pluck from it the card you picked.

How?

To make this simpler to write down, imagine we were doing the trick with just the diamonds. Here's how a riffle shuffle looks on the page. You start with the cards in order:

2, 3, 4, 5, 6, 7, 8, 9, 10, J, K, Q, A

you cut them into two piles, not necessarily the same size:

2, 3, 4, 5, 6   7, 8, 9, 10, J, Q, K, A

and then comes the brrrrrrrip:

2, 3, 7, 4, 8, 9, 10, 5, J, 6, Q, K, A

The cards are shuffled, but if you peer closely you can see they still have some "memory" of the order they started in. Start at 2 and then jump forward to where you see the next-highest card, 3; then jump forward to the 4; and keep jumping until you'd have to go *backward* to

* No, not a "sphere"; that's the set of points at distance exactly 1 from a given point. The Earth's surface is a sphere (okay, an ever-so-slightly oblate spheroid), but the Earth itself is a ball. The distinction is the same one we made earlier between a "circle" and a "disc."

get to the next-highest number, which happens when you get to the 6 of diamonds. I've boldfaced the ones that you land on.

2, 3, 7, **4**, 8, 9, 10, **5**, J, **6**, Q, K, A

Now go back to the first card you haven't used, which is 7, and repeat the process. This time you cover all the rest of the cards. In fact, the two sequences you've marked are the two piles you riffled together. No matter how you riffled, the deck will always split into two rising sequences like this.

Now suppose you cut the deck into two piles again,

2, 3, 7, 4, 8, 9   10, 5, J, 6, Q, K, A

move a card—the queen, say—from one pile to the other,

2, 3, 7, Q, 4, 8, 9   10, 5, J, 6, K, A

and mail one pile back to me, the mind reader.

Here's how the trick works. Whatever cards I get in the mail, I put in order and organize into sequences of consecutive cards. If you hadn't switched a card from pile to pile, there would be two of those sequences. As it is, there will probably be three. If one of the sequences consists of a single card, that's the card that moved. If not, but if there's a *missing* card whose presence would glue two of the sequences together, that's the card that's missing from the pile. Let's see how it goes in this case. If you mail the first pile, I put the cards in ascending order:

2, 3, 4, 7, 8, 9, Q

and notice that there are two strings of consecutive cards (2, 3, 4 and 7, 8, 9) and one card all alone—and that is the moved card, the out-of-place queen.

And if you send the other pile? In order, those cards are

5, 6, 10, J, K, A

If you group these into strings of consecutive cards, you get three pairs: but you can see that you could make it into just two consecutive strings if you only had the one card that separates 10, J from K, A: the missing queen.

Don't get me wrong, this might not work. What if you moved the 10 from the second pile to the first and gave me the pile you enlarged? Then you'd mail me 2, 3, 4, 7, 8, 9, 10, which splits into two perfectly good consecutive sequences: 2, 3, 4 and 7, 8, 9, 10. I would have no idea what's out of place. With just thirteen cards, this kind of thing happens a lot. But with a full deck of fifty-two, the trick almost always works.

Of course, Jordan didn't mail people a deck in its factory order; that would make the trick too obvious. Neither should you, if you're doing this at home. You *do* need to know what order the deck was originally in; so you might want to put it in an order you can easily remember. When you get the half deck back, put it in order according to whatever rule you chose, and the card that moved should jump right out at you.

What makes the trick possible is that a shuffled deck isn't in a random order. Or rather, to use the proper math terminology, it's not in a *uniformly* random order: that's how we convey that not every ordering is equally likely. Mathematicians like to use the word "random" in a more general way than this: if a coin is weighted to land heads two-thirds of the time, the result of the coin flip is still random! But it's not uniform, because one of the two possible outcomes is more likely than the other. Even a coin with two heads is random, by our lights! It just happens to be a random event where one of the outcomes, heads, occurs 100% of the time. You can insist that's not truly "random" because the outcome doesn't involve chance; but that, to me, is like declaring that zero is not a number because it doesn't refer to a quantity of something, but rather the absence of a quantity. (Even now this bad idea survives in the terminology "natural numbers" for the whole numbers starting from 1, a notation I hate; there's nothing more natural than zero. There are lots of things there's none of!)

The more you shuffle a deck, the more uniformly random it gets. This feels natural (and it would be pretty upsetting to blackjack dealers around the world were it to be found wrong) but isn't so easy to prove. One early justification is found in a book by good old Poincaré, taking a break from geometry to write a treatise on probability. The math involved here is much the same as what underlies Google PageRank; it's the Law of Long Walks again. As you wander randomly around the space of all orderings, memory of your original starting point starts to fade away. You might have started anywhere. What makes PageRank different from cards is that some web pages are just plain better than others, and the internet wanderer on average will spend more time at those pages, giving them a higher PageRank. The orderings of a deck of cards are all equally good, and if you shuffle between them long enough, the chance of winding up at one is the same as the chance of winding up at any other.

If the victim of Jordan's telepathy trick shuffled twice instead of once, the trick wouldn't work, or at least wouldn't work in exactly the same way. This inspired Diaconis and his collaborator Dave Bayer* to ask: Just how much *do* you have to shuffle the deck to make the ordering of the deck so close to perfectly uniform that you can't do card tricks with it?

It turns out shuffling six times is enough to make every ordering of the cards possible. You might say six is the "radius" of this geometry, the greatest distance you can travel from the center before running out of room to run. Just as 13 is the biggest Erdős number any mathematician has, 6 is the maximum shuffle number any permutation of the cards has. (As you might expect, the ordering where the cards are exactly reversed from their starting order is one of those that takes six full shuffles to get to.) So the geometry of card shuffling is big, but somehow, like a world with a lot of direct intercontinental flights, also small; there are a lot of different places, but it doesn't take you many steps to get from one position to any other.

---

* Who, as the hand double standing in for Russell Crowe in all the blackboard scenes in the John Nash biopic *A Beautiful Mind*, has a Bacon number of 2 via Ed Harris, and an Erdős number of 2 via Diaconis, whose paper with Erdős on the greatest common divisor was published eight years after Erdős's death. That paper, in turn, is the first link in the length-4 path from Erdős to Danica McKellar.

But even after six shuffles, some orderings are a lot more likely than others. No amount of shuffling makes every ordering *exactly* equally likely, it turns out; but pretty quickly the probabilities become so close to equal that there's no meaningful difference. No magician, however deft, could tell whether you'd moved a card from one spot in the deck to another. Diaconis and Bayer were able to quantify this convergence to uniformity just about exactly. Everybody in the math world calls this result the "seven shuffle theorem" because seven shuffles meet a reasonable benchmark of mixed-upedness.

Diaconis was interested in card shuffling because he was a magician. But why was Poincaré? In part, it came down to physics. Poincaré was puzzled, as were all scientists at the time, by the problem of entropy. Boltzmann's vision, that the behavior of matter could be derived from the aggregated physics of myriad individual molecules bouncing around subject to Newton's laws, was appealing and elegant. But Newton's laws are time-reversible; they work the same way backward and forward. So how can it be that, as the second law of thermodynamics demands, entropy always increases? Hot and cold soup swirled together quickly becomes lukewarm, but lukewarm soup never spontaneously organizes itself into hot and cold sides of the bowl.

One answer comes from probability. Maybe it's not that entropy *can't* increase, but that it is *incredibly unlikely* for it to increase. Shuffling a deck of cards is a time-reversible process, too. You have probably never shuffled a mixed-up deck of cards only to find it restored to perfect factory order. But that's not because it's impossible—it's not! It's merely improbable. In the same way, a long flexible string like your headphone cord will tend to get tangled up if you stuff it in your pocket—life experience and a peer-reviewed 2007 paper with the killer title "Spontaneous Knotting of an Agitated String" agree on this point—not because there's a universal law that tangledness must increase, but because, more or less, there are more ways for a string to be tangled than to be untangled, and so random jostles are unlikely to result in the rare untangled state.*

* A physicist reading this will understand this to be an oversimplification of the way modern people think about entropy. It's better, though still oversimplified, to think of entropy not as a measure of the state of the

We are back again to the St. Louis exposition of 1904, and Poincaré's lecture, where he addressed the multiple crises besetting physics. In the 1890s, Poincaré had stoutly opposed the incursion of probability into the subject. But he was no ideologue; he grappled with the theory he didn't like by teaching a course on it, and in so doing came to see that it had virtues. If the probabilistic view were right, he told his St. Louis audience, "Physical law would then assume an entirely new aspect; it would no longer be solely a differential equation, it would take the character of a statistical law."

THE ONLY KARDYHM IN THE WORLD

Shuffling a deck of cards is a lot like Ross's mosquito. In both cases, we take a sequence of steps, each one chosen at random from a menu of options. The mosquito, at each tick of the clock, chooses to flit north, east, west, or south; the deck gets shuffled via one of the riffle shuffles available to it.

But there the two geometries part ways. The mosquito, remember, wanders very slowly. If it starts at the center of a 20 x 20 grid, it takes twenty days for it to even have a chance of getting to the far corner; and, as we've seen, its random motion tends to diverge from its starting point much less quickly than that. It would take hundreds of moves for the mosquito's position on the grid to become more or less random. The deck of cards, even though the number of possible orderings is so much larger, explores its entire geometry in six steps and is pretty well uniform in seven.

One obvious difference is that there are four directions for the mosquito to move in, and four billion different riffle shuffles that cards can undertake. But that's not what makes the shuffles get around faster. If you choose four types of shuffle out of those four billion options, and force

soup, but as a measure of our *uncertainty* about the state of the soup—as time progresses, our uncertainty increases, and to say uncertainty is maximized is (very) roughly to say that all states are equally likely, and many more states of molecules correspond to lukewarm soup than to temperature-segregated soup. So soup is likely lukewarm in the long run.

your shuffler to choose one of those four shuffles at random each time they address the deck, the ordering will *still* get random extremely fast.

No, there is a true structural difference between the flight of the mosquito and the shuffling of the deck. The former is tied to the usual geometry of space. The latter is not. That makes the difference. Abstract geometries, like the geometry of the shuffled cards, are typically really fast to explore, much faster than geometries drawn from physical space. The number of places you can reach grows exponentially with the number of steps you take, following the terrifying law of geometric increase, which suggests that you can get almost anywhere in very short order. The Rubik's cube has 43 quintillion positions, but you can get from any of them back to the original setting in just twenty moves. The hundreds of thousands of published mathematicians are all (with the exception of the applied Ukrainians and other isolates) only thirteen collaborations away from Paul Erdős.

But math is a human activity, mathematicians are humans, and the network that captures our interest the most, if we're to be honest, is the network of people and their interactions. That's also the one that's relevant to the spread of a pandemic. So what kind of a network is it? More like card shuffling, or more like Ross's meandering anopheles?

It's a little of both. Most people you cough on live pretty near you. But there are long-distance connections: a businessperson from Wuhan flies to California, a skier in northern Italy flies home to Iceland. Those long-distance transmissions are rare, but they matter. In graph theory we call these networks mixing short and long connections "small worlds," a phrase that goes back to the 1960s and the social psychologist Stanley Milgram. Milgram is probably best known for convincing subjects to deliver fake electric shocks to actors under authoritarian suasion, but in his cheerier moments he studied more positive forms of human connection. In the geometry of human acquaintance, where two people are connected whenever they know each other, how likely is it, Milgram asked, that two people can be joined by a chain of connections, and, if so, how long a chain is needed? John Guare's play *Six Degrees of Separation* puts a summary of Milgram's results in the mouth of one of his brittle, upper-crust New York art world characters:

> I read somewhere that everybody on this planet is separated by only six other people. Six degrees of separation. Between us and everybody else on this planet. The President of the United States. A gondolier in Venice. Fill in the names. I find that a) tremendously comforting that we're so close and b) like Chinese water torture that we're so close. Because you have to find the right six people to make the connection. It's not big names. It's *anyone*. A native in a rain forest. A Tierra del Fuegan. An Eskimo.

That isn't quite the finding of Milgram, who studied only Americans, asking people in Omaha to find a chain of acquaintances ending with a particular stockbroker in Sharon, Massachusetts. And he didn't find that *everyone* was connected; on the contrary, only 21% of the Nebraskans were able to find a path to the Sharon stockbroker. The completed paths were typically four to six people long, but in at least one case, ten degrees of separation were needed. The Guare play tweaks the findings to make the study serve better as a metaphor for racial anxiety—the white characters in the play want to be able to say they're part of a diverse and modern world, but are physically pained by the awareness that the rain forest and its "natives" may not be as far from the Upper East Side as they imagine. (The "separation" Guare attaches to Milgram's six degrees surely carries a silent suffix, "but equality.") Milgram actually did a follow-up study in 1970 in which 540 white people in Los Angeles were asked to find chains to eighteen men in New York, half Black, half white. About a third of the white-white connections were successfully completed; but only one in six of the California whites was able to find their way to a Black man.

The phrase "Six Degrees of Separation" turned into "Six Degrees of Kevin Bacon," the common name for the process of drawing short paths to Kevin Bacon in the geometry of movie stars. To bring us full circle back to COVID-19, Bacon launched a "six degrees" public relations campaign in March 2020 asking his fans to keep up their social distancing. "I'm technically only six degrees away from you," he said in the video he shared. "I'm staying home because it saves lives and it is the only way we're gonna slow down the spread of the coronavirus."

Nowadays we can do degrees-of-separation experiments without relying on human beings sending postcards, as Milgram did. In 2011, Facebook had about 700 million active users, with an average of about 170 friends each, and mathematicians in the research arm of the company have access to that entire meganetwork. Pick two users at random anywhere on the globe: the average length of the shortest chain of Facebook friends between them, it turns out, is just 4.74 (that is, there are typically three or four intermediaries between the two users). Almost all the pairs —99.6% of the total—were within six degrees. Facebook is a small world. (And getting smaller even as the number of users gets bigger: by 2016, the average path length had dropped a bit, to 4.57.) The reach of Facebook is so great that its network dominates geography. The separation between two random users in the United States is 4.34; between two random Swedish Facebook accounts it's 3.9. To Facebook, the world is only a little bigger than Sweden.

Analyzing this gigantic graph is a serious computational effort. Facebook will tell you how many friends you have, but to carry out this path analysis you need to know how many friends of friends you have, and how many friends of *those* friends of friends there are, and so on for at least a few more iterations. This gets complicated: you can't just add up the number of friends each of your friends has, because you'll get a lot of repeated names! And searching that whole list for repeated names requires storing and accessing hundreds of thousands of records, which is going to slow you down too much.

The trick for doing this fast is a process called the Flajolet-Martin algorithm. Rather than explain exactly what goes into it, I'll tell you a simpler version. Facebook won't tell you how many friends of friends you have; but it will let you search your friends of friends for people named Constance. I have 25. Constance isn't a common name; in the age groups most of my social circle comes from, between 100 and 300 people per million born in the United States are called that. If my friends of friends are about as likely to be named Constance as the typical American, that means I have between 85,000 and 250,000 friends of friends. I tried this for a few more names, sticking to uncommon ones so I'd get a list short enough to count: 50 Geralds, 18 Charitys

(Charities?). I mostly got estimates around a quarter million, so that's my estimate.

The Flajolet-Martin algorithm isn't *quite* like that, though it operates on the same principle. It's more like running through the list of every friend of every friend, one by one, keeping track all the while of the rarest name you've seen so far. Every time you encounter a name that's rarer than the current champ, you throw out the name you were saving and replace it with the new one. No large-scale storage required! At the end of the process, you have a presumably very rare first name, and the bigger your list, the rarer that name is likely to be. So you can go backward; from the rarity of the rarest name, you can estimate how many different people there were among your friends of friends!

This won't always work. For example, I have a friend named Kardyhm, whose parents put together the initials of their seven best friends' names in the first pronounceable order they could find and attached the result to their baby. I believe my friend to be the only Kardyhm in the world. So the estimate for any of Kardyhm's friends of friends, thanks to the exceeding rarity of the name, will be unreasonably high. The real Flajolet-Martin algorithm doesn't use first names, but another kind of identifier called a *hash*, which you have enough control over to avoid Kardyhm problems.

One small warning about these calculations: if you do them for yourself, you are likely to be faced with the ego-wounding fact that your friends, on average, have more friends than you do. I'm not trying to insult the social skills of my typical reader here. A large-scale analysis of the Facebook network in 2011 found that 92.7% of users had fewer friends than their average friend did. It's utterly normal for your friends to have more friends than you do, because your friends, in real life or on the screen, are not a randomly selected sample of the population. By virtue of being friends with *you*, they're more likely to be the kind of people who have a lot of friends.

SIX DEGREES OF SELMA LAGERLÖF

To most people, it's pretty astonishing that a social network as vast as Facebook can be traversed end to end in so few steps. But we now know that small-world networks are common, thanks to foundational work in the late 1990s by Duncan Watts and Steven Strogatz that laid the mathematical groundwork. Watts and Strogatz ask you to contemplate the following kind of network. You start with a bunch of dots arrayed around a circle, each one connected to a small set of its nearest neighbors. That network is like the mosquito's walk; you can't move very fast, and if the circle is thousands of dots in circumference it's going to take you a long time to get all the way around. But what if you add some random long-distance connections to the network, to simulate the occasional connections we know exist between distant people?

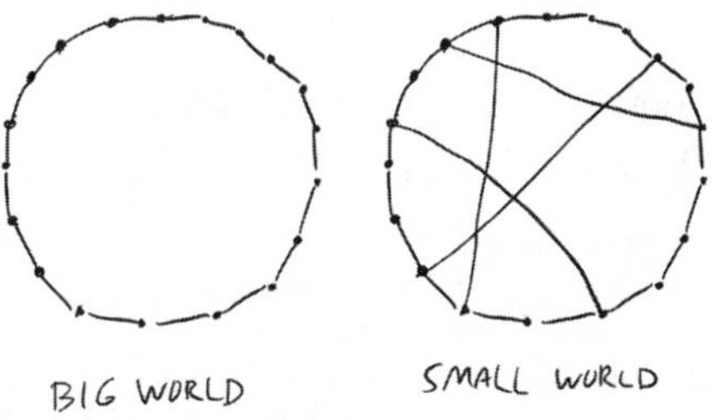

What Watts and Strogatz discovered is that it takes only a tiny number of these new links to make the network into a small world, where every individual is connected to every other via a short path. They write, in a passage that comes off now as unsettlingly Nostradamic, that "infectious diseases are predicted to spread much more easily and quickly in a small world; the alarming and less obvious point is how few short cuts are needed to make the world small." The developments in the math of small-world networks show that the initially surprising phenomenon found by Milgram ought not to have been surprising at all. This is the nature of good applied mathematics: it turns "How could it be?" into "How could it be otherwise?"

Stanley Milgram is the face of the "six degrees" theory partly because of the experiment he ran, and partly because he was such a skilled marketer of his own work. His first write-up of the postcard study, two years before any formal scientific publication, was in the popular magazine *Psychology Today*—in fact, it was a feature in the very first issue. But Milgram wasn't the first to contemplate the smallness of the networked world. His experiment was designed to test an existing theoretical prediction of small-worldness, produced but never published by Manfred Kochen and Ithiel de Sola Pool—the latter, in the spirit of short chains, my college roommate's grandfather. And before that, in the early 1950s, Ray Solomonoff and Anatol Rapoport, writing in biology journals, had understood the tipping point that Erdős and Rényi would later independently discover in the pure math context; once a certain density of connections has been reached, disease can start anywhere and get almost everywhere. And before *that*, in the late 1930s, the social psychologists Jacob Moreno and Helen Jennings were studying "chain-relations" in social networks at the New York State Training School for Girls.

But the very earliest appearance of the idea of the small world wasn't in biology or sociology, but in literature. Frigyes Karinthy was a Hungarian satirist* who in 1929 published a story called "Chain-links" (*"Láncszemek"*):

> Planet Earth has never been as *tiny* as it is now. It shrunk—relatively speaking of course—due to the quickening pulse of both physical and verbal communication. This topic has come up before, but we had never framed it quite this way. We never talked about the fact that anyone on Earth, at my or anyone's will, can now learn in just a few minutes what I think or do, and what I want or what I would like to do. . . . One of us suggested performing the following experiment to prove that the population of the Earth is closer together now than they have ever been before. We should select any person from the 1.5 billion inhabitants of the Earth—anyone, anywhere at all. He bet us that, using no more than five individuals, one of whom is a personal acquaintance, he could contact the se-

* Erdős and Rényi were Hungarian, too, as was Milgram's dad, and graph theory is considered a strongly Hungarian kind of subject to work on even today; make of this what you will.

> lected individual using nothing except the network of personal acquaintances. For example, "Look, you know Mr. X.Y., please ask him to contact his friend Mr. Q.Z., whom he knows, and so forth. "An interesting idea!"—someone said—"Let's give it a try. How would you contact Selma Lagerlöf?" "Well now, Selma Lagerlöf," the proponent of the game replied. "Nothing could be easier." And he reeled off a solution in two seconds: "Selma Lagerlöf just won the Nobel Prize for Literature, so she's bound to know King Gustav of Sweden, since, by rule, he's the one who would have handed her the Prize. And it's well known that King Gustav loves to play tennis and participates in international tennis tournaments. He has played Mr. Kehrling, so they must be acquainted. And as it happens I myself also know Mr. Kehrling quite well.

Apart from the low figure for the world's population, this could have been written in 2020. The anxiety and unsettledness the narrator feels is the same one we now feel, in the middle of a global pandemic, and the same one Guare's characters feel, holed up in their Upper East Side apartment. It is an anxiety about the geometry of the world we live in. We evolved to understand a world where what was near us was what we could see, hear, and touch. The geometry we inhabit now, and the one Karinthy in the 1920s was already having to get used to, is different. "The famous worldviews and thoughts that marked the end of the nineteenth century are to no avail today," Karinthy writes later in the story. "The order of the world has been destroyed."

The world's geometry is even smaller now, and more connected, and more prone to exponential spread. There are so many wrinkles in time it's almost all wrinkles. It's not easy to draw this on a map. The abstractions of geometry step in when our ability to draw gives out.

Chapter 14

How Math Broke Democracy (And Might Still Save It)

The night of November 6, 2018, was a joyous one for long-suffering Democrats in the state of Wisconsin. Republican governor Scott Walker, who had survived two general elections and a recall campaign, who had brought Washington-style polarization to the state during his eight years in Madison, and who had, for a little while, seemed poised to be his party's 2016 presidential nominee, had finally been brought down, edged out by Tony Evers, a gee-whiz-saying, euchre-playing ex-schoolteacher of a certain age whose highest previous position was state superintendent of public instruction. In fact, Democrats swept the statewide offices up for election that night. Their Senate candidate, Tammy Baldwin, was reelected by an 11-point margin, the biggest victory by a statewide candidate of either party since 2010. They took over the attorney general and state treasurer positions previously held by Republicans. And all this was in the context of a national wave of pro-Democratic sentiment that saw the party win the majority in the U.S. House of Representatives, gaining forty-one seats.

But not everything was beer and roses for Wisconsin Democrats. In the Wisconsin State Assembly, the lower house of the state legislature, Republicans lost only one seat, retaining a 63–36 majority. In the state senate, the GOP actually gained a seat.

Why would the legislative elections in 2018, a year of massive Democratic gains, come out pretty much the way they did in 2016, when Republican U.S. senator Ron Johnson cruised to reelection and a Republican presidential nominee won the state for the first time in decades? One might look for a political explanation; maybe Wisconsin voters think Republicans are better at legislating even if they prefer a Democratic executive? If that were the story, you'd expect there to be a bunch of assembly districts that voted for a GOP representative while supporting Evers for governor. But in fact, if you plot the share of the vote Scott Walker got in each assembly district against the share of the vote the Republican assembly candidate got there, it looks like this.*

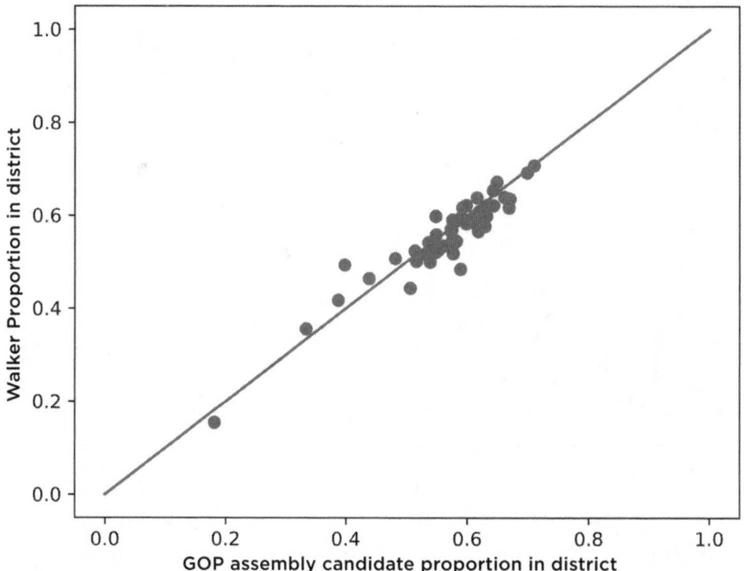

Districts liked Scott Walker just about exactly as much as they liked their Republican legislative candidate. Only two districts, both represented by Republican incumbents, voted for Evers but went for the GOP in the assembly. Walker lost the governorship while getting more votes in sixty-three out of ninety-nine assembly districts. Most of

* The very careful reader will note that there aren't ninety-nine dots on that chart, but only sixty-one; that's because we're only showing the sixty-one districts where both parties fielded candidates.

Wisconsin's *voters* in 2018 chose Democrats, but most of Wisconsin's *districts* chose Republicans.

This might seem like a funny accident, except that it's no accident and it's funny only in the hollow-laugh-with-your-head-in-your-hands kind of way. The districts in Wisconsin are Republican because the district lines were drawn by Republicans, and they were precisely engineered to produce that outcome. Here's a plot showing Walker's percentage of the vote in each assembly district, where I've ordered the districts by increasing Republican-ness:

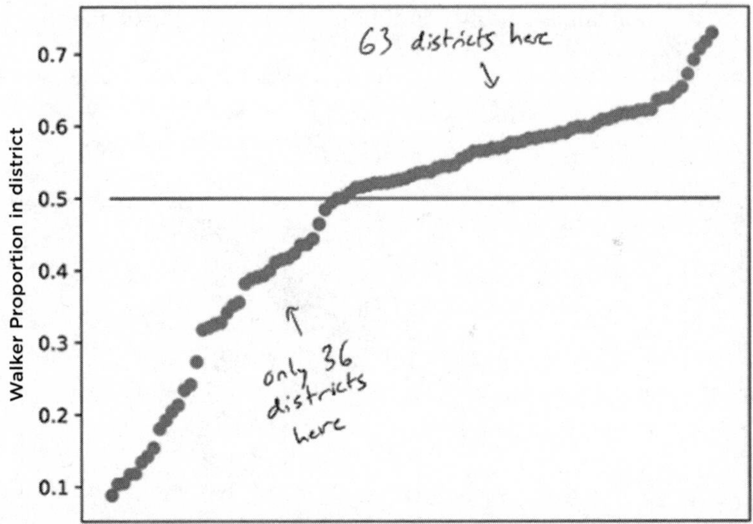

There's a clear asymmetry here. Notice the dominance of districts where Walker just barely clears 50% of the vote. In thirty-eight out of ninety-nine districts, Walker's share was between 50% and 60%. His opponent, Tony Evers, got between 50% and 60% in just eleven districts. Tony Evers's tiny lead in the statewide race is a combination of winning big in about a third of the districts and losing, mostly narrowly, in the rest.

There are a few ways to read this graph. You could say that Democratic strength in Wisconsin is driven by a small, politically inflamed region that doesn't truly represent the politics of the state. That is, naturally, the view of the Wisconsin Republican Party, one of whose leaders, Robin Vos, remarked after the election, "If you took Madison and Milwaukee out of the state election formula, we would have a clear majority."* A more Democratic spin on state politics would be to observe that there are eighteen districts where Scott Walker got less than a third of the vote, against only five districts where Evers did that badly. In other words (still spinning here), Republicans have written off whole regions making up a fifth of the state, while there are substantial numbers of Democrats just about everywhere, including GOP-majority districts. Seventy-eight percent of Wisconsinites who voted for Scott Walker have a Republican representative in the assembly, but only 48% of Evers voters are represented by a Democrat.

Both these accounts treat the asymmetry of the curve as a peculiar natural feature of Wisconsin's political geography. It isn't. In fact, that curve was *built*, in the spring of 2011, in a locked room at a politically connected Madison law firm, by a group of aides and consultants working for Republican legislators. The project was part of a national effort by the Republican Party to translate its 2010 electoral gains into favorable district lines. That last digit, *0*, at the end of 2010 is important; it's in the years divisible by 10 that the United States conducts a census, which generates new official population statistics, which, given the natural slosh of population from place to place, tends to make some of the existing districts bigger than others. That means new districts need to be made, and partisan actors jostle to be the ones to make them. In previous census years, both Democrats and Republicans had controlled either a house of the Wisconsin legislature or the governor's mansion, so any map that could be passed into law had to satisfy both parties. In practice that meant no map could be passed into law, and the courts had to do the job. In 2010, Republicans had majorities in both houses, and a brand-new Republican governor: Scott Walker, eager to set the rules for ten years of Wisconsin elections before he even finished measuring for

* As the old Yiddish proverb goes, "If my grandmother had wheels, she'd be a wagon."

drapes. There was nothing but their own sense of decorum to hold them back from angling for maximal political advantage.

This is not going to be a story about the triumph of decorum.

THE LIFE AND TIMES OF JOE AGGRESSIVE

The Wisconsin mapmakers were bound to icy secrecy. Even Republican legislators were shown only their own proposed district, and forbidden from discussing what they'd seen with their colleagues. Democrats saw nothing at all. The map as a whole remained under wraps until the week before the legislature voted it into state law, along party lines,* as Act 43.

The mapmakers in their locked chamber had worked for months to build a map that was maximally advantageous to Republican interests. Among them was Joseph Handrick, no newcomer to this game. Since his teenage years, he told an interviewer, "every big decision in my life was made with the backdrop of wanting to run for the state assembly." He made his first run for the assembly seat in his up-north district as a twenty-year-old college junior. In an unusually data-driven campaign for the mid-1980s, he worked out a precinct-by-precinct chart to identify where the popular Democratic incumbent had been overperforming the local partisan lean, and targeted those voters with a strong ideological campaign on taxes and Native American fishing rights. (He was against both.) The conventional wisdom was that the popular incumbent couldn't be beaten by a college kid with a spreadsheet, and the conventional wisdom was right. But the race made Handrick an up-and-comer in state Republican politics, and he later served three terms in the assembly. By 2011 he was out of elected office and consulting for Wisconsin legislators. "What I like about campaigns more than anything," Handrick once said, "is the planning of strategy and development of the game plan." In the back room of the law firm, he was deep in the part of politics he liked best.

The map team classified maps as "assertive" when they helped Re-

* Or almost along party lines: Samantha Kerkman of Randall was the lone Republican in the state legislature to vote against it.

publicans a lot, and "aggressive" when they helped Republicans even more than that. They named each map by combining that adjective with the name of the person who drew it. The map they finally went with, the one still in use in 2018, was one of Joseph Handrick's. They called it "Joe Aggressive."

Here's how aggressive Joe Aggressive was. Keith Gaddie, an Oklahoma political science professor brought in to consult, estimated that Republicans would typically maintain a 54–45 majority in the assembly even in an election where their statewide share of the vote dipped to 48%. The Republicans would have to be losing statewide by a 54–46 margin before Democrats would pick up a majority of seats.

There's a back-of-the-envelope way to check how Gaddie's work held up, seven years on. If you rank the ninety-nine Wisconsin districts by how well Scott Walker did there in 2018, the one in the middle is Assembly District 55, in Winnebago County, about halfway between Madison and Green Bay. Walker got 54.5% of the vote there,* about four points ahead of his popular vote share. Forty-nine districts were better for Walker, and forty-nine were worse; in the language of statistics, we say that District 55 is the *median* among the districts. If a Democrat wins District 55, there's a pretty good chance the party wins the forty-nine districts more Democratic than that, and thus secures a majority; and the same goes for the Republican. The bellwether status of District 55 isn't just hypothetical; in the statewide elections held in Wisconsin since this map was drawn, the candidate who won District 55 has won a majority of districts in every single case.

How good a year would Democrats need to have in order to scrape out a victory in AD55? In 2018, a year where the two gubernatorial candidates got almost the same number of votes, Scott Walker won that district by 9 points. So you might estimate that for Democrats to break even in District 55, they'd need to come out ahead by 9 percentage points statewide, winning 54.5–45.5—just about the same figure Gaddie came up with when the maps were drawn. This is just a rule of thumb, not a precise prediction about future elections, but it gives some

* 54.5% of the combined Republican and Democratic vote, that is; for simplicity I'm going to throw out votes for smaller parties (sorry, Libertarians) and use the two-party share.

sense of the headwind Democrats face in their quest for an assembly majority with the current district boundaries.

Another way to see the effect of the Act 43 map is to compare it to the one that came before it, which was drawn by an exasperated federal district court in 2002 after it found "unredeemable flaws" in all sixteen of the maps proposed to it by interested actors on both sides.

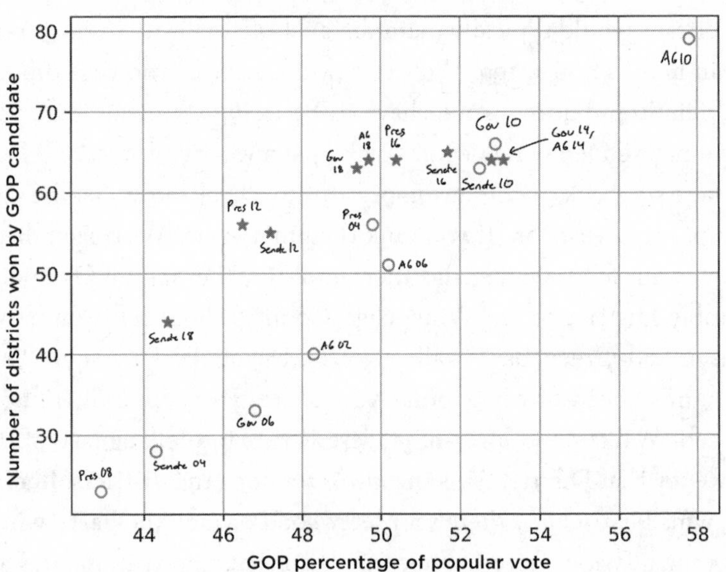

What you're looking at here is a list of November statewide elections held in Wisconsin between 2002 and 2018.* The horizontal axis shows the share of the statewide vote the Republican candidate got, and the vertical axis shows how many of the ninety-nine assembly districts gave the Republican more votes than the Democrat.

The circles are elections held under the court-drawn 2002 map, and the stars are Joe Aggressive elections. Notice anything? In 2004, John Kerry barely edged George W. Bush in Wisconsin's presidential race, getting 50.2% of the two-party vote; Bush won in fifty-six assembly districts. In a similarly close election in 2006, Republican J. B. Van Hol-

* Not every single one; I took out a 2006 Senate blowout and the somewhat weird 2002 gubernatorial election, in which the younger brother of the former Republican governor ran as a Libertarian and got 10% of the vote.

len beat out Kathleen Falk to become Wisconsin's attorney general; he won fifty-one of the assembly districts. Those are the two circles near the middle of the plot. Republican Ron Johnson, in his 2010 senate race, did better, getting 52.4% of the vote against incumbent Russ Feingold and winning in sixty-three assembly districts.

Starting in 2012, things look different. Donald Trump in 2016 and Scott Walker in 2018 were in nearly tied elections, just like Bush and van Hollen; but where those two Republicans came out ahead in fifty-six and fifty-one districts, Trump and Walker both won in sixty-three out of ninety-nine, the same number Ron Johnson got under the court-drawn map while solidly beating his Democratic opponent. In 2012, the first year of the Act 43 maps, Republican Mitt Romney won 46.5% of the two-party vote but won fifty-six of the ninety-nine districts; Democrat Tammy Baldwin got 52.9% of the vote in her Senate race and won just forty-four districts. When Baldwin ran for reelection in 2018, she did a lot better, thumping challenger Leah Vukmir by 11 points. She won in fifty-five districts; a majority, to be sure, but in 2004, when Russ Feingold won his Senate race for the Democrats by the same margin, he won in seventy-one out of ninety-nine districts on the old map.

That's a lot of words to make a point the picture already does. The stars float high above the circles, which means the same electoral facts now translate into many more Republican seats than they did ten years ago. Nothing suddenly changed in Wisconsin politics between 2010 and 2012. The difference is the maps.

Tad Ottman, another of the locked-away mapmakers, told the Republican caucus, "The maps we pass will determine who's here 10 years from now. . . . We have an opportunity and an obligation to draw these maps that Republicans haven't had in decades." The language is telling. Not just an opportunity, but an *obligation*. The implication here is that the first duty of a political party is to protect its own interests against the whims of a potentially hostile electorate. The practice of drawing district lines to secure advantage for yourself or your co-partisans is called *gerrymandering*, and it's the way you end up with a swing state like Wisconsin where Republicans enjoy a bigger legislative lower-house majority than they do in more conservative states like Iowa and Kentucky.

Is that fair?

Short answer: no.

The long answer is going to require some geometry.

ARTIFICIAL DISTINCTIONS AND SYLLOGISTIC SUBTLETIES

Democratic governments are founded on the principle that every citizen's views are to be represented in the decision-making of the state. Like all good principles, this is easy to state, difficult to make precise, and almost impossible to implement in a fully satisfying way.

For one thing, modern governments are *big*. Even a modestly sized city is large enough that it would be impractical for every decision about zoning, school curriculum, public transport, and taxes to be put to a public plebiscite. There are workarounds. For criminal cases, we pick twelve people's names out of a hat and let them decide. For much of the day-to-day management of cities and states, decisions are made within government agencies with only occasional and indirect input from voters. But when it comes to legislation, the basic infrastructure of government action, we use the system of elected representatives, in which a small group of legislators is elected by the people at large and commissioned to speak on their behalf.

How to choose those representatives? That's where the details start to matter. And there are a lot of ways the details can look. Voters in the Philippines cast a vote for as many as twelve candidates, and the top twelve vote-getters overall join the Senate. In Israel, each political party makes a list of proposed legislators, and voters choose a party, not an individual candidate. Then each party occupies seats in the Knesset according to its proportion of the popular vote, going down the party list until they hit their appointed number. But the most common way to set this up is the way the United States does; you divide the population into predefined districts, and each district chooses a representative.

In the U.S., districts are drawn geographically. But it doesn't have to be that way. In New Zealand, Māori people have their own electoral districts, which are superimposed on the general districts; Māori voters have the choice in each election whether to vote in the Māori or the general district containing their residence. Or the partition might have no geographic aspect at all. In Hong Kong, there's a seat in the legislative council only teachers and school administrators can vote for, one of thirty-five seats elected by so-called functional constituencies. The Centuriate Assembly of the Roman Republic had constituencies separated by wealth bracket. In the upper house of the Oireachtas in Ireland, there's a three-seat constituency consisting of students and graduates of Trinity College Dublin, and another for alumni of the National University of Ireland. Jews have their own seat in the Parliament of Iran.

As an American, and thus someone trained to think of the American way as the one and only way, I find it agreeably freeing to think about the different ways we could divvy up the U.S. voting public. What if our state legislative districts, instead of geographic regions, were age bands of equal size? With whom do I have more political priorities and values in common: an elderly retiree who lives ten miles from me, or a fellow forty-nine-year-old, who has about as much life left to plan for as I do, who likely has kids around the same age, but who happens to live across the state? Would legislators have to "live" in their chronological district? (If so, that would neatly solve the problem of lazy incumbents staying in office forever by dint of inertia; unless representatives were spaced extremely evenly by birth date, the progression of time would regularly pit incumbents against each other as they aged through the brackets.)

The U.S. states are, at least formally, semiautonomous governments, each with its particular interests. The districts within states, on the other hand, are patches of land without much meaning. Nobody in the Second Congressional District of Wisconsin, where I live, wears a WI-2 sweatshirt, or could recognize the district from its silhouette. As for my state legislative district, I had to look it up to be sure I had the number right. Those districts have to be determined somehow, despite lacking any inherent political identity. Somebody has to cut up the state into segments. This process, called *districting*, is technical and time consuming, and involves spreadsheets and maps. It does not make good television and it has not traditionally drawn much public attention.

That has now changed. It has changed because we now understand something we didn't really grasp before, a fact that is both mathematical and political: the way you cut up the state into districts has an enormous effect on who ends up in the statehouse making laws. Which means the people with the scissors have enormous power over who gets elected. And who wields the power scissors? In most states, it's the legislators themselves. The voters are supposed to choose their representatives, but in many cases, the representatives are choosing their voters.

To some extent, it's obvious that the district-drawers have a lot of power. If I'm in complete control of the districting of Wisconsin, with the power to partition the population any way I wish, I can just find a cabal of like-minded people, declare each one of them to be their own district, and then create one more district consisting of everybody else. My handpicked candidates vote for themselves and then rule the legislature with at most one potential voice of opposition. Democracy!

That's plainly not fair. Certainly the people of Wisconsin, with the exception of the cabal itself, would be right to feel themselves unrepresented in the decision making of the state. It's also ridiculous; no democratic government would ever be run this way! Except, of course, the ones that are. In England, for example, there were "rotten boroughs" that persisted for centuries, duly electing members to Parliament despite having dwindled to near emptiness. The town of Dunwich, once as big as London, fell into the North Sea bit by bit and was largely abandoned by the seventeenth century, but continued to send two members to the House of Commons until it was dissolved by Whig prime minister Earl Grey (admit it, you thought he invented tea) in the Reform Act of 1832. By that time Dunwich was down to thirty-two voters. And that wasn't the rottenest of the rotten boroughs! Old Sarum had once been a thriving cathedral town, but lost its reason for being when the new Salisbury Cathedral was erected; the town was emptied out and its buildings demolished for scrap in 1322. And yet, for five hundred years,

Old Sarum had two MPs, chosen by whatever wealthy family held title to the stony, unpopulated hill. Even Edmund Burke, generally a friend to tradition, complained of the need for reform: "The representatives, more in number than the constituents, only serve to inform us that this was once a place of trade . . . though now you can only trace the streets by the color of the corn, and its sole manufacture is in members of Parliament."

Things were more rational over here in the colonies, but only just. There were no rotten boroughs, but nonetheless, some Americans were more represented than others. Thomas Jefferson complained about the unequal sizes of legislative districts in Virginia, insisting that "a government is republican in proportion as every member composing it has his equal voice in the direction of its concerns." Well into the twentieth century, the city of Baltimore was limited to 24 of the 101 seats in the Maryland House of Delegates, even though Baltimoreans made up half the population of the state. Maryland attorney general (and Baltimore native) Isaac Lobe Straus begged for a change in the constitution that would give Baltimore equal representation, quoting Jefferson and Burke and then really going for it: "Will some one explain, upon what principle of justice or ethics or law or politics or philosophy or literature or religion or medicine or physics or anatomy or aesthetics or art, a man in Kent County is entitled to twenty-nine times the representation that a man in Baltimore City is entitled to?"

(Lest I leave you with the impression that Straus was a principled tribune of democracy, he went on in the very same 1907 speech to recommend a further amendment that would require a literacy test for voting, with the goal of mitigating "the evil of an unthinking suffrage wielded by a large body of illiterate and irresponsible voters in this State, who became voters as a consequence of the war between the Northern and Southern States, and not only not through any act of the people of Maryland, but in the teeth of their solemn rejection of the amendment to the Federal Constitution under which the persons in question vote." For any readers unfamiliar with the customary code words of American politics used here, he means Black people.)

The era of unequal representation came to an end in America only in 1964, when the Supreme Court threw out the Alabama state legislative districts, in the case of *Reynolds v. Sims*. Alabama law apportioned representatives by county; the formula in force awarded a single state senator to the 15,417 residents of Lowndes County, and the same to Jefferson County, which contained the city of Birmingham and had a population of more than 600,000. W. McLean Pitts, arguing in Alabama's defense, warned that overturning the district maps would mean "the larger, dense-populated counties would have a stranglehold on the Alabama Legislature on a one man, one vote basis, and the people in the rural areas would not have any say so in their own government." The court saw it differently, writing in an 8–1 decision that Alabama had violated the Fourteenth Amendment by depriving voters in the larger counties of the "equal protection" of the laws governing the vote.

The requirement of equal representation means we can't stop gerrymandering by forbidding governments from futzing with the district boundaries. The futzing is mandatory. People move from place to place, the old die and the young reproduce, some regions bulge as others wither, and so the boundaries that are constitutional when drawn become unconstitutional when the next census rolls in. That's why the years that end in *0* matter so very much.

The W. McLean Pitts principle—"Why should the people of Birmingham have more power over the law just because there are more of them?"—sounds funny to modern ears, but in a real way Americans still live by it. Each state has two senators, whether it's tiny Wyoming or vast California. This has been controversial from the beginning. Alexander Hamilton complained in *Federalist* #22:

> Every idea of proportion and every rule of fair representation conspire to condemn a principle, which gives to Rhode Island an equal weight in the scale of power with Massachusetts, or Connecticut, or New York; and to Delaware an equal voice in the national deliberations with Pennsylvania, or Virginia, or North Carolina. Its operation contradicts the fundamental maxim of republican government, which requires that the sense of the majority

> should prevail. . . . It may happen that this majority of States is a small minority of the people of America; and two thirds of the people of America could not long be persuaded, upon the credit of artificial distinctions and syllogistic subtleties, to submit their interests to the management and disposal of one third.*

History has dehypotheticalized Hamilton's angry worry; the twenty-six smallest states, whose fifty-two representatives make up a majority of the Senate, speak for just 18% of the population.†

It's not just the Senate. Each state, however small, gets at least three votes in the Electoral College, which ultimately decides the presidency. Wyoming's 579,000 people—about as many as live in greater Chattanooga—share three electoral votes among them, which means each electoral vote represents about 193,000 Wyomingites. California has almost 40 million people, so each of its fifty-five electoral votes stands for more than 700,000 Californians.

This is, as your constitutional-originalist friends probably often remind you, by design. The idea that the president should be chosen by the majority of the national vote seems rather natural to most Americans today, even to those who see reasons to support the Electoral College system. But there was little appetite among the founders for the idea. James Madison was a notable exception, and even he supported a national popular vote only because he thought all the other options were worse. Small states worried that only a candidate from a populous state would have a chance. Southerners (Madison excepted) didn't like the fact that a national election would blunt their hard-won "three-fifths compromise," which allowed them to derive extra representation in Congress from the large, enslaved, and disenfranchised Black population. In a national popular majority system, your state gains no power from people unless you let them vote.

---

* Technically, Hamilton is venting about the Articles of Confederation here, not the senatorial allocation in the just-drafted Constitution, but he spoke similarly during the arguments over that document, asking: "[I]s it our interest in modifying this general government to sacrifice individual rights to the preservation of the rights of an artificial being, called states?"

† Eighteen percent of the population of the fifty states, that is. The percentage is even lower if you count Americans residing in Washington, D.C., Puerto Rico, or other U.S. territorial possessions, who have no representation in the Senate at all.

The manner of electing the president was a source of rancorous division, and it dragged on and on through the long constitutional summer of 1787. Plan after plan was brought up and voted down. Elbridge Gerry of Massachusetts suggested that the governors should choose, each with a vote weighted by their state's population; that idea was soundly rejected. So were proposals that the president be selected by state legislatures, or by Congress, or by a committee of fifteen members of Congress chosen at random. The main body of the group was unable to agree, finally punting the decision about election of the president and some other persistent points of dissent to a group of eleven unlucky members called the Committee on Unfinished Parts. The system we eventually arrived at shouldn't be thought of as a brilliant encapsulation of the founders' wisdom; it was a compromise reluctantly and wearily arrived at, nobody having been able to come up with anything better. If you have ever sat in a long meeting as day care pickup got nearer and nearer, knowing you couldn't go home until the meeting produced a policy document everyone there could make themselves grumblingly sign, you have a pretty good idea of how the Electoral College came to be.

Even if you're on board with the representational inequities baked into the Electoral College, you'd better be aware that they've gotten a lot more intense since the framers' time. In the 1790 census, the largest state, Virginia, had eleven times the population of the smallest, Rhode Island. Right now, the ratio between Wyoming's population and California's is about 68. Would the constitutional convention have been game to assign Rhode Island so much power to appoint senators and electors if it had been six times smaller than it was?

Perhaps the simplest way of diluting the inequality of the Electoral College would be to increase the size of the House of Representatives. There were 435 representatives in 1912 and there are 435 representatives today, in a country more than three times as large. The number of electors in each state is the number of representatives *and* senators from each state. If the House had 1,000 members, 120 of them would be from California, and 2 from Wyoming. So California would have 122 electoral votes, one for every 324,000 Californians, while Wyoming would have 4, one for every 144,500 people in Wyoming; still unequal, but not as unequal as before. A bigger House would mean a more representative House of Representatives, and an Electoral College that better represented the people voting, without changing a single jot of the founders' plan.

As extreme as electoral inequity is now, it's been even worse. When Nevada was admitted to the Union in 1864, it had only forty thousand or so inhabitants; the state of New York was more than a hundred times as large! That vast difference didn't happen by chance. Abraham Lincoln and the Republicans had hustled the Nevada territory into statehood, despite its meager population, in the run-up to the 1864 election; concerned that three major candidates might split the vote and throw the election to the House of Representatives, they needed the reliably Republican Nevadans to have a voice there, disproportionate though it was to their actual numbers. Nevada became a state with weeks to spare before the election, and dutifully cast its votes for Honest But Also Shrewd When He Needed To Be Abe. Nevada eventually got bigger, but it took a while. In 1900, it was still just 1/171 the size of New York, and its Senate delegation had sent just one Democrat to Washington for just one term in the state's thirty-six years of existence.

Disproportions like this can be obscured by the fact that some small states look big. Politicians of a GOP bent are fond of displaying maps of the United States that show a sea of Republican red from almost-coast to almost-coast, with the Democratic strongholds of California and the Northeast a minor blue fringe along the shoreline. From this point of view, it hardly seems unfair that Wyoming has two senators—look how *much* Wyoming there is!

But this, of course, is an artifact of the way we draw the map. Senators represent people, not acres. We've already encountered the "too much Greenland" problem—standard maps like the Mercator projection distort areas, making some regions look bigger than the space they actually take up on the globe. What if there was a map that assigned each state an amount of space according to its *population* instead of its area, more accurately rendering the people the Senate is supposed to represent? Geometry can do that. This kind of map is called a cartogram:

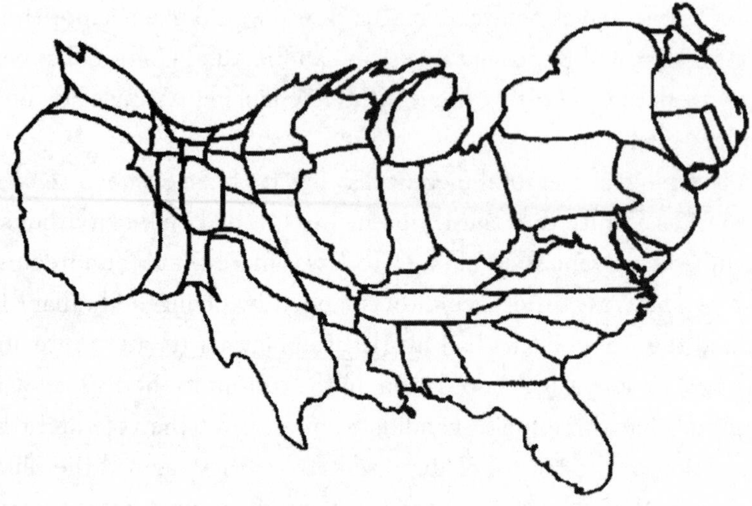

The cartogram makes plain how much of the U.S. population, even today, is in the original thirteen colonies of the East, and how narrow a wasp waist the Great Plains really is.

A voter in Pennsylvania may have less influence in the presidential election than one in New Hampshire, but they have infinity times as much as an American who lives in Puerto Rico or the Northern Marianas or Guam. (The civic-minded Guamanians, despite not being granted any electoral votes, hold presidential primaries and a presidential election every year anyway; in 2016, they got a turnout of 69%, better than all but three U.S. states.)

You might think of the Senate and Electoral College as a kind of standardized test, a quantifiable proxy for whatever we think of as the popular will. Like any standardized test, it roughly measures the thing it's supposed to; but it can be gamed, and the longer it persists in fixed form, the better people get at gaming it, getting more and more accustomed all the while to thinking of the test itself as the thing that really matters. Sometimes I imagine a distant future where whole regions of the United States, ravaged by climate change and unchecked pollution, are inhabited solely by a handful of cyborg-people aged a hundred and up, kept in stasis in purified-air boxes and roused into consciousness by their machine parts once every even-numbered year, just long enough to mark a ballot for the congressional representatives the Constitution guarantees them. And there will still be opinion pieces in the newspapers praising the founders' keen insight in designing a system of self-government that has served us so well, and for so long.

The states are mostly fixed in place now; we are not ever going to replace them with some highly rationalized machine-drawn division of the nation into equal-sized chunks where Wyoming and greater Chattanooga have the same say in lawmaking. There will continue to be some much smaller than the others. By contrast, legislative districts, post-*Reynolds*, are all roughly the same size. That blunts the power of the district-drawers, preventing them from brazenly creating rotten boroughs to preserve their power. But it does not eliminate it. Chief Justice Earl Warren, in his majority opinion in *Reynolds*, wrote: "Indiscriminate districting, without any regard for political subdivision or natural or historical boundary lines, may be little more than an open invitation to partisan gerrymandering."

So it has proved. There are many flavors of mischief available to legislators who—if you'll allow me a redundancy—are strongly motivated to advance the interests of their political faction. Let's see how this works in the state of Crayola.

WHO RULES CRAYOLA?

In the great state of Crayola two parties vie for power, the Oranges and the Purples. The state has a lean: 60% of the one million voters there support the Purple Party. Crayola has ten legislative districts, each one of which sends a senator to fulfill government's solemn duties at the statehouse in Chromopolis.

Here are four ways the voters could be divided into those ten districts:

	OPTION 1		OPTION 2	
	Purple	*Orange*	*Purple*	*Orange*
District 1	75,000	25,000	45,000	55,000
District 2	75,000	25,000	45,000	55,000
District 3	75,000	25,000	45,000	55,000
District 4	75,000	25,000	45,000	55,000
District 5	75,000	25,000	45,000	55,000
District 6	75,000	25,000	45,000	55,000
District 7	35,000	65,000	85,000	15,000
District 8	35,000	65,000	85,000	15,000
District 9	40,000	60,000	80,000	20,000
District 10	40,000	60,000	80,000	20,000

All four of these districtings split Crayola up into equal-sized districts with 100,000 voters each. In all four, the columns add up to 600,000 Purple voters and 400,000 supporters of Orange. But the legislatures they produce are wildly different. In the first map, Purple wins six seats and Orange four. In the second, Orange takes over the majority, with six out of the ten seats. In the third, Purple holds the majority seven seats to three. And in the final map, Orange is completely shut out, and Purple makes the laws without a dissenting voice to be heard in the chamber.

Which one is fair?

That's not a rhetorical question. Actually think about it for a minute! There's no sense in reading dozens of pages about a difficult social problem until you've reflected on what goal you think we're trying to achieve.

A minute goes by . . .

There's no obvious answer, as I hope you can see. I give a lot of talks about districting, and I always ask this question, and I get all kinds of answers. Almost always, a majority of people like Option 1 the best.

OPTION 3		OPTION 4	
Purple	*Orange*	*Purple*	*Orange*
80,000	20,000	60,000	40,000
70,000	30,000	60,000	40,000
70,000	30,000	60,000	40,000
70,000	30,000	60,000	40,000
65,000	35,000	60,000	40,000
65,000	35,000	60,000	40,000
55,000	45,000	60,000	40,000
45,000	55,000	60,000	40,000
40,000	60,000	60,000	40,000
40,000	60,000	60,000	40,000

Most people pick as the unfairest choice Option 2, where Orange holds the majority despite being decidedly in the minority of the population. But I once talked to a group of Unitarians who thought Option 4 was clearly the worst, because one party was deprived entirely of its right to participate. And the Unitarians are far from alone in this view.

Is this even a math question? It's not *not* a math question. But there's a legal strand, a political strand, and a philosophical strand, too, and there's no way to unwind these from each other. There's a long and unimpressive tradition of mathematicians approaching the problem of districting as an exercise in pure geometry, asking questions like "How can we cut up Wisconsin along perfectly straight lines such that the resulting polygonal regions have equal population?" You can do that—but you *shouldn't,* because you'll get districts that have nothing to do with the real political facts on the ground. Those districts might have agreeable geometric properties, but they'll cut cities and neighborhoods in half, and cross county lines, which, in Wisconsin and many other states, is a constitutional no-no unless you have to do it to make the districts equal in population.

On the other hand, when lawyers and politicians think about redistricting while neglecting the mathematical strand, the result of their work will be no better; and that, by and large, is exactly how these issues have been addressed until recently. To get districting right, there's no alternative to digging into the numbers and shapes.

Looking at those numbers for the four districtings of Crayola makes the basic quantitative principle of gerrymandering perfectly clear. If you get to draw the district lines, you want your opponents' voters packed into a few districts where they predominate. Best of all is if you can get that done by drawing those enemy voters out of formerly competitive adjacent districts, giving your party the advantage there. As for your own voters, you want them carefully allocated in a larger number of districts where they make up a reasonably safe majority. That's what happens in Option 2: the majority of the Purple vote is crunched into the four districts where Orange doesn't have a chance, while the other six districts lean Orange by a solid margin of 55–45.

It's also what happens in Wisconsin. The line between Waukesha County and Milwaukee County is one of the state's sturdiest political boundaries. When you drive east from Madison to see a Brewers game in an election year, the signs in the yards instantly snap from Republican red to Democratic blue as you cross 124th Street. Until 2010, the assembly district lines largely stopped where the counties did, with reliably Republican districts to the west in Waukesha County and Democratic-leaning ones making up Milwaukee County. The map enacted in 2011 changes all that.

The 13th, 14th, 15th, 22nd, and 84th Districts, among others, now dip across the county line to mix Democratic voters—but not *too* many—in with the Republican wards of Waukesha.* Those five districts have been represented by Republicans from the time of their creation through 2018, when Robyn Vining, a former pastor and official 2017 Wisconsin Mother of the Year, won the 14th District for Democrats by

* Didn't I just say the Wisconsin constitution doesn't allow you to break county lines? Well, yes; but so far, court challenges to this map have gone through the federal courts, which don't address potential violations of the state constitution.

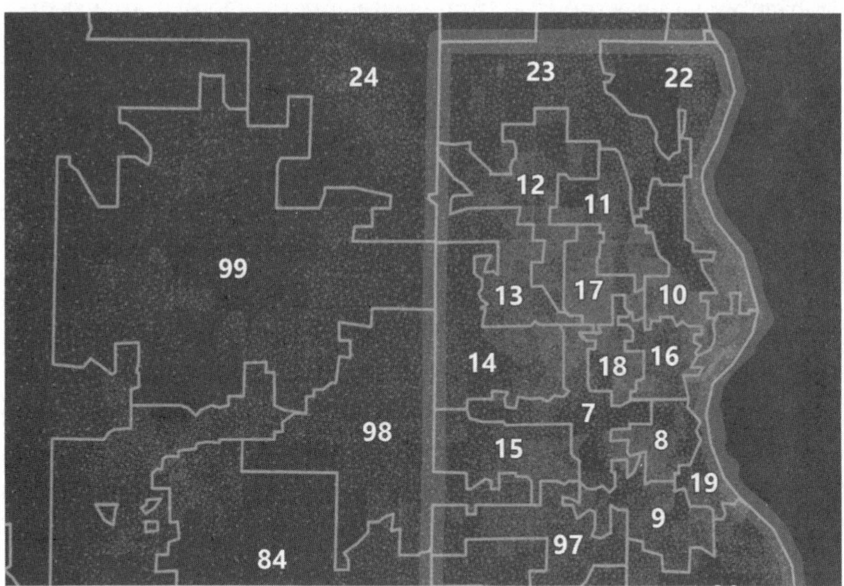

2001 Map

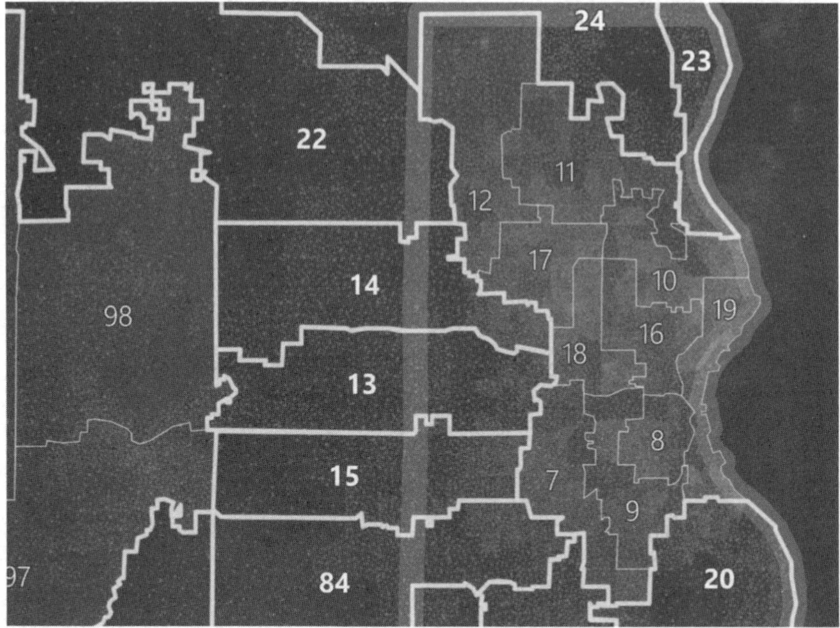

2011 Map

less than half a percent.* The number of districts located entirely within Milwaukee County dropped from eighteen in the old map to just thirteen. Democrats hold eleven of those assembly districts, and ten of those are so uncompetitive that Republicans didn't even field a candidate in 2018.

Politics, as the saying goes, ain't beanbag, and from one point of view, there's no unfairness to be found here. Legislating is a game where whoever's ahead gets to change the rules on the fly, and there's no right or wrong, just winning and losing. But most people see something to be leery of in the practice of gerrymandering, and some of those people are federal judges. Wisconsin's districts were the subject of court challenges almost from the moment Scott Walker signed them into law. Two of the districts were modified by judges in 2012 to make the map less hostile to Hispanic voters in Milwaukee, in a decision that starts out, "There was once a time when Wisconsin was famous for its courtesy and its tradition of good government," and which describes the map-drawers' claims that they worked without partisan bias as "almost laughable." Then, in 2016, a three-judge panel of the U.S. District Court for the Western District of Wisconsin threw out the whole map as a specimen of political gerrymandering in violation of the U.S. Constitution. That decision was appealed and made its way to the U.S. Supreme Court, which had long labored to find a reasonable legal standard for how much partisan gerrymandering was too much. What happened after that was a collision of math, politics, law, and motivated reasoning whose implications American politics is still absorbing.

"A MINORITY RULE IS ESTABLISHED"

If you know anything about gerrymandering, it is probably this pair of facts, which are glommed together right there in the name: first, that it was invented by Elbridge Gerry, who as governor participated in a dis-

* Added in proof: One more, the 13th, was flipped by Democrats in November 2020. The statewide presidential vote was almost exactly even, as the Walker-Evers race had been, and Repubicans maintained a 61–39 assembly majority.

tricting of Massachusetts designed to assist the Democratic-Republicans in fending off the Federalists in the 1812 election; and second, that it involves districts drawn with bizarrely sinuous boundaries, like the "salamander"-shaped district in Massachusetts that a cartoonist immortalized as the "Gerry-mander."

Both these facts are wrong. First of all, gerrymandering in America goes back well before the word and well before Gerry. According to Elmer Cummings Griffith's definitive study, his 1907 PhD dissertation in history at the University of Chicago, the practice dates back at least as far as the colonial assembly of Pennsylvania in 1709. And in early America, the most notorious example of politically motivated district making was carried out by Patrick Henry—"Give me liberty or give me death" Patrick Henry, whose pro-liberty attitude was tempered by his desire to maintain iron control over the Virginia legislature. Henry was a bitter opponent of the new U.S. Constitution, and was determined to keep one of its chief architects, James Madison, out of Congress in the 1788 election. At Henry's direction, Madison's home county was placed in a district with five counties that were seen as anticonstitutional, which Henry hoped would vote for Madison's opponent, James Monroe. Just how unfair this district was is disputed to the present day, but there's no question Madison and his allies felt Henry was playing dirty. Madison didn't get the easy path to Congress he'd hoped for, instead having to return home from New York to campaign for weeks throughout the district. He had a bad case of hemorrhoids that made travel difficult, and picked up frostbite on his face debating Monroe outdoors in January in front of a crowd of Lutherans. Gerrymander or no gerrymander, Madison prevailed, in part by winning his home base of Orange County by 216 votes to 9.

So by the time Gerry gerrymandered, it was no innovation, but an established political technology. (Stigler's Law strikes again!) By 1891, the practice, intertwined with other colors of electoral shenanigans, was so severe as to move President Benjamin Harrison* to warn in his State of the Union address:

* Who won a decisive majority of the Electoral College in 1888 while losing the popular vote, for what it's worth.

> If I were called upon to declare wherein our chief national danger lies, I should say without hesitation in the overthrow of majority control by the suppression or perversion of the popular suffrage. That there is a real danger here all must agree; but the energies of those who see it have been chiefly expended in trying to fix responsibility upon the opposite party rather than in efforts to make such practices impossible by either party. Is it not possible now to adjourn that interminable and inconclusive debate while we take by consent one step in the direction of reform by eliminating the gerrymander, which has been denounced by all parties as an influence in the selection of electors of President and members of Congress?

Harrison's description of democracy under the gerrymander has lost none of its aptness: "A minority rule is established that only a political convulsion can overthrow."

This raises a question. If legislators have been drawing district boundaries to suit their partisan interests for three hundred years, and democracy has more or less persisted, why, now, is urgent reform suddenly required?

That's partly a technology story. An old Wisconsin election hand once told me how redistricting used to be done. There was a person who, over decades of experience in Wisconsin politics, had memorized the idiosyncrasies of every voting ward in the state, from Kenosha to Superior. And the districting savant would have a big paper map spread out on a giant conference table, and would gaze at it, move a chunk here and a chunk there, denote changes with a marker, and the thing was done.

Gerrymandering used to be an art; advanced computation has made it a science. Joe "Aggressive" Handrick and his team of mapmakers tried out map after map, tweak after tweak, not on a wooden table but on a screen. They ran each potential districting through simulations that tested its performance in a wide range of political climates, until they converged on a map optimized to preserve Republican control in all but the most extreme circumstances. That process isn't just faster, it's better. A lawyer involved in the suits against the state told me that the effectiveness of the Act 43 gerrymander went far beyond what any old-fashioned map master was able to accomplish.

What's more, a gerrymander that works well in the initial elections after the cycle creates a population of incumbents for the gerrymandering party, adding even more advantage to what the gerrymander provides. Opposition donors, assessing the map as too tilted to be overcome, allocate contributions elsewhere. And so the gerrymander feeds itself.

Justice Sandra Day O'Connor, writing in dissent in the 1986 case *Davis v. Bandemer*, argued that courts didn't need to intervene in redistricting cases. Remember, making a good gerrymandered map involves constructing a lot of districts where your party has a moderate advantage, set against a few where your opponent predominates. Doesn't that mean, O'Connor asks, that gerrymandering is an inherently risky strategy? In her account, parties will refrain from gerrymandering a state unreasonably strongly, because it puts their incumbents at too much risk of being knocked over by an unanticipated political gust. "[T]here is good reason to think," she wrote, "that political gerrymandering is a self-limiting enterprise."

Back then, she may have been right. But today's computational power has blown away the self-limiting nature of the enterprise, as it has with so many other limits. (Ask Marion Tinsley.) Just as the maps can be tuned to produce substantial partisan advantage, they can be jiggered to reduce risk to incumbents at the same time. And that's not just because a souped-up modern computer is faster than an Apple II. Voters have changed, too! We Americans like to think of ourselves as dispassionately assessing the ballot without prejudice, making a study of each candidate's policy platform and temperamental fitness for the office, and choosing the one who makes the best case for our vote. But actually we're pretty predictable, and getting more so. Most of us just vote for the person with the right letter after their name. The proportion of "floating voters" who switch parties from one presidential election to the next, which hovered around 10% from the mid-twentieth century through the 1980s, has dropped to half that. The more stable and predictable voters are in their choices, the greater the ability a party has to draw a map that preserves its majority *and* protects incumbents *and* persists long enough in its effect to get to the next census, and thus to a brand-new map drawn by the same old legislative majority in the same old locked room.*

STOP KICKING DONALD DUCK

A traditional point of view on gerrymandering is that, as Justice Potter Stewart said in a very different judicial context, you know it when you see it. And yes, there are some weird-shaped legislative districts out there, like the Fourth Congressional District of Illinois, the "earmuffs," which consists of two distinct regions connected by a mile or two of freeway, or this beauty in Pennsylvania, colloquially known as "Goofy Kicking Donald Duck":

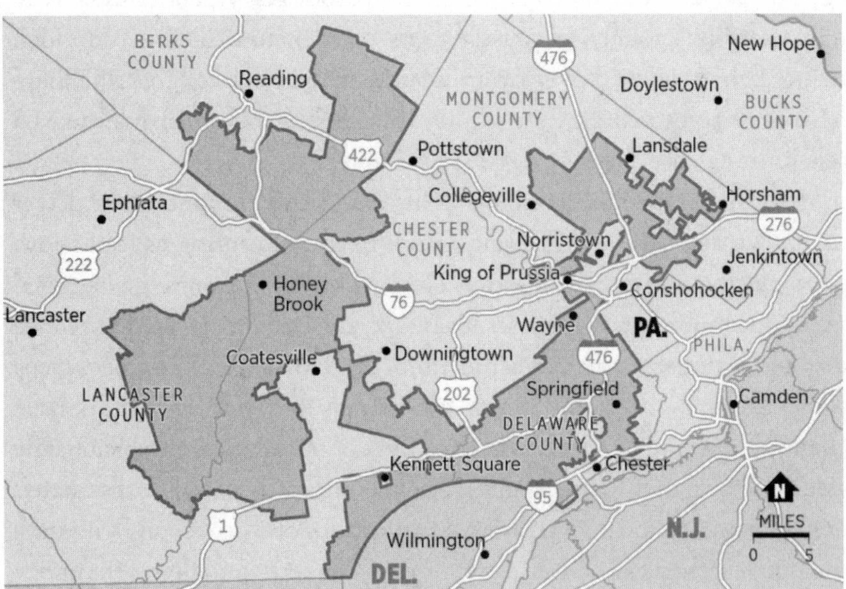

Pennsylvania's Seventh District was drawn this way in order to capture enough scattered Republicans to form a GOP-favoring district. The

* Compare what Elmer Cummings Griffith has to say about a similarly polarized era: "By 1840 two great parties had settled down for a steady and continuous struggle for political supremacy. With general political stability, elections could be predicted with greater promise of success. When parties become established, the change of voters from one to the other is a very small per cent. And the results of an election can be safely predicted within certain definite limits."

two main figures are connected only by means of the grounds of a hospital located at the tip of Goofy's business foot. Goofy's neck is a single parking lot.

This district was thrown out by the Pennsylvania Supreme Court in 2018 as an example of partisan gerrymandering gone too far: a victory for fair elections and roughly round shapes. A commonplace belief in the history of redistricting reform is that we can prevent the excesses of gerrymandering if we require districts to have "reasonable" shapes, thus limiting the ability of legislators to make mischief. Many state constitutions even have provisions directing mapmakers to avoid districts shaped like Disney melees; Wisconsin's, for instance, requires districts to be "in as compact form as practicable." But what exactly does this mean? Lawmakers have never settled on a consensus standard. And attempts to specify what shapes are "compact" sometimes make things even more muddled. In 2018, Missouri voters approved a referendum amending their constitution to require forevermore that "compact districts are those which are square, rectangular, or hexagonal in shape to the extent permitted by natural or political boundaries." First of all, a square is a kind of rectangle. And what has Missouri got against triangles, pentagons, and nonrectangular quadrilaterals? (My personal theory is that Missouri is self-conscious about being a trapezoid and is overcompensating.)

The discipline of geometry does offer some options for measuring the "compactness" of a shape. Your intuition is probably that a very complicated shape, like the erstwhile Pennsylvania Seventh, encloses its area very inefficiently, using a long, complicated, jiggery boundary. So maybe we want shapes whose perimeter isn't too large compared to their area.

Your first thought might be to use a ratio: How much area is there per mile of perimeter? A higher score would be better. Here's the problem with that idea. A tiny square district four miles north to south and four miles east to west will have a perimeter of 16 and an area of 4 × 4 = 16. So your area/perimeter score would be 16/16 = 1. But what if you magnified the square district so that it was 40 miles on a side? Now the perimeter would be 160 and the area 1,600. The score has improved to 1,600/160, or 10.

This is an unpleasant state of things. How "compact" a square is shouldn't depend on its size! Nor should it depend on whether we denominate its size in kilometers, miles, or furlongs! Whatever measure of "compactness" we use should be what geometers call an *invariant*;* it shouldn't change when the region is moved, or rotated, or enlarged, or shrunk. When we move or rotate a region, its perimeter and area don't change, so that's no problem. When we magnify it by a factor of 10, though, its perimeter grows by that same factor of 10 while its area grows by a factor of 100. This suggests that a better ratio to use is

$$\text{area}/\text{perimeter}^2$$

which doesn't change when you magnify or shrink the district. By the way, a very handy way to keep track of this sort of thing is to write everything with units of measurements attached! The perimeter of our 40-mile square is 160 miles, while its area is 1,600 *square* miles; so the area divided by the perimeter isn't 10, it's 10 *miles*, a length, not a number.

Redistricting types call the ratio above the Polsby-Popper score, after two lawyers who realized its relevance in the 1990s, but the notion is older than that. For a circle of radius r, the perimeter is $2\pi r$ and the area is πr^2, so the score is

$$(\pi r^2)/(2\pi r)^2 = \pi r^2/4\pi^2 r^2 = 0.079\ldots$$

and note that the answer doesn't depend on the radius of the circle at all! The r's cancel out. That's the invariance at work. It's the same deal for squares; if the length of a side is d, the perimeter is 4d and the area is d^2, so the Polsby-Popper score is

$$d^2/(4d)^2 = d^2/16d^2 = 1/16 = 0.0625$$

which doesn't depend on the length of the side. The square's score isn't quite as good as $1/4\pi$. In fact, it turns out that $1/4\pi$ is the very highest

* In particular, an invariant for the *similarities* discussed in chapter 3.

score any shape can possibly have! This jibes with what our intuition tells us about how large the area of a figure can be if you fix its perimeter. Put a loop of string on the table and try to "inflate" it by packing as much material inside the loop as possible; don't you feel like it would take on circular form? This fact was known and proved by Zenodorus, in the somewhat casual sense that most ancient mathematicians proved things, a century or so past Euclid. Mathematicians call it "the isoperimetric equality." It didn't get a proof up to the standards of modern geometers until the nineteenth century.

So you can think of the Polsby-Popper score as measuring how "circle-like" a district is, at which point you should start to wonder whether it's actually a good idea. Is a circular district actually better than a square? Is a longer rectangle like this one

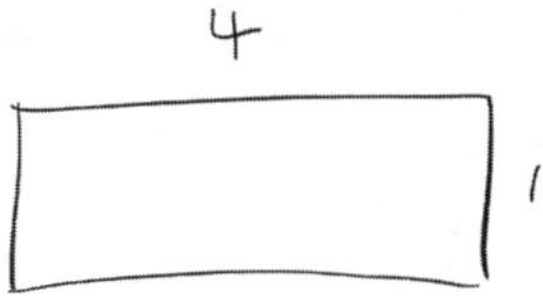

with a score of 4/100 = 0.04, really that much worse?

For that matter, what do we actually mean by perimeter? Real-life boundaries of districts are partly straight surveyor's lines but partly things like seacoasts, which are fractally wiggly at every scale, so they get longer and longer the more closely you measure every miniscule jig and jag. The quality of a district shouldn't depend on the size of your ruler!

Let's take another tack. In many ways, the most manageable geometric figures are those that are *convex*. Speaking vaguely, a convex figure is one that only bends outward,

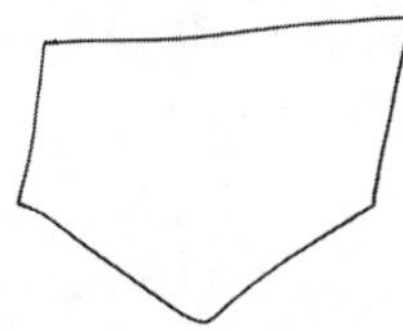

never inward:

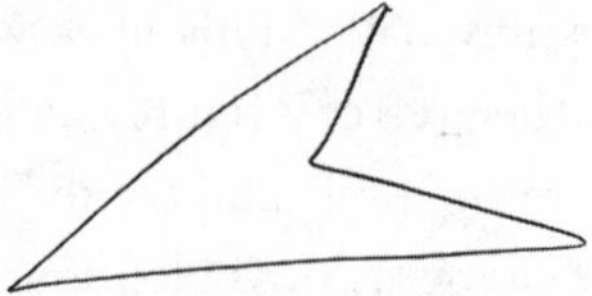

But there's a lovely official definition: a shape is convex when the line segment between any two points in the shape is entirely contained within the shape. (This definition makes sense in two dimensions, or three, or even some much larger number of dimensions well beyond your capacity to visualize "inward" or "outward" bending.) You can see how the latter shape above fails that line segment test:

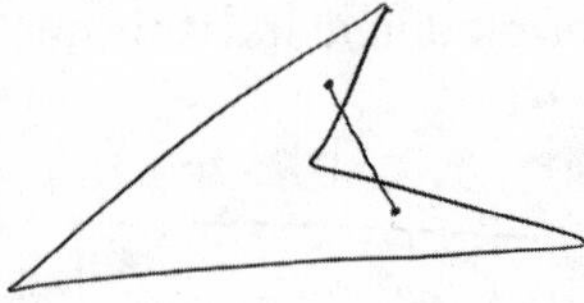

The *convex hull* of a shape is the union of every line segment joining every pair of points in the shape:

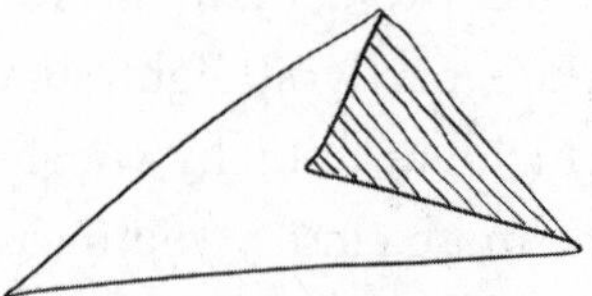

You can think of it as "filling in all the non-convex places," or, more physically, as wrapping plastic wrap around your shape as tightly as you can. The convex hull of a golf ball is a sphere; the dimples get filled in. Your own convex hull is very tightly wrapped around you if you press your legs together and your arms to your sides, but much greater in extent if you extend your limbs in all directions.

Anyway: the "Population Polygon" score of a district is the ratio between the number of people who live in the district and the number of people who live in its convex hull. The convex hull of Goofy and Donald Duck contains all those people *between* Goofy and Donald Duck, so the score of that district is going to be pretty bad.

Population Polygon is an improvement over Polsby-Popper, since it takes into account where people actually live. But there's a deeper problem with enforcing compactness as a brake on gerrymandering, which is that it doesn't work. Maybe in the days of paper maps people had to resort to wacky shapes to get just the portfolio of voters they wanted, but no longer. The map software that allows you to assess a million maps in an afternoon simultaneously lets you pick the ones that are nicely shaped *and* achieve your objectives. Those gerrymandered districts around Milwaukee are innocent-looking quasi-rectangles, which would get acceptable scores on any quantitative measures of compactness.

Sandra Day O'Connor once wrote* that when it comes to legislative districts, "appearances do matter": a salamandery district creates the *perception* that something other than democratic ideals is at work. If you ask me, those ideals aren't much shored up by replacing Goofy and Donald with a map that's just as partisan but less offensive to the eye. There are some good reasons to want a district to be compact, I guess—a shorter average drive to your state representative's office, a modestly increased alignment of political priorities among the constituents. But to the extent that compactness constrains blunt gerrymandering at all, it's just because they're constraints. The fewer choices map-drawers have, the less likely they are to be able to find options that are grievously rigged. It's not that there's anything inherently more fair about roughly round districts; it's that there are many fewer ways to break a state into districts if you have to make them roughly round.

What we now know is that traditional compactness measures just aren't enough to keep parties from stacking the deck in their favor, any more than the equal-population requirement of *Reynolds v. Sims* was. Of course, you could be stricter about the compactness measures to make the constraints more severe, or enforce state laws about breaking county boundaries, or just invent purely arbitrary rules ("the number of registered voters in each district must be a prime number") to limit the

* In the racial gerrymandering case *Shaw v. Reno*—the drawing of districts to guarantee, or prevent, minority representation is a whole other aspect of the districting story that this chapter isn't big enough to contain.

wiggle room available to legislators caught up in their decennial gerrymandering estrus. But arbitrary rules like this aren't really politically feasible. If the goal is to stop gerrymandering, the strategy is going to have to target gerrymandering directly. That means we need a measure of a map that tells us not how equally populated the districts are, or how round and plump they are, but *how gerrymandered* they are. That's a harder problem still. But geometry can get us there.

HIVE THE GRITS!

To paraphrase H. L. Mencken, almost every interesting question of applied mathematics has an answer which is simple, mathematically elegant, and incorrect. For districting, that answer is *proportional representation*: the principle that a party ought to get a share of seats in the legislature equal to the proportion of the popular vote drawn by its candidates. That's one straightforward quantitative answer to what it would mean for a district map to be "fair," and it's really popular. *The Washington Post* reported that, in the 2016 Wisconsin State Assembly elections, 52% of votes went to Republican assembly candidates, but the Republicans won 65% of the seats; the GOP, the paper wrote, "seems to have benefited from gerrymandering, given that discrepancy between votes won and seats held." The implicit suggestion is that something's smelly when those numbers don't match.

Proportional representation is the reason people tend to like Option 1 of the Crayola maps. The Purple party got 60% of the vote, and it gets 60% of the seats.

But would proportional representation actually be the outcome if maps were drawn fairly? Almost surely not! Take the Wyoming State Senate. Wyoming is by some measures the most strongly Republican state in America. Two-thirds of its voters picked Donald Trump in 2016, and the same proportion voted Republican in the governor's race in 2018. But the state senate isn't two-thirds Republican; there are twenty-seven GOP senators and only three Democrats. Should we really see that as unfair? When a state's population is two-thirds Republican, it's pretty likely that almost every geographic chunk of the state is pretty Republican. In the extreme case of a state that's totally homogeneous politically, every neighborhood in every town having the same proportion of Democrats and Republicans, the party with the popular vote majority would win every single seat in the legislature. That's the scenario of Crayola Option 4. And the single-party legislature wouldn't be the result of gerrymandering, but of the state's strangely consistent voter distribution.

Idaho has two representatives in Congress, and so does Hawaii, and we don't think it's strange that Idaho's delegation has been fully Republican and Hawaii's fully Democratic for the past decade,* even though the proportion of voters who back the majority party in each state is closer to 50% than it is to 100%. I don't think a fair splitting of Idaho into two districts would return one Democrat and one Republican. I don't even think you *could* draw a non-ridiculous district that covered half of Idaho and had a Democratic majority.

And what about the plight of the Libertarians? The proportion of Americans voting for Libertarian candidates for the House of Representatives consistently hovers around 1%; but there has never been a U.S. representative elected from that party, let alone the three to five proportional representation would recommend, because there is no such thing as a libertarian city or even neighborhood (though it's fun to imagine!). In Canada, whose elections are structured very similarly to those in the U.S., the deviations are even starker; in the 2019 federal elections there, the New Democratic Party drew 16% of the vote against only 8% for the Bloc Québécois, but the Bloc, whose voters are concentrated in a single province, won substantially more seats in Parliament.

Canada, by the way, doesn't have a gerrymandering problem, despite having a U.S.-style legislature. That's not because Canadians are nicer than Americans. It's because Canada has assigned the drawing of districts (called *ridings* up north) to nonpartisan commissions since 1964.

* Each state had one representative from the minority party for one term after the 2008 election, to be fair.

Before that, district drawing was as politically motivated and dirty as it is in the United States. Canada's very first prime minister, the Conservative Sir John Macdonald, wielded the districting pen in a ruthless effort to diminish the power of his opponents in the Liberal Party, the so-called "Grits." The map used for the election of 1882 was so brazen as to inspire this poem published in the *Toronto Globe*, surely the clearest explanation of the principles of gerrymandering ever set to trochaic tetrameter:

> Therefore let us re-distribute
> What constituencies are doubtful
> So as to enhance our prospects;
> Hive the Grits where they already
> Are too strong to be defeated;
> Strengthen up our weaker quarters
> With detachments from these strongholds
> Truly this is true to nature
> In a mighty Tory chieftain!

Proportional representation is a perfectly reasonable system; many countries build it into their method of assembling legislatures. But it's not *our* system, and it's unreasonable to expect proportional representation to be the outcome of a U.S. election. Nonetheless, the specter of proportional representation still haunts the gerrymandering discourse. In a closed-door seminar advising Republicans on how to draw maps in their favor without running afoul of judges, secretly taped by a participant, the GOP election lawyer Hans von Spakovsky warned his audience about those who would try to overturn the maps in court:

> What they were arguing was that, if for example, the Democratic Party has a presidential candidate who gets 60% of the votes statewide in the presidential election, why then they're entitled to 60% of the state legislative seats, and 60% of the congressional seats.

This is false, though it's not clear to me whether Spakovsky knows it's false. Proportional representation isn't the standard reformers are arguing for. So what is?

MIND THE GAP

The 2004 Supreme Court case of *Vieth v. Jubelirer* put the problem of partisan gerrymandering into a curious legal limbo. Four justices felt that the practice of gerrymandering for partisan gain was entirely nonjusticiable; that is, that it was a purely political matter the federal courts were forbidden to interfere with. Four felt that the map insulted the right to representation so grievously as to constitute a constitutional violation.

Justice Anthony Kennedy, in this case as in many others the fulcrum of the court, joined the majority in upholding the gerrymandered map at issue, but disagreed with the authors of the majority opinion on the critical matter of justiciability. Courts *did* have the power and duty to stop partisan gerrymandering, he wrote, if only there were a reasonable standard judges could use to determine when a map is so bad it makes the Constitution puke.

We've seen that proportional representation isn't that standard, and neither are measures of geometric compactness. So a new idea was needed. Reformers got one from political scientist Eric McGhee and law professor Nicholas Stephanopoulos in the form of the "efficiency gap."

Remember: what makes a gerrymander work is that your party wins a lot of districts by a little and a few districts by a lot. You can think of that as an "efficient" allocation of your party's voters. Looking at Crayola Option 2 through that lens, we can see a massive failure of efficiency on Purple's part. What good does their 85,000–15,000 win in the 7th District do them? They would have been better off swapping 10,000 of those voters into District 6 and bringing 10,000 Orangists into District 7 in exchange; they'd still win the 7th by a commanding 75,000–25,000 margin, but now they'd win the 6th District 55,000–45,000 instead of losing it by the same amount.

Those extra Purples in District 7 are, from the point of view of their party, wasted. In Stephanopolous and McGhee, "wasted votes" are votes which are either

- votes cast in a district where your party loses; or
- votes above the 50% threshold in a district where your party wins.

In Option 2, the Purple party wastes a *lot* of votes. Here's a chart:

WASTED	Purple votes	Orange votes	WASTED
45,000	45,000	55,000	5,000
45,000	45,000	55,000	5,000
45,000	45,000	55,000	5,000
45,000	45,000	55,000	5,000
45,000	45,000	55,000	5,000
45,000	45,000	55,000	5,000
35,000	85,000	15,000	15,000
35,000	85,000	15,000	15,000
30,000	80,000	20,000	20,000
30,000	80,000	20,000	20,000

There are 45,000 votes wasted in each of the six losing districts; in each of the 7th and 8th, the 35,000 votes exceeding what Purple needs for a majority are wasted, too; and in the 9th and 10th Districts, 60,000 of the 160,000 Purple votes are wasted. That adds up to 6 × 45,000 + 70,000 + 60,000 or 400,000 wasted votes.

Orange, by contrast, is incredibly efficient. Only 5,000 of its votes are wasted in each of the first six districts; and in the districts where it loses, it loses *badly,* wasting only 30,000 votes in Districts 7 and 8 and 40,000 in Districts 9 and 10. That's 100,000 votes wasted in all, 300,000 fewer than Purple.

The *efficiency gap* is the difference between the number of votes wasted by the two parties* expressed as a percentage of the total num-

* The *two* parties? But what if . . . yes, I know, I know. Quantifying gerrymandering when more than two parties are involved is a largely unexplored topic, and I encourage you to think about it!

ber of votes cast. In the case of Option 2, that gap is 300,000 out of a million, or 30%.

That's a *huge* efficiency gap. In real elections, the figure is typically in the single digits. Some lawyers have suggested that any figure over 7% should be enough to induce a court to have a careful look-see.

Not all the options we laid out for Crayola are that gappy. Here's a chart for Option 1, the map that satisfies proportional representation:

WASTED	Purple votes	Orange votes	WASTED
25,000	75,000	25,000	25,000
25,000	75,000	25,000	25,000
25,000	75,000	25,000	25,000
25,000	75,000	25,000	25,000
25,000	75,000	25,000	25,000
25,000	75,000	25,000	25,000
35,000	35,000	65,000	15,000
35,000	35,000	65,000	15,000
40,000	40,000	60,000	10,000
40,000	40,000	60,000	10,000

Purple wastes 25,000 votes in each of the first six districts, 35,000 each in Districts 7 and 8, and 40,000 in Districts 9 and 10, for a total of 300,000. Orange also wastes 150,000 in the first six districts, but only 15,000 each in Districts 7 and 8 and 10,000 each in Districts 9 and 10, for a total of 200,000. So the efficiency gap drops to 100,000 out of a million, or 10%, still in favor of Orange. In Option 4, the map where Purple holds all the seats, the Orange Party wastes 40,000 votes in every single district, while Purple wastes only 10,000; so we get another huge efficiency gap of 30%, but this time favoring Purple. What about Option 3?

WASTED	Purple Votes	Orange Votes	WASTED
30,000	80,000	20,000	20,000
20,000	70,000	30,000	30,000
20,000	70,000	30,000	30,000
20,000	70,000	30,000	30,000
15,000	65,000	35,000	35,000
15,000	65,000	35,000	35,000
5,000	55,000	45,000	45,000
45,000	45,000	55,000	5,000
40,000	40,000	60,000	10,000
40,000	40,000	60,000	10,000

Now, each party wastes the same number of votes: 250,000. This map has efficiency gap *zero*; from the point of view of this measure, it's as fair as can be, despite its departure from proportional representation.

As far as I'm concerned, that's good! In practice, maps that are drawn by neutral arbiters rarely approach proportional representation, except for those cases where both seat share and popular vote share are close to 50–50. Instead, the seat share is usually farther from 50–50 than the vote share is. By the efficiency gap standard, an election where one party got 60% of the votes and seated 60% of the legislature might be evidence *for* gerrymandering, not against it.

The efficiency gap is an objective measure, it's easy to calculate, and reams of empirical evidence show that it jumps in maps we know were gerrymandered, like Wisconsin's. So it became a fast favorite of plaintiffs. It played a major role in the court case that threw out Wisconsin's maps in 2016, after years of legal dispute.

And here is where I pull the football away once again. The efficiency gap started to come under fire almost as soon as it got popular. It has flaws, severe ones. For one thing, it's badly discontinuous. Whether votes are wasted depends on who wins the district, which means that the efficiency gap can change wildly under very small changes in election outcome. If Purple wins a district by 50,100–49,900, Orange has wasted 49,900 votes there and Purple only a hundred. A tiny shift in the vote gives Orange a 50,100–49,900 victory, and flips the wasted votes; now it's Purple that racks up nearly 50,000 more. That changes the efficiency gap by almost 10% all by itself! A good measure shouldn't be this brittle.

Another problem with the efficiency gap has more to do with law than math. To get a court to throw out a map, or even to have the case heard, the person bringing the case has to have standing; that is, the plaintiff has to show that they, personally and individually, were denied some portion of their constitutional rights by the state's map. When districts are wildly different in size, it's plain who was harmed: the person in the gigantic district whose vote counts for less. The standing claims in a gerrymandering case are much murkier. And the efficiency gap doesn't help much. Whose rights were denied, or at least meaningfully shaved down? It can't be everyone whose vote counts as "wasted"—that, for instance, would include everyone on the losing side of a close district election, and the voters in the most competitive districts certainly don't seem like the ones whose right to vote is being abridged. This issue of standing was precisely where the Wisconsin case foundered at the U.S. Supreme Court, which found unanimously that the plaintiffs had not done enough to establish that they, personally, were harmed by the gerrymander. The case was returned to Wisconsin for repair, but never made it back to the Supreme Court, which decided to use cases from North Carolina and Maryland to render its judgment on gerrymandering.

The efficiency gap also suffers from a certain over-rigidity. If the number of votes in every district is the same, as in our Crayola examples, then it turns out that the efficiency gap is simply the difference between

> the winning party's victory margin in the popular vote

and

> *half* the winning party's victory margin in seats.

So you get efficiency gap zero when the victory margin in seats is exactly twice the victory margin in the popular vote, and the closer you come to that standard, the smaller the efficiency gap is. In Crayola, the Purple party won the popular vote by a 20-point margin. So, as far as efficiency gap is concerned, the "right" margin of victory in legislative seats is twice that, or 40 points. That's exactly what happens in the efficiency-gap-zero Option 3, where Purple wins 70% of the seats. In Option 1, where Purple wins both the popular vote and the race for seats by a 20-point margin, the efficiency gap is 20% – 10% = 10%.

Courts don't like systems where there's is a single "correct" number of seats assigned to a given vote share. They smack of proportional representation, even when, as here, the formula is usually incompatible with proportional representation.

I say "usually" incompatible because there's one situation in which efficiency gap and proportional representation (and probably you, too) agree about what's fair. That's the scenario where each party gets exactly half the overall votes. Then there's a basic symmetry you might expect any map called "fair" to satisfy. If the population of the state is exactly evenly split, shouldn't the two parties share the legislature equally as well?

The Republican Party of Wisconsin would say no. And however I may feel about their districting skullduggery in the spring of 2011, I have to admit they have a point.

Crayola map 2 awards a majority of seats to the Orange Party, even as they get thumped by the Purples in the popular vote. But what if the Purples of the state congregate in a couple of dark-Purple metro areas, set against the background of a countryside that leans orange? It's possible you'd see results a lot like this, without any scale-thumbing by the map-drawers. Is this kind of asymmetry actually unfair if the Purple people are gerrymandering themselves?

Wisconsin's Republican attorney general Brad Schimel argued in an amicus brief to the Supreme Court that this scenario is exactly what's happening in Wisconsin. The assembly district in Madison where I live, AD77, delivered 28,660 votes to the Democrat Tony Evers. The GOP's man Scott Walker got only 3,935. In Milwaukee's District 10, Evers was even more dominant, winning 20,621 to 2,428. There were no districts where Republicans won by even close to that much. That's not because gerrymandering packed those districts full of Democrats. Madison *just is* full of Democrats.

The superficially fair criterion that a 50–50 split of votes should result in a roughly 50–50 split of seats, Schimel argued, would actually be biased *against* Republicans, not just in Wisconsin but in every state whose dense cities were dominated by Democratic voters—which is to say, just about every state.

BIPARTISAN STATISTICAL MALFEASANCE

Not all the points in that legal brief are good ones. The Act 43 map was designed to be voter-proof by building in resistance to a uniform change in voter mood; if every district moves toward Democrats by the same fixed percentage, it takes quite a big shift to dissolve the GOP's engineered advantage. Schimel, whose task was to deny the effectiveness of the map his own party had worked so hard to create, points out that all ninety-nine districts do not, in fact, swing exactly in tandem.

There are many statewide statistical measures one could compute in order to test just how uniform those swings tend to be, and just how well the Act 43 map resists Democratic gains under a more realistic model of year-to-year variation. That would be an interesting and useful analysis. It's not what Schimel did.

Rather, he picked out one district, State Senate District 10, which had given the Republican candidate 63% in one election and 44% in the next, a 19-point swing. Is it really plausible that the whole state got that much more Democratic in such a short time? If it had, Schimel estimates, the Democrats were en route to "*winning 77 out of 99 districts*" (breathless italics Schimel's). I guess the gerrymander wasn't so bad after all!

What Schimel doesn't say is that the District 10 race won by Democrat Patty Schachtner (a bear-hunting grandmother of nine whose highest previous office was county medical examiner) was a special election for an open seat, held in January, with turnout about a quarter of what you get in a normal election year. And in the election before that, the Republican candidate who got 63% was a popular incumbent who'd held the seat for sixteen years. You can't make the case that Wisconsin politics swings by 19 points in eighteen months unless you choose your data points very carefully to get you to that conclusion, which is exactly what Schimel did.*

Statistical malfeasance of this kind isn't limited to Wisconsin Republicans. The total number of votes cast for Democratic Wisconsin assembly candidates in 2018 was 1,306,878. Republican assembly candidates got just 1,103,505. So Democratic candidates got 53% of the popular vote for assembly—but only won thirty-six of the ninety-nine seats. That's not just a departure from proportional representation, which would be forgivable; it's an almost veto-proof majority seized by a party that got a minority of the votes. This statistic was shared everywhere. It appeared on Rachel Maddow's popular liberal TV show and was tweeted out by the head of the state Democratic Party, as proof positive of the rigging of Wisconsin's district map.

But *I* haven't mentioned it. Here's why. One of the gerrymander's chief effects is to pack Democrats into districts so homogeneous that Republicans don't have a speck of a chance there. In a year with a Democratic mood, like 2018, it's not even worth a Republican candidate's time to run. So there were thirty out of ninety-nine districts with no 2018 Republican candidate at all—my district in Madison was one of them, natch—against only eight districts lacking a Democrat. Each of those thirty uncontested races *would* have given some votes to a Republican candidate, if any had been willing to run. But that 53% number treats them as if they had no Republican sentiment at all.

Both Schimel's number and Maddow's number were correct. That somehow makes it worse! A false figure can be corrected. A true one chosen to make the wrong impression is a much harder poodle to muzzle. People often complain that no one likes facts and numbers and reason and science anymore, but as someone who talks about those things

* Added in proof: and indeed, in November 2020, her first try in a regularly scheduled election, Schachtner lost her bid for reelection by a 19-point margin.

in public, I can tell you that's not true. People *love* numbers, and are impressed by them, sometimes more than they should be. An argument dressed up in math carries with it a certain authority. If you're the one who outfitted it that way, you have a special responsibility to get it right.

WRONG QUESTIONS

If even the basic principle that even-steven votes should lead to even-steven legislatures is suspect, what hope is left of defining fairness? How are we to judge which of the four maps of Crayola is the right one? Is it Option 1, the one that satisfies proportional representation, where Purple gets six seats and Orange four? Is it Option 3, the one with zero efficiency gap, where Purple has a 7–3 edge? What about Option 4? It feels wrong for Purple to hold all the seats, but as we've seen, that's just what happens if the state of Crayola happens to be politically homogeneous, with the same 60–40 ratio of Purple to Orange in all four compass directions, in the cities and the hamlets. In that case, no matter how you draw the lines, each district will tilt 60–40 in Purple's favor and you get a monochrome legislature.

The Wisconsin GOP would suggest even Option 2 shouldn't be ruled out; if the concentration of Purples in Purpleopolis is strong enough, it may well be that any reasonable map will produce four highly Purple districts and six modestly Orange ones.

We seem to have hit an impasse; there is no clear way to look at those numbers and agree on which map is fair. This feeling of futility is welcomed by gerrymanderers, who prefer to do their dark work unconstrained. It is at the center of every argument put to the court in defense of the practice: maybe it's fair, maybe it's not, but tragically, Your Honor, there is just no way to judge.

Maybe not. But you and I are not judges. We are, for the moment, mathematicians. We're not bound by the limits of the law; we can use every tool at hand to try to figure out what's really going on. And if we're very lucky, we'll come up with something that will stand up in court.

The legal battles over gerrymandering came to a climax in March 2019, when the U.S. Supreme Court heard oral argument in two cases with the potential to finally open or close the constitutional door Justice Kennedy had left tantalizingly ajar. Kennedy himself wasn't there to hear the case; he had retired the year before, replaced on the Court by Neil Gorsuch. One case, *Rucho v. Common Cause,* came from North Carolina; the other, *Lamone v. Benisek,* was from Maryland. Both the disputed maps concerned districts for the U.S. House of Representatives; the North Carolina map, gerrymandered by Republicans, had arranged for ten of the state's thirteen seats to be reliably Republican, while in Maryland, the state's all-Democratic government had cut the number of plausible Republican seats in the state to just one out of eight. The Maryland mapmakers were advised by veteran Democratic congressman and House Majority Leader Steny Hoyer, who once told an interviewer, "Now let me make it clear, I am a serial gerrymanderer." Ironically, Hoyer's career in politics started when, as a twenty-seven-year-old political rookie, he won the 1966 race for Maryland State Senate seat 4C, a seat that had come into existence just that year after the Supreme Court threw out Maryland's unequally sized senate districts in the wake of *Reynolds v. Sims.* (Isaac Lobe Straus, sadly, didn't live to see it.)

The twin set of cases presented a perfect opportunity for the court to rule on gerrymandering without appearing to take partisan sides. The most high-profile gerrymanders in the country, in places like North Carolina, Virginia, and Wisconsin, had been drawn by Republicans, so the fight to reform redistricting was usually seen as a Democratic struggle; but high-profile Republican officials like Ohio governor John Kasich and Arizona senator John McCain also weighed in against gerrymandering, submitting amicus briefs to the court presenting their own woeful experiences with the effect of motivated mapmaking on democracy. Experts from around the country submitted their own briefs. There was a historians' brief, which quoted no fewer than eleven different *Federalist Papers*; there was a brief from a team of civil rights organizations addressing the impact on minority rights; there was a political scientists' brief disputing Justice O'Connor's view that the problem of gerrymandering would take care of itself; and, for the first time in the history of the Supreme Court, there was a mathematicians' brief.* I signed it. In a few pages we'll get to what's in it.

Mathematicians are like Ents, the sentient trees in *The Lord of the Rings*—we don't like to get involved in the mundane conflicts of state, which are out of sync with our slow time scale. But sometimes (and I'm still inside this Ent simile, by the way), events in the world so offend our particular interests that we have to lumber in. Our intervention was necessary here because of some fundamental misapprehensions about the nature of the problem, which we hoped by means of our brief to correct. From the very beginning of the oral arguments, it was clear we hadn't fully succeeded. Justice Gorsuch, questioning the North Carolina plaintiff's lawyer Emmet Bondurant, cut to what he believed to be the chase: "How much deviation from proportional representation is enough to dictate an outcome?"

In math, wrong answers are bad, but wrong questions are worse. And this is the wrong question. Proportional representation, as we've seen, isn't what usually happens when districts are neutrally drawn. Yes, more than three-quarters of North Carolina's districts were firmly in Republican hands, even though Republican voters in North Carolina don't make up anything like three-quarters of the electorate. But that wasn't the actual problem the plaintiffs were asking the court to remedy.

It's easy to see why the judges would *want* that to be the request, because that would make their job easy; they could just say no. The case of *Davis v. Bandemer* had already established that a lack of proportional representation didn't make a map unconstitutional. But the actual issue in *Rucho* was subtler. To explain it, as is so often the case in math when we get really stuck, we need to go back to the beginning of the problem and start over.

* Technically, the "Amicus brief of Mathematicians, Law Professors, and Students," but most of us were mathematicians.

DRUNK DISTRICTING

We have tried to find a numerical standard for "fairness" and we have failed. The reason is that we've made a basic philosophical mistake. The opposite of gerrymandering isn't proportional representation, or efficiency gap zero, or adherence to any particular numerical formula. The opposite of gerrymandering is *not gerrymandering*. When we ask whether a district map is fair, what we really want to ask is:

> Does this districting tend to produce maps similar to the ones a neutral party would have drawn?

Already we are in a realm that makes lawyers tug nervously at their chins, because we're asking questions about a *counterfactual*: what would have happened in a different, fairer world? To be honest, it doesn't sound much like math, either. The question requires knowledge of the desires of the map-drawer. What does math know about desires?

A path out of this thicket was first chopped by the political scientists Jowei Chen and Jonathan Rodden. They were troubled by the problems with traditional measures of gerrymandering, especially the principle that 50% of the votes should yield 50% of the seats in the legislature. It was clear to them that concentration of one party in city districts was likely to produce what they called *unintentional gerrymandering* favoring the more rural party, even in maps drawn by disinterested actors. That's what we saw in Crayola; the party whose voters are crammed into a few districts is at an asymmetrical disadvantage when it comes to winning seats. But would that asymmetry be big enough to explain the disparities we observe? In order to find out, you need to get some neutral parties to draw maps for you. And if you don't know any neutral parties, you can just program a computer to act like one. The idea of Chen and Rodden, which is now central to the way we think about gerrymandering, is to generate maps *automatically*, and in large number, by a mechanical process that has no preference between the parties because we didn't code it to have one. So we can rephrase our initial question:

> Does this districting tend to produce outcomes similar to those a computer would have drawn?

But of course there are lots of different ways a computer might have drawn a map; so why not leverage the computer's power to let us look at *all* the possibilities? That lets us rephrase the question in a way that starts to sound more like math:

> Does this districting tend to produce outcomes similar to a map *randomly selected* from the set of all legally permissible maps?

This suits our intuition, at least at first; one might imagine that a mapmaker truly indifferent to how many seats each party gets would be equally happy with any of the ways of scissoring up Wisconsin. If there were a million ways to do it, you could roll a million-sided die, read the tiny number off the top, pick your map, and relax until the next census.

But that's not quite right. Some maps are better than others. Some are downright illegal—if the districts are noncontiguous,* for instance, or if they violate the Voting Rights Act requirement of districts where racial minorities are likely to be able to elect representatives, or if the populations of the districts differ from each other by more than the rules allow.

And even among the maps that don't run afoul of statute, we have preferences. States want to reflect natural political divisions, avoid cutting up counties, cities, and neighborhoods. We want our districts to be reasonably compact, and in the same vein we want their boundaries not to be too snaky. You can imagine giving each district map a score that measures just how well it does with respect to these measures, which in legal terminology are called *traditional districting criteria*, but which I will call *handsomeness*. And now you choose a district at random from

* Except in Nevada, the one state with no contiguity requirement—hold that thought, we'll need it later!

among the lawful options, but in a way that's biased to favor the handsomest maps.

So let's try one more time:

> Does this districting tend to produce outcomes similar to a map randomly selected, with a bias toward handsomeness but no bias regarding partisan outcome, from the set of all legally permissible maps?

A question now presents itself. Why don't we just let our computer search and search until it finds the very handsomest of all maps, the one that best respects county boundaries and does the least non-convex squirming along its trim perimeter?

There are two reasons. One is political. People who actually work with state governments, in my experience, are unanimous in the opinion that elected officials and their voters *hate* the idea of a computer-drawn map. Districting is a task given to the people of the state, through some official body that's supposed to represent our interests. Delegating that task to an unauditable algorithm is not going to be acceptable.

If you don't like that reason, here's another one: it would be absolutely and definitively impossible to do that. A computer can pick the best map out of a hundred. It can pick the best map out of a million. The number of possible districtings is . . . a lot more than that. Remember 52 factorial, the astronomical number of orders you can put a deck of cards in? That number is like a tiny shriveled bean next to the colossal quantity of ways you can divide the state of Wisconsin into ninety-nine contiguous regions of roughly equal population.* Which means you can't simply ask your computer to assess each map's handsomeness and pick out the best one.

Instead, we may look at just a few possible maps, where by "just a few" I mean 19,184. You get a picture like this:

* No exact formula for this number, or even a decent approximation, is known. The number of ways to divide the 81 little boxes in a 9 x 9 square into nine equally sized regions, each one connected—the number of possible "Jigsaw Sudoku" pictures, if you're into that—is already 706,152,947,468,301. Wisconsin has 6,672 wards that you have to divide into 99 regions.

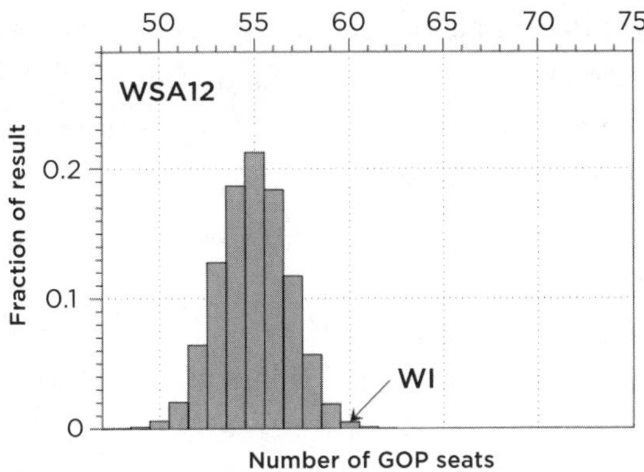

What you're looking at is what's called an *ensemble,* a set of maps generated at random by a computer. This particular computer was managed by Gregory Herschlag, Robert Ravier, and Jonathan Mattingly at Duke University. For each of those nineteen-thousand-some randomly generated maps, they take the Democratic and Republican votes cast in the actual 2012 Wisconsin State Assembly election and assign them to their new autogenerated district.* For each map, you count the number of districts in which Republicans got more votes. That's what you see in the bar chart above. The most common outcome, which happens in more than a fifth of the machine-generated maps, is that Republicans win fifty-five of the seats. Slightly less frequently, Republicans win either fifty-four or fifty-six. Together, these three possibilities cover more than half of the simulations. As you get further in either direction from that most frequent outcome† of fifty-five seats, the bar chart tails off; like so many random processes, it forms something vaguely bell-curvish, and outcomes very far from fifty-five are very unlikely. They are, to use a statistical term of art, *outliers.*

A districting that separates the 2012 voters into sixty districts with more Republicans and only thirty-nine with more Democrats is one of those outliers. That a map would yield such a good outcome for the

* There are wrinkles here, like: What do you do with people who live in wards where the real-life election was uncontested? You have to make your best guess as to how those voters *would* have voted, given a candidate in each party; you can do this by extrapolating from how the ward voted in the contested races for president, senator, and House representative going on at the same time.

† The value of a variable that occurs most frequently—the place where a bar graph peaks—is usually called the *mode,* which is yet another word invented by Karl Pearson.

GOP is highly unlikely, happening in less than one in two hundred of the computer's trials. Or rather—that kind of map is highly unlikely *if the map is chosen randomly by a person or machine without a partisan interest.* If the map is, instead, chosen by a group of consultants in a locked room with the explicit mission of maximizing GOP seats, it's the opposite of unlikely.

The ensemble also shows you the truth and the lie in the Wisconsin legislature's defense of its map. We can't help it, they say, if Democrats choose to congregate in cities among their own effete kale-eating liberal kind; that makes the legislature skew Republican even when the popular vote is split.

And that's true! But with the ensemble, we can estimate *how* true. In 2012, a typical neutrally drawn map, under conditions of near parity between Democratic and Republican assembly votes statewide, would have given the GOP a 55–44 majority. That's a lot less than the 60–39 majority they actually acquired. Six years later, in the 2018 election, Scott Walker won just under half the popular vote; nonetheless, a typical neutral map would have him coming out ahead in fifty-seven assembly districts. But the GOP-drawn map manages to create sixty-three Walker-favoring districts! The political geography of Wisconsin helps Republicans; the turbo-boost they get from the gerrymander goes above and beyond that.

At least, sometimes it does. In 2014, a midterm election year where the whole country was in a somewhat Republican mood, the GOP did well in Wisconsin, getting almost 52% of the statewide assembly vote. But they increased their assembly majority by only three, winning sixty-three out of ninety-nine seats. Filter that same election through the 19,184 random maps and it doesn't look like an outlier at all; in the 2014 election, it turns out, sixty-three Republican seats is just about what a random neutral map would likely have delivered.

What happened? Did the gerrymander lose its mojo in just two years? That would be evidence that gerrymandering doesn't need judicial intervention, but wears off on its own like a hangover. But it's not quite like that. It's more like Volkswagen. A few years ago it was revealed that the auto company had been systematically evading pollution tests, installing software in its diesel cars to fool regulators into thinking the engines were meeting emissions standards. Here's how it worked: the software detected when the car was being tested, and *only then* did it turn on the antipollution system. The rest of the time, the car just sailed down the highway spewing particulate matter.

The Wisconsin map is a similarly audacious piece of engineering. And the ensemble method reveals it; because it gives you information not just about what happened in the state's elections, but what *might* have happened if the elections had gone a bit differently. What if we take the assembly election of 2012 and shift each of the 6,672 wards 1% in the direction of the Democrats or the Republicans? Does the gerrymander bend or does it break? This is the same flight into the counterfactual Keith Gaddie used when Republicans were designing the map in the first place. And it reveals something startling. In electoral environments where Republicans get a majority of the statewide vote, the gerrymander doesn't have much effect; those are elections where the GOP would get an assembly majority anyway. It's only in Democratic-leaning environments that the gerrymander really kicks in, acting as a firewall to maintain the Republican majority against prevailing popular sentiment. You can see the firewall in the plot on page 346: in years when the Republicans did well, the circles and the stars aren't far apart, but as the Republican popular vote share gets lower, the stars separate from the circles, staying stubbornly above the fifty-seat line that gives the GOP the majority.

The Duke team estimates through its ensembles that the Act 43 map does exactly what Gaddie predicted it would do. The map keeps the assembly in Republican hands unless Democrats win the statewide vote by 8 to 12 points, a margin rarely achieved in this evenly split state. As a mathematician, I'm impressed. As a Wisconsin voter, I feel a little ill.

I've left something out. There are six hundred kajillion possible maps; that's why we can't just pick the very best one. So how is it that we can pick nineteen thousand of them at random?

For that, we need a geometer. Moon Duchin is a geometric group theorist and a math professor at Tufts University in Massachusetts. Her Chicago PhD thesis was about a random walk on Teichmüller space.

Don't worry about what Teichmüller space is,* just focus on the random walk; that's the key. We saw it with Go positions and we saw it with card shuffling and in a minor way we even saw it with mosquitoes: a random walk, our old friend the Markov chain, is the way to explore an unmanageably large set of options.

Remember—to walk randomly among the district maps, you need to know which map you can stumble into from which other map, which is to say you need to know which maps are *near* which other maps. We are back to geometry, but geometry of a very high and conceptual kind: not the geometry of the state of Wisconsin, but the geometry of the collection of all ways to break up that geometry into ninety-nine pieces. That is the geometry the mapmakers explored in order to find their gerrymander, and it's the geometry mathematicians have to map out in order to show what a gruesome outlier the gerrymander is.

There's no controversy about what geometry to use on the state of Wisconsin itself. Madison is close to Mount Horeb, Mequon is close to Brown Deer. For the higher geometry of the space of all districtings, you have a lot of choices; and it turns out these choices matter. My favorite is the one that Duchin developed together with Daryl DeFord and Justin Solomon, called the *ReCom* geometry, short for "recombination." The random walk on that geometry works like this.

1. Randomly choose two districts in your map that border each other.
2. Combine those two districts into one double-sized district.
3. Make a random choice among the ways of splitting that double-sized district in half, yielding a new map.
4. Check if the map you made violates any legal constraints; if it does, go back to 3 and choose a new splitting.
5. Go back to 1 and start again.

The "split and recombine" (or "ReCom") move of steps 2 and 3 is to districting what shuffling is to a deck of cards. And, as with the cards,

* If you must know, it's a kind of geometry of all two-dimensional geometries named after, though by no means entirely developed by, one of early-twentieth-century math's most fervent Nazis.

you can explore lots and lots and lots of different configurations with just a few moves. It's a small world. You can randomize a deck with seven shuffles. Seven ReComs is not, unfortunately, quite enough to explore the space of districtings. A hundred thousand ReComs seems to do the trick; that sounds like a lot, but it's tiny compared to the problem of sorting through *all* the districtings one by one. You can ReCom a hundred thousand times on your laptop in an hour. That gives you a sizable ensemble of neutrally drawn maps, to which you can compare the map you suspect of being gerrymandered.

The point of the ensemble method isn't to eliminate partisan gerrymandering entirely, any more than the point of *Reynolds v. Sims* was to require districts to have equal population down to the very last voter. Every decision made by a map-drawer, from incumbent protection to promotion of competitive races, might have partisan impact. The goal isn't to enforce an impossible absolute neutrality, but to block the very worst offenses.

Think back to Tad Ottman's speech to the Republican legislators, about the "obligation" the party had to seize the opportunity to cement control. If your job is to get and hold a majority of the legislature, and the law allows you to play as dirty as you like, then dirty is your duty. Blunting gerrymandering's power, establishing that there's some level of unfairness democracy won't tolerate, would have a healthy effect on the whole process. Politicians would be more likely to make reasonable compromises if the rewards of gerrymandering weren't so very great. If you don't want kids to shoplift, maybe don't leave *so* many candy bars *so* close to the front door.

THE TRIUMPHANT RETURN OF GRAPHS, TREES, AND HOLES

I could gloss over the part of ReCom where you split the double-sized district in two, but I won't, because talking about it lets me bring back two characters from earlier in the book. First of all, the voting wards in a district, like stars in movies or atoms in a hydrocarbon, form a network, or as James Joseph Sylvester would have called it, a graph: the vertices are the districts, and two vertices are connected just when the corresponding districts border each other. If the wards look like this:

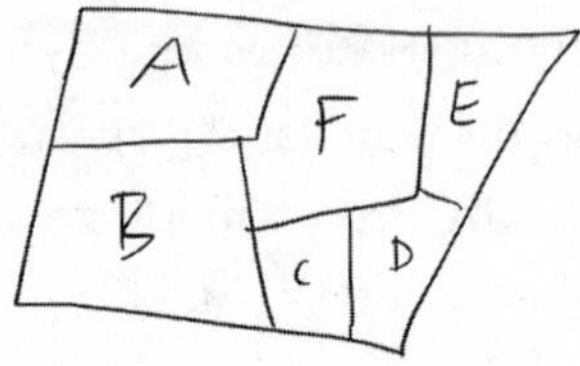

the graph looks like this:

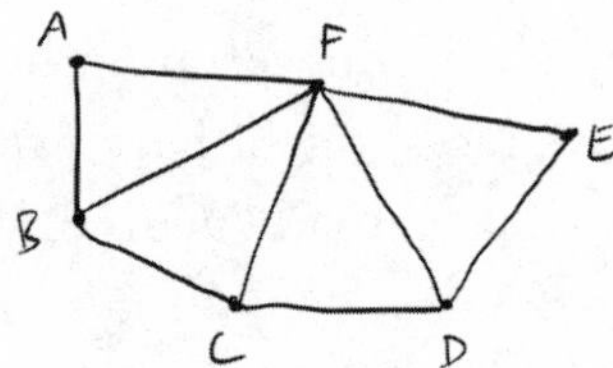

We need to find a way to split up the wards into two groups, and we need to make sure that each group of wards forms a connected network on its own.

Putting A, B, and C in one group and D, E, and F in the other works fine:

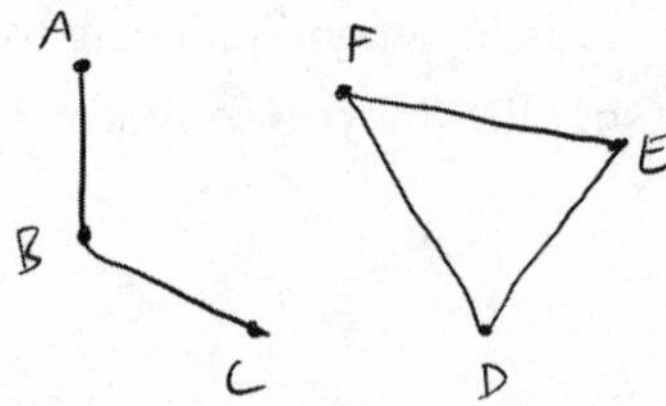

But grouping C, D, and F leaves you with A, B, and E, which don't form a connected district.

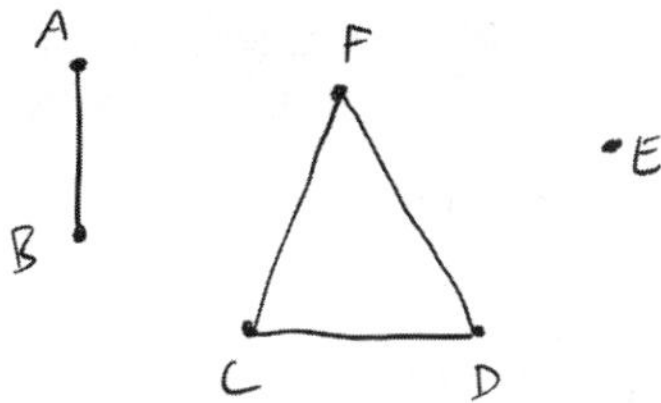

We're standing on the lip of a whole bubbling caldera of graph theory here. John Urschel, an offensive lineman for the Baltimore Ravens, dumped his professional football career in 2017 to work on this stuff, which is what he'd really wanted to do all along. One of his very first papers after leaving football was about how to split graphs into connected chunks using the theory of eigenvalues we met in chapter 12.

There are a lot of ways to split a graph. When the graph is as small as this one, you can list all the splittings, and choose one from the list at random. But listing all the possible splittings gets complicated as soon as the graph gets even a little bigger. There's a trick to picking one at random, and it involves more old friends. Suppose Akbar and Jeff play a game; they take turns removing an edge from the network, and whoever breaks it into disconnected pieces loses. In the graph above, Akbar might remove the edge AF, and then Jeff could remove DF, and then Akbar could remove EF (but not AB, because then he'd disconnect A and lose!) and Jeff could remove BF, and now Akbar is stuck; any edge he erases is going to break the graph into two disconnected pieces.

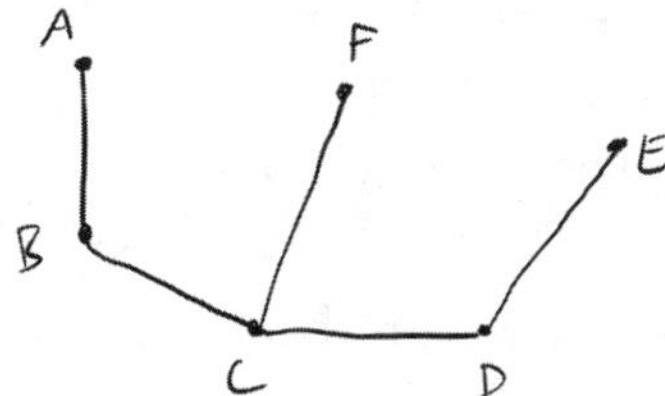

Could Akbar have played smarter and won? No, because this game has a secret feature: as long as neither player blunders and disconnects the network unnecessarily, it doesn't matter which moves you make; the game will always end after four turns, with Akbar the loser. In fact, no matter how big the network, the number of moves in this game is fixed. There's even a nice formula for it: it's number of edges – number of vertices + 1

At the beginning of the game, there are nine edges joining the six districts, so you get 9 – 6 + 1 = 4. When the game ends, with only five edges left, that number has dropped to 0. And what remains of the network has a very special form; there's no way to trace out a closed loop in the graph, as you could in the original graph by traveling in a cycle from A to B to F back to A. If there were a loop like that, you could remove one of the edges in that cycle without disconnecting the graph. What's left is a graph with no cycles at all; and a graph with no cycles is a tree.

How many holes are there in a network? That is, in its way, a confusing question, just like the question of how many holes there are in a straw or a pair of pants. But for this question I've already told you the answer; it's that very number just above, edges minus vertices plus one. Every time you snip an edge out of a cycle, you're getting rid of a hole. When you can't snip anymore, you've ended up with a graph with no holes in it at all: a tree. This is not just a metaphor; there is a fundamental invariant of any kind of space, called its *Euler characteristic,* which very, very, very roughly tells you the number of holes in it.* We've seen it once before, when we were counting holes in straws and pants. Straws have Euler characteristics and so do networks and so do string-theoretic models of twenty-six-dimensional spacetime; it's a unified theory that covers geometries from the modest to the cosmic.

So we're back to the geometry of trees. A tree like the one at the end of the edge-cutting game, which touches every vertex of our network, is called a *spanning tree.* These things appear all over math. A spanning tree of a square grid network like Manhattan streets is something you've seen before: it's called a maze. (In this picture, the white lines are the edges. If you've got a pencil you can convince yourself the maze is connected; you can draw a path from any point to any other without leaving the white lines. In fact, there's only *one* route you can take without backtracking. It's my book, I give you permission to write in it.)

* Only slightly less roughly, it's more like "the number of even-dimensional holes minus the number of odd-dimensional holes." If you have the appetite to learn what it *really* is, see Dave Richeson's book *Euler's Gem.*

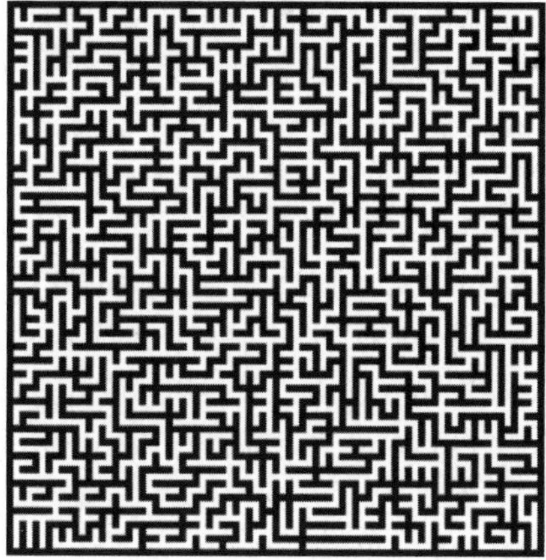

Or you can draw the spanning tree with vertices as dots and edges as line segments, more like the way we drew the districting graph:

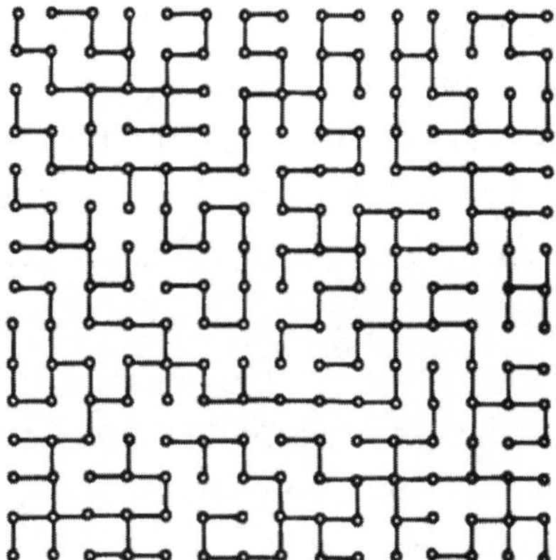

Most graphs of any decent size have a lot of spanning trees. The nineteenth-century physicist Gustav Kirchhoff worked out a formula that can tell you exactly how many, but that doesn't come close to answering all the questions they present, and a century later these trees are still an active research area. There is regularity and structure. For instance: How many dead ends does a random maze have? Of course there will be more and more dead ends the bigger the maze is, but what if we ask what proportion of locations in the maze are cul-de-sacs? A very cool theorem of Manna, Dhar, and Majumdar from 1992 shows that this proportion doesn't grow to 1 or drop to 0 as the maze gets large—instead, it just gets closer and closer to, for some reason, $(8/\pi^2)(1 - 2/\pi)$, just under 30%. You might think that the number of spanning trees of a random graph would be a more or less random number. Not so. My colleague Melanie Matchett Wood proved in 2017 that, if your graph is chosen at random,* the number of spanning trees is slightly more likely to be even than to be odd. To be precise, the chance the number of spanning trees is odd is the infinite product

$$(1 - 1/2)(1 - 1/8)(1 - 1/32)(1 - 1/128) \ldots$$

where the denominator of each fraction is four times that of the previous one. (Geometric progressions again!) The product comes to about 41.9%, quite a distance from 50/50. This asymmetry is the signature of a deeper geometric structure on the collection of all spanning trees; it turns out, for example, that there is a meaningful way of saying when a sequence of spanning trees forms an arithmetic progression!

But to explain this I would have to delve into the fascinating details of the "rotor-router process," and we still haven't saved democracy. So let's get back to districts.

Once you have a spanning tree in hand, there's an easy way to cut the network into two parts; just make the game-losing move and cut an edge, disconnecting the graph. Any choice you make will break the graph in two; with a little work, you can usually find an edge that makes the two pieces roughly equal in size. (If you can't, pick another tree and start over.) It looks like this, where the two sections are the part of the tree I've scribbled over and the part I haven't.

* In the sense of Erdős and Rényi from chapter 13.

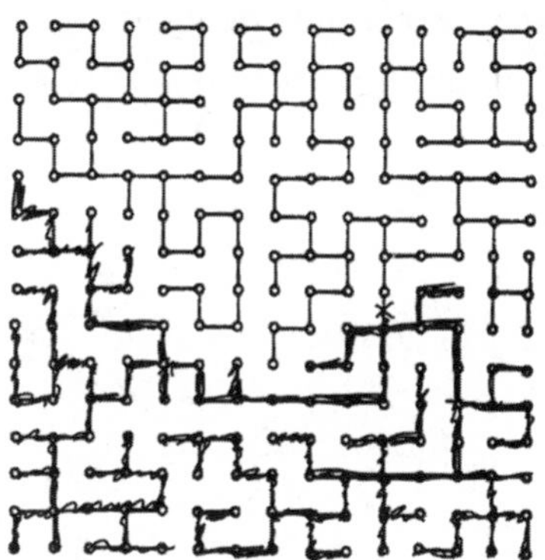

And now you know more or less how ReCom works.* You take your double-sized district, you choose a spanning tree at random—for example, by playing the edge-cutting game with randomly chosen moves†—pick a random edge in that tree, cut it, and the graph cleaves neatly into your two new districts.

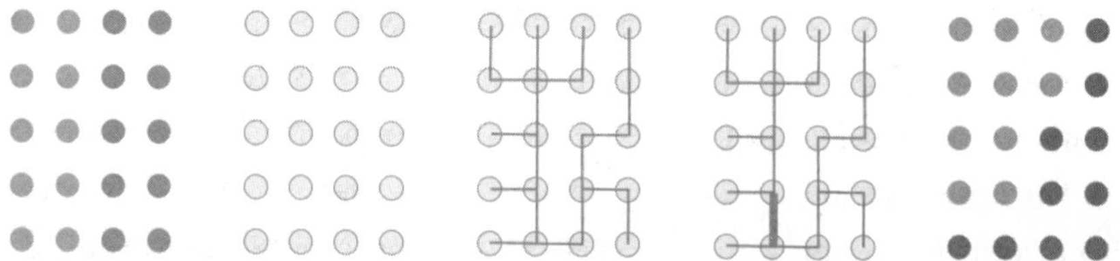

I'd better stop here for a caveat. There's a big difference between a random walk by ReCom on the space of maps and the random walk by shuffling on the space of ways to order a deck of cards. In the latter case, we have the seven-shuffle theorem, where by theorem I mean *theorem*; there is a mathematical proof that a certain number of shuffles (six!) is sufficient to explore every possible ordering, and what's more, that given just a few shuffles (seven!) every single order is just about equally likely.

When it comes to districtings, there are no theorems. We know

* And if you want to know more rather than less, see the 2021 book *Political Geometry*, edited by Duchin along with Ari Nieh and Olivia Walch.

† Though if you really want to get each spanning tree with equal probability, you have to be a little more intentional about how you choose which edges to cut; or you can follow DeFord, Duchin, and Solomon and use *Wilson's algorithm*, which is also a bit faster.

much less about the geometry of districtings than we do about the geometry of shufflings. The space of all districtings might, for instance, look like this:

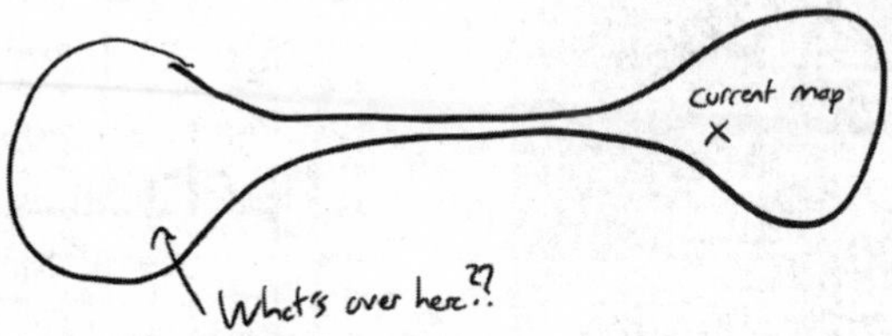

in which case, if you started at one end, you could conceivably randomly wander around for a long time before ever starting to explore what's on the other side of the isthmus. Or, for all we know, the space of all districtings could split into two separate connected pieces, or more than that. There might be an undiscovered country of possible maps of North Carolina, radically different from any ever contemplated by mathematicians, computers, or unscrupulous politicians, and perhaps among *those* maps, having ten out of thirteen Republican seats is quite typical. If we can't rule that out, do we still have the right to say the current gerrymandered map is an outlier?

Yes, as far as I'm concerned. We may not know with absolute certainty whether or not some secret reservoir of alternate maps exists; but we know that, in practice, if you start with the map the North Carolina legislature made, and mess with it, it gets less Republican no matter what you do to it. That experiment provides a strong indication, in any meaningful statistical sense, that the map was cooked. It is not a *proof* that the map-drawers set up to rig the game. For that matter, neither are the emails and memos from the map-drawers directly asserting they were trying to rig the game; after all, there is no Euclidean demonstration that they didn't actually mean to type "Let's get down to business and draw district maps that impartially capture the will of the people" but their fingers slipped and "Let's gerrymander this state so bad we can't possibly lose" came out instead. It's a proof in the sense of law, not in the sense of geometry.

UNITED STATES V. TUNA MELT

The ensemble of maps produced by random walks was at the very heart of the gerrymandering cases the Supreme Court heard in the spring of 2019. The point wasn't to prove that the maps had been drawn with partisan intent; that question wasn't in dispute. Thomas Hofeller, the North Carolina map's architect, had already testified that his aim had been "to create as many districts as possible in which GOP candidates would be . . . successful" and "to minimize the number of districts in which Democrats . . . [could] elect a Democratic candidate." The question was: Had the plan worked? You can't throw out a map for merely *trying* to be unfair. You have to prove it actually was.

The ensemble method is the best tool we have for that. Older ideas, like the efficiency gap, were largely absent from the plaintiffs' request. What they were asking the court was to recognize that the North Carolina map was an outlier, as out of place among its neutrally drawn counterparts as a warthog in a litter of piglets. This outlier analysis, they argued, is the "manageable standard" the court had been seeking. Jonathan Mattingly, a Duke mathematician and a member of the team that made the ensemble of Wisconsin assembly maps, had done the same for North Carolina's congressional districts; he testified that in his ensemble of 24,518 maps, there were just 162 in which Republicans won ten districts. The existing map packed North Carolina's Democrats so efficiently into three districts that the Democratic vote shares there were 74%, 76%, and 79%; not a single one of the 24,518 simulated maps created any districts that lopsided.

The mathematicians' brief made a similar point, though ours had prettier graphs.

And then came the oral argument, which for everyone who had been watching the case through a mathematical lens was a monumental bummer. It was as if years of progress and research on districting had never happened, and we were back to the stale question of whether 55% of the statewide vote should guarantee 55% of the seats in the legislature. Paul Clement, defending the North Carolina maps, got it started, telling Sonia Sotomayor: "I think you've put your finger on what my friends on the other side perceive to be the problem, which is a lack of proportional representation." Justice Sotomayor tried to tell Clement that her finger was elsewhere, but he pressed on, saying to Stephen Breyer, "[Y]ou can't talk even generally about outliers or extremity unless you know what it is you're deviating from. And I take it, implicit in your question and implicit in Justice Sotomayor's question, that what's bothering people is a deviation from a principle of proportional representation." "Actually—" Sonia Sotomayor broke in. "You keep saying that, but I don't quite think that's right," Elena Kagan objected. That didn't stop Clement, who had a point to make about Massachusetts. Republicans never have a congressional representative in Massachusetts, he observed, though Republicans make up a third of the state's population. "Nobody thinks that's unfair, because you really can't draw districts to do it because they're evenly distributed. It might be unfortunate for them, but I don't think it's unfair."

The plight of the Massachusetts Republican, as it happens, is also discussed in the mathematicians' brief. Our account of it basically matches Clement's, apart from one important detail; what he says the plaintiffs are asking to enforce is in fact what the plaintiffs are asking to forbid. It's *not* unfair that Republicans in Massachusetts don't get proportional representation. You can make an ensemble of maps, thousands of them, drawn with no nefarious partisan intent, and *every single one of them* will send nine Democrats and zero Republicans to Congress. That's why Common Cause wasn't asking the Supreme Court to guarantee proportional representation. Proportional representation is a lousy criterion for fairness. A map in Massachusetts that resulted in proportional representation wouldn't be immune from accusations of gerrymandering; it would, in fact, be a gerrymander just as bad as Joe Aggressive.

But many of the justices persisted in treating proportional representation as the issue they were being called on to decide. Neil Gorsuch worried that, if he decided against North Carolina, "We're going to have to, as part of our mandatory jurisdiction, in every single redistricting case, look at the evidence to see why there was a deviation from the norm of proportional representation. That's—that's—that's the ask?"

It was not the ask. This seemed hard for Gorsuch to accept. There's a really astonishing exchange near the end of the oral argument between Gorsuch and Allison Riggs, representing the League of Women Voters in opposition to the gerrymandered map. Riggs is explaining that her client is asking the court only to throw out the most freakishly outlying gerrymanders, whose partisan performance sets them apart from all but a handful of neutral alternatives. States would then still have a lot of breathing room to choose from the other 99% of all maps with a free hand, taking whatever nonpartisan criteria they liked into account. Gorsuch breaks in:

Justice Gorsuch: —but with—with respect, counsel, and I'm sorry to interrupt, but breathing room from what?

Ms. Riggs: Breathing room to—

Justice Gorsuch: From—how much breathing room, from what standard? And isn't the—isn't the answer that you just—I understand you don't want to give it, but isn't the real answer here breathing room from proportional representation up to maybe 7 percent?

Ms. Riggs: No.

After some more tussling, Gorsuch seems to concede that Riggs is not going to accept his enforced paraphrase. "We need a baseline," he says. "And the baseline, I still think, if it's not proportional representation, what is the baseline that you would have us use?"

He was asking a question that had been answered, a moment before, by Elena Kagan: "[W]hat's not allowed is deviation from whatever the state would have come up with, absent these partisan considerations."

Reading through the transcript of the argument, for a mathematician, feels like teaching a small seminar where only one student has done the reading. Justice Kagan gets it. She delivers a clear and succinct paraphrase of the quantitative argument she's being asked to consider. And then . . . everybody carries on as if she hadn't said a word. Sonia Sotomayor and John Roberts don't say much, but what they say is mostly right. Stephen Breyer has his own gerrymandering test, which neither side likes much. And Gorsuch, Samuel Alito, and to some extent Brett Kavanaugh, with help from Paul Clement, collaborate on building a fictional version of the case in which the plaintiffs are asking the court to impose some form of proportional representation on the states.

If you need a break from ensembles, random walks, and outliers, here's how the oral argument would have gone if it had been about ordering a sandwich instead:

> ***Ms. Riggs:*** I'd like a grilled cheese.
>
> ***Justice Alito:*** Okay, one tuna melt.
>
> ***Ms. Riggs:*** No, I said grilled cheese.
>
> ***Justice Kavanaugh:*** I hear the tuna melt's good.
>
> ***Justice Gorsuch:*** You want that tuna melt open face or closed?
>
> ***Ms. Riggs:*** I don't want a tuna melt, I want a—
>
> ***Justice Gorsuch:*** It seems like you don't want to just come out and say it, but don't you want a tuna melt?
>
> ***Ms. Riggs:*** No.
>
> ***Justice Kagan:*** She asked for a grilled cheese. That's not a tuna melt because there's no tuna in it.
>
> ***Justice Gorsuch:*** But if, as you say, you don't want a tuna melt, what do you want? Are we supposed to just make up a sandwich for you?
>
> ***Justice Alito:*** You come in here, you ask for a hot sandwich on toasted bread with cheese on it. That, to me, is gonna be a tuna melt.
>
> ***Justice Breyer:*** Nobody ever orders the chopped liver, but have they really given it a chance?
>
> ***Mr. Clement:*** The framers had every opportunity to make you a tuna melt, but they chose not to.

Maybe you already know how this ended up, and if you don't know, you can probably guess. On June 27, 2019, the Supreme Court ruled in a 5–4 decision that it was outside the scope of the federal courts to decide whether a partisan gerrymander was constitutional or not; in technical terms, the matter is "nonjusticiable." In lay terms, states can gerrymander their legislative maps with unlimited wanton abandon. Chief Justice Roberts, writing for the majority, explained:

> Partisan gerrymandering claims invariably sound in* a desire for proportional representation. As Justice O'Connor put it, such claims are based on "a conviction that the greater the departure from proportionality, the more suspect an apportionment plan becomes."

Roberts does concede, far down in the decision, that proportional representation isn't what the *Rucho* plaintiffs are asking for, but much of what he writes is devoted to reiterating his opposition to that unasked-for requirement. The constitutional constraint that no one's voting power be diluted, he insists against no opposition, "does not mean that each party must be influential in proportion to the number of its supporters."

No, I won't make you a tuna melt. You know we don't serve tuna melt here!

I'm not a lawyer, and won't pretend to be one. Nor will I pretend the constitutional issues in this case were easy. Easy cases don't make it to the Supreme Court. So I'm not going to tell you that the majority ruled wrongly here as a matter of law. If that's what you're looking for, I recommend Justice Kagan's dissent, which is so sardonic and bleak it seems at times about to break out in bitter laughter.

For Roberts, it's crucial that a certain amount of partisan bias in redistricting has been explicitly allowed by courts in the past. The question before the court was whether, at some point, there is such a thing as *too much*. The majority in *Rucho* said no: if it's constitutional to do it, it's constitutional to overdo it. Or, more precisely, if the court can't find a clear, agreed-on universal line to draw between permissible and forbidden, then the court can't consider the matter at all. It's a legal version of the *sorites* paradox, which goes all the way back to Eubulides, an old sparring partner of Aristotle. The *sorites* paradox asks us to work out

* As best I can tell from my lawyer friends, "sound in" here means something in between "derives from" and "amounts to"—and people say *mathematicians* talk in impenetrable jargon!

how many grains of wheat it takes to make a heap (in Greek, a *soros*). A single grain isn't a heap, and neither is two grains, surely. In fact, no matter how much wheat is on the table, it's impossible to imagine a situation where adding one grain of wheat to something that's not a heap of wheat results in something that is. So three grains of wheat isn't a heap, and neither is four, and so on. . . . Carrying this argument to its limit shows there's no such thing as a heap of wheat; yet somehow heaps of wheat exist.*

Roberts sees gerrymandering as unavoidably soritical. Any line drawn between "acceptable gerrymandering" and "I'm sorry, but it's just too much" will inevitably be arbitrary, he says, and will eventually be complicated and case-dependent (he might be satisfied with a rule that ninety-nine grains isn't a heap but a hundred is, but not if the threshold depended on whether the grains were wheat or sand).

I take his point. And yet I keep thinking about Nevada. Alone among the fifty states, it has no contiguity requirement for its legislative districts at all. In principle, the legislature in this Democratic-leaning state could fill three of the twenty-one state senate districts entirely with registered Republicans, and balance the party composition of the remaining districts to be just about exactly 60% Democratic, making them almost certainly safe seats and locking in a veto-proof 18–3 supermajority in the upper chamber, which would persist even if the state swung quite a bit to the right and elected a Republican governor. By the reasoning of *Rucho,* there is no clear-cut way to identify such a plan as "too much." Sometimes legal reasoning—even if it's sound as a matter of law—parts ways from common sense.

The majority's ruling, in the end, turns on a technical point—that partisan gerrymandering is a "political question," which means that, even if the Constitution *has* been violated, the Supreme Court is forbidden to intervene. That the results of gerrymandering "reasonably seem unjust"—that they are, in fact, "incompatible with democratic principles"—isn't in dispute; those are in fact quotes from the majority decision! And the gerrymanderers' unconvincing protests that their

* The reader is invited to note the resemblance to the idea of proof by induction that showed up when we talked about Nim.

maps weren't really *that* good at locking in electoral advantage are dismissed almost without comment. But just because something is unjust and incompatible with democratic principles and fiendishly effective, Justice Roberts writes, doesn't mean it's within the purview of the court to find a constitutional violation. Gerrymandering stinks, but not so badly the Constitution can smell it.

You can feel the discomfort in the decision: not just in the concessions that gerrymandering impedes democracy, but in the fervently expressed wish that somebody other than the justices of the Supreme Court will do something about it. Maybe something will be discovered in state constitutions that forbids gerrymandering, Roberts writes. Or, if not, maybe voters in afflicted states will rise up and change the system by ballot referendum, if they happen to live in a state where the legislature can't immediately reverse the outcome of such a vote. Maybe the U.S. Congress will do something about it, who knows?

I imagine Roberts as a worker who, on his way out of the factory at 5:05, notices that the building has caught fire. There's a fire extinguisher right there on the wall—he *could* grab it and spray foam all over the problem, but wait a minute, there's a principle at stake here. It's after five and he's not on the clock. Union rules are pretty clear that he's not supposed to put in unpaid overtime. If he puts out *this* fire, he sets a precedent; now is he on the hook every time the building catches fire after the whistle blows? There's probably somebody around working late who can put it out. And there is, after all, a fire department—that's who's supposed to put out fires! Admittedly, there's no telling how long it'll take them to show up, and the truth is, this town's fire department is known to be pretty lax about showing up at all. But still—it's officially their job, not his.

"THAT ONLY A POLITICAL CONVULSION CAN OVERTHROW"

For opponents of gerrymandering, the Supreme Court's decision was not the hoped-for happy ending. But it could be a happy beginning.

Justice Roberts, after all, wasn't wrong that there are other possible avenues for reform. Within a year of the decision in *Rucho,* the North Carolina congressional districts were thrown out by a panel of state judges for violating the North Carolina constitution. The Pennsylvania Supreme Court had done the same in 2018 (at which point the governor brought in Duchin, of the ReCom algorithm, to help build new and fairer maps). The House has passed a bill, currently blocked by Senate leadership, that would create nonpartisan commissions for the drawing of U.S. House districts (but not of state legislative districts, over which Congress has no power).*

And the mere existence of a high-profile case brought gerrymandering much more sharply into the public's view. The HBO news-comedy show *Last Week Tonight* ran a full twenty-minute segment on redistricting. Three siblings in Texas's badly malformed Tenth Congressional District made their own gerrymandering board game, Mapmaker, which sold thousands of copies after gerrymandering archenemy Arnold Schwarzenegger boosted it on social media. More people know about gerrymandering than used to, and when people know about it, they don't like it. Fifty-five out of Wisconsin's seventy-two counties, some that lean Democratic, some Republican, have passed resolutions asking for nonpartisan districting.

Voters in Michigan and Utah approved new nonpartisan districting commissions by popular referendum. In Virginia, which had a legislative map gerrymandered by Republicans, a bipartisan group in the legislature managed to pass a constitutional amendment handing control of redistricting to an independent commission. But the state shifted so far left so rapidly that, in 2019, Democrats broke the gerrymander and took over both houses of the legislature. Many members of the newly minted majority, now in the driver's seat for the next census, were suddenly less gung ho for reform.

Justice Kagan's dissent argues that one can't expect too much from the political process. The political process is exactly what gerrymander-

* At least that's the conventional wisdom; but some lawyers have lately argued that Congress can regulate state legislative districts under its Fourteenth Amendment power.

ing seeks to constrict. Maryland's congressional gerrymander, for instance, is still firmly in place, even with a Republican governor; Democrats hold a veto-proof majority in the state legislature and are expected to keep the map as it is.

How would Wisconsin get fairer maps? Wisconsin's state constitution has so little to say about district boundaries (and what it does say is so routinely ignored already) that it's hard to see a court challenge to the current maps succeeding.* Wisconsinites have no way to put a ballot initiative to a popular vote, unless the legislature initiates it, and the legislature likes things the way they are. Wisconsin could elect a new governor who would be sure to veto a GOP-gerrymandered map—in fact, it did just that in 2018. There are rumors the legislature plans to ask state courts to declare that redistricting is the job of the legislature and the legislature only, not requiring the governor's signature, and they may find sympathy in the state judiciary. If that were to happen, it's hard to see how the people of Wisconsin could gain any say in the matter.

In Michigan, the independent redistricting commission has been mired in legal challenges from the state's Republicans since the day it was voted into law by 61% of the state's voters. In Arkansas, redistricting reform group Arkansas Voters First gathered more than a hundred thousand signatures in the middle of a pandemic to get a constitutional amendment on the November ballot; the secretary of state declared the petitions void, because the certification that paid canvassers had undergone criminal background checks was misworded on the relevant form. State politics is full of veto points, so political factions with turf to protect have many ways to shield themselves from the public.

For all that, I'm an optimist. Americans used to shrug at legislative districts of wildly different sizes, saying that's just the way the game is played; now most people I talk to are shocked the practice was ever allowed. We are inclined to dislike what's unfair, and our ideas about

* Though when I once told a retired Wisconsin judge I didn't see how the text of the Wisconsin constitution was enough to support a legal challenge, he fixed me with a world-weary eye and said, "I see you are not a litigator."

fairness are never fully separated from mathematical thinking. Talking to people about the dark art of gerrymandering is a form of teaching math, and math has intrinsic purchase on the human mind, especially when it's twined up with *other* things we care deeply about: power, politics, and representation. Gerrymandering was a huge success when it was done behind locked doors. I'd like to believe it can't persist in an open, well-lighted classroom.

CONCLUSION

I Prove a Theorem and the House Expands

The British architect Herbert Baker, one of the designers of the Indian colonial capital of New Delhi, argued that the new city ought to be built on a neoclassical plan. Architecture of a more indigenous flavor wouldn't suit the empire's goals: "While in this style we may have the means to express the charm and fascination of India," he wrote, "yet it has not the constructive and geometrical qualities necessary to embody the idea of law and order which has been produced out of chaos by the British administration." Geometry can be marshaled as a metaphor for unquestioned-because-unquestionable authority, the mathematical analogue of a natural order centered on the king, or the father, or the colonial administrator. The monarchs of France spent untold *pistoles* laying out formal gardens whose perfect lines converging on the palace represented the unchanging order they took to be axiomatic.

Maybe the purest example of this point of view is the short novel *Flatland*, written by English schoolmaster Edwin Abbott in 1884. It is a story told by a square. (The first editions were published pseudonymously, with "A Square"* as the author.) The book takes place in a two-dimensional world, whose inhabitants, like Sylvester's bookworms, are

* Possibly a math pun—Abbott's full name was Edwin Abbott Abbott, so his monogram EAA could be styled algebraically as "E A squared."

unable to conceive any direction not spanned by the four compass points they know. The people in the plane are geometric figures, whose shape determines their position in society. The more sides a person has, the higher their station, the most exalted of all being polygons so multisided as to be indistinguishable from circles. In the other direction, isosceles triangles constitute the masses, with social standing proportional to the size of their central angle. The really narrow triangles, with their dangerously sharp angles, are soldiers; the only people below them are women, who are mere line segments, and who are presented in the novel as terrifying creatures, nearly mindless, lethally pointy, and invisible when viewed head-on. (What about non-isosceles triangles? They're considered grotesquely defective, and are institutionalized away from proper society, or, if sufficiently scalene, mercifully euthanized.)

The Square travels in a dream to the world of Lineland, whose proud one-dimensional King is unable to comprehend his visitor's explanations of the planar universe beyond his domain. Upon awakening, the Square is startled by a disembodied voice, which reveals itself as coming from a tiny circle that has somehow gotten inside his house. The circle grows and shrinks inexplicably; but of course that's because the circle is not a circle but a sphere, whose cross-section inside our narrator's universe grows and shrinks as the sphere moves up and down in the third dimension. The Sphere tries to explain itself to the Square, but after words fail, the Sphere lifts the narrator out of his home plane and tilts him so that he can see for himself the shape of the world he had previously only inferred. Returning to his plane after this revelation, the Square tries to spread the news about what he's seen. Predictably, he's imprisoned, and that's where the novel leaves him: locked up, his revelation ignored.

Flatland was received at the time of publication with a mixture of bemusement and dismissal. *The New York Times* said: "It's a very puzzling book and a very distressing one, and to be enjoyed by about six, or at the outside seven, persons in the whole of the United States and Canada." But it's become a favorite of young people with a taste for geometry, adapted for film several times and continually in print. I read it again and again as a kid.

But I didn't understand as a child that the book was a satire, lampooning rather than embracing the already old-fashioned view of social hierarchy that holds sway in Flatland. Far from seeing women as empty-headed death needles, Abbott was an advocate of equality in education. He served on the council of the Girls' Public Day School Company, which financed secondary education for women. And since I didn't know that Abbott was an Anglican priest whose publication record, other than this novel, was primarily of a theological nature, I *certainly* didn't grasp the Christian allegory that animates the story: the principles of geometry, far from enforcing an oppressive social order, are a way out of it, for those able to accept the reality of the world beyond.

The power of geometry, in this telling, is that a two-dimensional being can infer, by pure thought, the properties of a higher world he can't directly observe. By analogy with the squares he knows about, he is able to work out that a cube *must* have eight corners, and six faces, each of those faces a square like himself. At this point, the analogy with Christianity either breaks down or becomes extremely subversive, because the Square goes further, and asks the Sphere what he knows about the *fourth* dimension, which can be reasoned about in just the same way. That's ridiculous, the Sphere tells him, there's no such thing as the fourth dimension, what would make you bring up such a stupid thing?

The geometry we know can be used to endorse conventional ways. But the geometry we don't yet know is a threat. In seventeenth-century Italy, the Jesuits stamped out mathematicians' attempts to develop a rigorous theory of infinitesimals and compute the areas and volumes of previously inaccessible figures; if it went beyond Euclid, it was suspect. In England, Newton's theory of the calculus underwent fervent ecclesiastical attacks, and had to be defended by books like James Jurin's *Geometry No Friend to Infidelity*. But it sort of *is* a friend to infidelity, if your faith is badly placed! Geometry, especially new geometry, offers a locus of authority that rivals the established order. In this way it can be a destabilizing force and a radical measure.

THE SOUL OF THE FACT

Rita Dove is a Pulitzer Prize–winning poet, a former poet laureate of the United States, and the Commonwealth Professor of English at the University of Virginia, where Thomas Jefferson and James Joseph Sylvester both thought deep mathematical thoughts in their day; but in the early 1960s she was a nerdy kid in Akron, Ohio. Her father was an industrial chemist, the first Black research chemist at Goodyear Tire. Dove remembers:

> My brother and I would get together to figure out our math homework. We would spend hours trying to solve a difficult problem on our own before giving up and approaching my father because, well, he was a *real* math whiz, and if we had a question about algebra he would say, "Well, it would be easier if we used logarithms." We would protest, "But we don't know about logarithms!" But out would come the slide rule nevertheless, and two hours later we'd learned logarithms, but the whole evening was gone.

That memory turns into a poem, "Flash Cards":

Flash Cards

In math I was the whiz kid, keeper
of oranges and apples. *What you don't understand,*
master, my father said; the faster
I answered, the faster they came.

I could see one bud on the teacher's geranium,
one clear bee sputtering at the wet pane.
The tulip tree always dragged after heavy rain
so I tucked my head as my boots slapped home.

My father put up his feet after work
and relaxed with a highball and *The Life of Lincoln*.
After supper we drilled and I climbed the dark

before sleep, before a thin voice hissed
numbers as I spun on a wheel. I had to guess.
Ten, I kept saying, *I'm only ten*.

"Flash Cards" depicts the facts of arithmetic as an authority imposed from above. (Doubly so—there's the tough father, and the math-loving Abraham Lincoln appearing in book form.) There's affection in this poem: as Dove says, "you also realize that they love you, since they are spending all this time on you. My father was very stern in those days; right before bedtime those flash cards had to come out. I hated them then, but I'm glad now." But in the end, you're running on a wheel in the dark, producing answers as quickly and correctly as you can. This is mathematics the way a lot of schoolchildren experience it.

Most great poets never write even one math poem, but Dove wrote two, and here's the other one:

Geometry

I prove a theorem and the house expands:*
the windows jerk free to hover near the ceiling,
the ceiling floats away with a sigh.

As the walls clear themselves of everything
but transparency, the scent of carnations
leaves with them. I am out in the open

And above the windows have hinged into butterflies,
sunlight glinting where they've intersected.
They are going to some point true and unproven.

* I don't *think* Dove is making veiled reference to expanding the House of Representatives to bring about more equal representation in the Electoral College, but a poetic line contains many overlapping meanings, so if you want that to be there, let's say it's there.

What a difference! Where arithmetic is a slog, geometry is a kind of liberation. Insight so powerful it blows the walls out sideways (or makes them invisible—this is poetry, I don't think we have to be too clear about the precise physics of the scenario). The intersections of planes in space become living things that might flutter away, beautifully *visible* even if you can't quite pin them to the two-dimensional page. What happens in the mind, when a proof reveals itself like this, is anything but a logical trudge.

There's something special about geometry, something that makes it worth writing poems about. Everywhere else in the school curriculum, you must, in the end, defer to the teacher's authority, or a textbook's, when it comes to who fought in the French and Indian Wars, or what the principal products of Portugal are. In geometry you make your own knowledge. The power is in your hands.

That, of course, is exactly why geometry was correctly seen as dangerous by the Flatlanders and the Italian Jesuits. It represents an alternative source of authority. The Pythagorean Theorem isn't true because Pythagoras said it was; it's true because we can, ourselves, *prove* that it's true. Behold!

But truth and proof are not the same thing. That's where Dove's poem ends; with the "point true but unproven." Poincaré ends up in the same place, when he insists on the necessary role of intuition. He writes:

> What I have just said is sufficient to show how vain it would be to attempt to replace the mathematician's free initiative by a mechanical process of any kind. In order to obtain a result having any real value, it is not enough to grind out calculations, or to have a machine for putting things in order: it is not order only, but unexpected order, that has a value. A machine can take hold of the bare fact, but the soul of the fact will always escape it.

We use formal proof as a scaffold, to extend our intuition's reach, but it would be useless, a ladder to nowhere, if we weren't using it to get to a point we could somehow, inexplicably, see.

We mathematicians present ourselves to the world as the people whose knowledge is eternal and unassailable, because we have proved it all. Proof is an essential tool for us—the measure of our certainty, just as it was for Lincoln. But it is not the point. The point is to understand things. We want not just the facts but the souls of the facts. It's at the moment of understanding that the walls go transparent, the ceiling flies off, and we're doing geometry.

A few years ago, a Russian mathematician named Grigori Perelman proved the Poincaré Conjecture. It wasn't Poincaré's only conjecture, but it was the one that stayed attached to his name, because it was difficult and because attempts to solve it tended to produce interesting new ideas; that's how a really good conjecture, so to speak, proves itself.

I am not going to state Poincaré's Conjecture exactly. It concerns three-dimensional spaces, but not necessarily the one we inhabit; rather, Poincaré is asking about three-dimensional spaces with a little more geometric richness to them, spaces that might be curved and bent in on themselves.* Imagine if the Square, lifted out of Flatland by his three-dimensional visitor, found that the plane he thought he lived on was actually the surface of a sphere,† or for that matter the surface of some kind of complicated doughnut, and then he said to his new higher-dimensional friend, what if *your* three-dimensional world actually has some kind of complicated shape visible only from dimension four? How would you even be able to tell?

Here's one way to tell whether you live on a doughnut or a sphere. On the doughnut's surface, you can make a closed loop of mildly elastic string

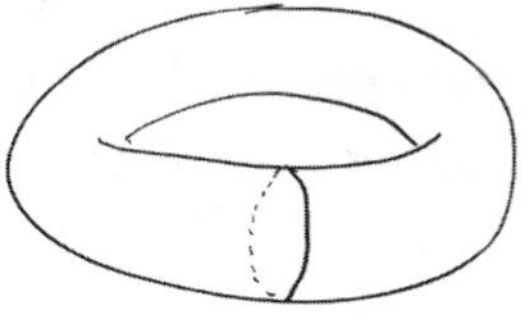

* Wait, isn't that what Einstein showed our space actually *is* like? Kind of, but in relativity we are doing the kind of geometry where angles are invariant, while Poincaré's question involves the looser flavor of geometry, the kind we've been calling topology (because that's what it's called), where a circle and a square and a triangle are all the same thing.

† This is in fact the plot of a sequel to *Flatland*, written by Dutch schoolteacher Dionys Burger in the 1950s, appropriately titled *Sphereland*, in which society is shaken by the discovery that very large triangles have angles that sum to more than 180 degrees.

that, no matter how much you zhuzh it around on the doughnut's surface, cannot be pulled shut. The sphere is a different story; any loop of string on its surface can be contracted to a point.

It's a little hard to imagine this in our own three-dimensional world, but why not try? A loop of string you can hold in your hand can surely be contracted without leaving the universe. But what about a spacecraft that travels gigaparsecs from the Earth only to find itself home again? If you think of its path as a long, long loop in space, is it clear you can pull it shut? The large-scale geometry of the universe is, in its way, as inaccessible to our direct observations as is the small-scale weirdness inside an electron.

Poincaré saw that this notion of closable and unclosable loops was a really fundamental one. His conjecture was that there was only *one* kind of three-dimensional space with no unclosable loops, and it's the one we're familiar with. Check that all loops can be pulled shut, and you know everything there is to know about the shape of space.

To be honest, Poincaré didn't exactly conjecture this. He merely asked, in a paper in the exposition year of 1904, whether it was so, without staking out one side or the other. Maybe it was his conservative temperament that kept him from committing; or maybe it was that, four years earlier, he had published a different conjecture in the same vein, which in the 1904 paper he showed to be completely wrong. This is more common than you might think. Even great mathematicians make lots of wrong guesses. If you never guess wrong, you're not guessing about hard enough things.

Perelman answered Poincaré's question, a hundred years later, using methods the older mathematician could barely have imagined. His proof goes a level higher, exploiting the geometry of all geometries, letting a mysterious loopless three-dimensional space flow through the space of all spaces until it becomes the standard three-dimensional space we know and love.

It is not an easy proof.

But the new ideas in Perelman's work touched off a huge wave of work on these abstract flows—they enlarged mathematicians' understanding of what geometry could be. Perelman himself was not a part of it. Having dropped his bomb of knowledge, he retreated into seclusion in his small St. Petersburg apartment, turning down both the Fields Medal and a $1 million bounty attached to the problem by the Clay Foundation.

Let me now propose a thought experiment. What if the Poincaré Conjecture had been proved, not by an introverted Russian geometer, but by a machine? Say, a grandchild of a grandchild of Chinook, which instead of solving checkers had managed to solve this part of three-dimensional geometry. And suppose the proof, like Chinook's perfect strategy for checkers, was something illegible to the human mind, a string of numbers or formal symbols that we can verify is correct but that we cannot, in any meaningful sense, understand.

Then, I would say, despite the fact that one of geometry's most famous conjectures had been settled, definitively proven to be now and forever correct, I would not care. Not one fig would I give! Because the point is not to know what's true and false. Truth and falsehood are just not that interesting. They are facts without souls. Bill Thurston, the modern era's great navigator of non-Euclidean three-dimensional geometries, and the designer of a grand strategy of classifying all such geometries that Perelman's work successfully completed, had no time for an industrial view of math as truth factory: "We are not trying to meet some abstract production quota of definitions, theorems and proofs. The measure of our success is whether what we do enables *people* to understand and think more clearly and effectively about mathematics." The mathematician David Blackwell put it more bluntly: "Basically, I'm not interested in doing research and I never have been," he said. "I'm interested in understanding, which is quite a different thing."

Geometry is made of people. It feels universal and eternal, manifesting in much the same form in every human community that's ever existed; but it is also right here, located in a time and space and among human beings. It's here to teach us things—to make the house expand.

Blackwell was a probabilist, who did lots of work on Markov chains, but he, like Lincoln, like Dove, like Ronald Ross, found inspiration in Euclid's plane. Geometry, he said, was "the only course I had that made me see that mathematics is really beautiful and full of ideas." He recalls a proof, maybe even the proof of the pons asinorum: "I still remember the concept of a *helping line*. You have a proposition that looks quite mysterious. Someone draws a line and suddenly it becomes obvious. That's beautiful stuff."

MY CHILDREN HAVE DEFEATED ME!

Famous Talmudic story: the oven of Akhnai. A group of rabbis is fervently arguing, as groups of rabbis do. The matter under dispute is whether an oven cut into pieces and then mortared back together is subject to the exact same laws of ritual cleanliness that would govern an oven made of an uncut stone. It doesn't really matter what they were arguing about, just that one rabbi, Eliezer ben Hyrcanus, is holding fast to a minority opinion counter to the view of everyone else in the room. Things get heated. Rabbi Eliezer, the Talmud says, presents "all the proofs in the world," but his opponents aren't moved. Eliezer turns to more dramatic forms of demonstration. "If my interpretation of the Torah is correct," he says, "let the carob tree prove it!" And a nearby tree uproots itself and leaps a hundred cubits away. Doesn't matter, says Rabbi Joshua, the leader of the opposition, a carob tree isn't a proof. Okay, says Rabbi Eliezer, if I'm right, let the stream prove it! And the stream begins to flow backward in its course. Who cares, say the rabbis, a stream isn't a proof. If I'm right, Eliezer says, let the walls of the academy prove it! And the walls begin to bend. Even this does not impress the rabbis on the other side.

But Eliezer has one more card to play. "If I am right about the Torah's law," he says, "let heaven prove me right." And the voice of God echoes from above, saying, "Why are you all giving Rabbi Eliezer so much trouble? You know he is always right in such matters."

That's when Rabbi Joshua stands up and says, "The voice of God isn't a proof! The Torah isn't in heaven any more, it's here on earth, written down, and the rules we've been given are clear; the ruling is determined by the opinion of the majority, and the majority opposes Rabbi Eliezer's view."

And God laughs. "My children have defeated me! My children have defeated me!" the heavenly voice delightedly declares, and goes silent.

This story about disagreement generates a lot of disagreement. Some people see Joshua as the hero, for his Prometheus-like snatching of authority from God. He's the country lawyer in this story, and I think Abraham Lincoln would have taken his side. As Lincoln's partner Herndon describes him: "He was remorseless in his analysis of facts and principles. When all these exhaustive processes had been gone through with he could form an idea and express it; but no sooner. He had no faith, and no respect for 'say so's,' come though they might from tradition or authority."

Others prefer Eliezer, for standing up for his beliefs against unified opposition. Elie Wiesel says of his rabbinical namesake, "I also like Eliezer because of his solitude. . . . He was who he was, he never tried to surrender, he remained faithful to his ideas, no matter what others said. He was ready to be alone." This echoes Alexander Grothendieck, who remade geometry from scratch in the 1960s, and whose work we've come to the end of the book without touching—oh well, maybe next time—who recalled about his early student days in Paris:

> [I]n those critical years I learned how to be alone . . . to reach out in my own way to the things I wished to learn, rather than relying on the notions of the consensus, overt or tacit, coming from a more or less extended clan of which I found myself a member, or which for any other reason laid claim to be taken as an authority. This silent consensus had informed me, both at the lycée and at the university, that one shouldn't bother worrying about what was really meant when using a term like "volume," which was "obviously self-evident," "generally known," "unproblematic," etc. I'd gone over their heads. . . . It is in this gesture of "going beyond," to be something in oneself rather than the pawn of a consensus, the refusal to stay within a rigid circle that others have drawn around one—it is in this solitary act that one finds true creativity. All other things follow as a matter of course.

And yet Grothendieck only became Grothendieck because of the fertile soil of French geometry that nourished his ideas, and the instant uptake of his innovations by dozens of other mathematicians of the Paris circle.

When we think, truly and deeply think, about geometric things—whether we are trying to chart the course of a pandemic, or walking through the tree of strategies that governs a game, or developing a working protocol for democratic representation, or understanding which things feel near to which other things, or trying to visualize the outside of the house from the inside, or, like Lincoln, rigorously criticizing our own beliefs and assumptions—we are, in a way, alone. But we're alone together with everyone on earth. Everyone does geometry differently, but everyone does it. It is, just as its name says, the way we measure the world, and therefore (only in geometry do we get to say "therefore") the way we measure ourselves.

Acknowledgments

My agent, Jay Mandel, his assistant, Sian-Ashleigh Edwards, and everyone at William Morris Endeavor have been tireless in their encouragement and support of the book from its earliest stages. And it was a pleasure to work again with my editor, Scott Moyers, at Penguin Press. They're consistently dedicated to publishing the books authors feel moved to write, not making authors write the books the house wants to sell. Thanks to the whole team there, especially Mia Council, Liz Calamari, and Shina Patel, and to Laura Stickney at Penguin UK, and to Stephanie Ross, who made the amazing cover image.

Thanks to Riley Malone, who wrote me out of the blue last summer asking if I needed a research assistant—I did! The book has benefited greatly from the hours she spent tracking down answers to my weird questions, checking my facts, and questioning my phrasing. Copy editor Greg Villepique expertly went through the whole manuscript with a nanodental comb, and saved me from several embarrassing factual errors, including the year of my own bar mitzvah.

I am fortunate to be able to rely on friends, acquaintances, and strangers to answer questions, workshop ideas, and patiently explain constitutional law and quantum physics to me. Some of the people who helped were Amir Alexander, Martha Alibali, David Bailey, Tom Banchoff, Mira Bernstein, Ben Blum-Smith, Barry Burden, David Carlton, Rita Dove, Charles Franklin, Andrew Gelman, Lisa Goldberg, Margaret Graver, Elisenda Grigsby, Patrick Honner, Katherine Horgan, Mark Hughes,

Patrick Iber, Lalit Jain, Kellie Jeffris, John Johnson, Malia Jones, Derek Kaufman, Emmanuel Kowalski, Adam Kucharski, Greg Kuperberg, Justin Levitt, Wanlin Li, the archivists at the London School of Hygiene and Tropical Medicine, Jeff Mandell, Jonathan Mattingly, Ken Mayer, Lorenzo Najt, Jennifer Nelson, Rob Nowak, Cathy O'Neil, Ben Orlin, Charles Pence, Wes Pegden, Douglas Poland, Ben Recht, Jonathan Schaeffer, Tom Scocca, Ajay Sethi, Lior Silberman, Jim Stein, Steve Strogatz, Jean-Luc Thiffeault, Charles Walker, Travis Warwick, Amie Wilkinson, Rob Yablon, Tehshik Yoon, Tim Yu, and Ajai Zutshi.

Special thanks are due to those who actually read through sections of the book in its unfinished and unlovely condition, and made it lovelier: Carl Bergstrom, Meredith Broussard, Stephanie Burt, Alec Davies, Lalit Jain, Adam Kucharski, Greg Kuperberg, Douglas Poland, Ben Recht, Lior Silberman, Steve Strogatz, and, most of all, supereditor Michelle Shih, who read most of the book and helped me believe it all made sense.

I am grateful to Moon Duchin for showing me that gerrymandering was not only a politically important problem but one with deep and interesting mathematics in it, and to Gregory Herschlag for carrying out extra analysis on data from the 2018 elections in Wisconsin.

As always, I am lucky to work at the University of Wisconsin-Madison, which has been unfailingly supportive of my work as a writer. A campus like ours is just about the maximally congenial environment for writing a wide-ranging book like this one; there is an expert in *everything* within walking distance. Also lots of places to get coffee.

The first person to teach me geometry was Eric Walstein, who died of COVID-19 in November 2020. I wish he could have taught more kids math.

One more thing I'd like to acknowledge is all the things it would have been great to write about in a book about geometry, but aren't here, because I ran out of time and space. I had meant to write about "lumpers and splitters" and the theory of clustering; Judea Pearl and the use of directed acyclic graphs in the study of causality; navigation charts of the Marshall Islands; exploitation versus exploration and multi-armed bandits; binocular vision in praying mantis larvae; the maximum size of a subset of the N by N grid with no three points forming an isosceles triangle (let me know if you solve this actually); much more on dynamics, starting from Poincaré and going through billiards, Sinai, and Mirzakhani; much more on Descartes, who launched the unification of algebra and geometry and somehow ended up nearly absent here; and much more on Grothendieck, who took that unification much farther than Descartes could have dreamed and somehow ditto; catastrophe theory; the tree of life. Doing geometry in the real world always involves holding the actual and the ideal up to the eye at once, and writing books is much the same; the ideal of this book is something you and I will just have to imagine, and I hope you've found the actual thing in your hands a good enough sketch.

A book is a family effort. My son, CJ, cleaned and analyzed many years of Wisconsin election data, my daughter, AB, drew some of the pictures, and everybody was patient with me when I went on a jag about why did I think it was a good idea to write an entire book about a subject some substantial proportion of people think they hate? And Tanya Schlam, as always, has been the wall I lean on, and the first and last reader of everything you see on the page, making rough sentences smooth, crooked passages straight, and obscure explanations clear. Without her, none of this exists.

Notes

Introduction: Where Things Are and What They Look Like

2 **When South American:** From p. 10 of Benny Shanon's "Ayahuasca Visualizations: A Structural Typology," *Journal of Consciousness Studies* 9, no. 2 (2002): 3–30, just in case you thought I was speaking from personal experience.

2 **You can give:** Jillian E. Lauer and Stella F. Lourenco, "Spatial Processing in Infancy Predicts Both Spatial and Mathematical Aptitude in Childhood," *Psychological Science* 27, no. 10 (2016): 1291–98.

3 **"The guys all":** Margalit Fox, "Katherine Johnson Dies at 101; Mathematician Broke Barriers at NASA," *New York Times*, Feb. 24, 2020, sourced to a 2010 interview with the *Fayetteville (NC) Observer.*

3 **"And haunted by itself":** As quoted, for instance, in Newton P. Stallknecht, "On Poetry and Geometric Truth," *The Kenyon Review* 18, no. 1 (1956): 2. Wordsworth revised *The Prelude* a lot, and some versions of the poem have "herself" instead of "itself" here.

4 **"Thus I often":** John Newton, *An Authentic Narrative of Some Remarkable and Interesting Particulars in the Life of John Newton*, 4th ed. (Printed for J. Johnson, 1775), 75–82.

4 **"Wordsworth was a profound":** Thomas De Quincey, *The Works of Thomas De Quincey*, vols. 3–4 (Cambridge, MA: Houghton, Mifflin, and Co.; The Riverside Press, 1881), 325.

4 **Wordsworth had done:** See the letter of June 26, 1791, from the poet's sister Dorothy Wordsworth to Jane Pollard (*Letters of the Wordsworth Family From 1787 to 1855*, vol. 1, ed. William Knight (Cambridge: Ginn and Company, 1907), 28, which recounts Wordsworth's failure to win a fellowship to Cambridge on account of not being able to force himself to study math. "He reads Italian, Spanish, French, Greek, Latin, and English, but never opens a mathematical book."

4 **who some believe:** Joan Baum, "On the Importance of Mathematics to Wordsworth," *Modern Language Quarterly* 46, no. 4 (1985): 392.

4 **Hamilton was fascinated:** Letter from Hamilton to his cousin Arthur, Sep. 4, 1822, reproduced in Robert Perceval Graves, *Life of Sir William Rowan Hamilton*, vol. 1 (Dublin: Hodges Figgis, 1882), 111.

5 **"came off with honor":** At least, so says Robert Perceval Graves in his profile of Hamilton in *Dublin University Magazine* 19 (1842): 95, written while his friend Hamilton was still alive, and so he repeats on p. 78 of his *Life of Sir William Rowan Hamilton*, ibid. The story appears in just about every more recent biography of Hamilton, all, as far as I can see, relying on Graves as a source. Hamilton's letters describe a meeting with Colburn in 1820, and having "seen" Colburn perform feats of computation, but I have not found any letters that refer to a competition; nor does Colburn mention such a contest, or in fact meeting Hamilton at all, in his own memoir, cited below, though he does rather proudly tell of *other* child prodigies he met and felt himself superior to. Did the competition really happen?

5 **Colburn had had it:** Zerah Colburn, *A Memoir of Zerah Colburn: Written by Himself. Containing an Account of the First Discovery of His Remarkable Powers; His Travels in America and Residence in Europe; a History of the Various Plans Devised for His Patronage; His Return to this*

Country, and the Causes which Led Him to His Present Profession; with His Peculiar Methods of Calculation (Springfield, MA: G. and C. Merriam, 1833), 72.

5 **Colburn had little insight:** Graves, *Life of Sir William Rowan Hamilton*, 78–79.

5 **"a *midnight walk*":** Letter from WRH to Eliza Hamilton, Sep. 16, 1827, quoted in Graves, *Life of Sir William Rowan Hamilton*, 261.

6 **At a dinner party:** Tom Taylor, *The Life of Benjamin Robert Haydon*, vol. 1 (London: Longman, Brown, Green, and Longmans, 1853), 385.

Chapter 1: "I Vote for Euclid"

9 **The source, Lincoln said:** "Mr. Lincoln's Early Life: How He Educated Himself," *New York Times*, Sep. 4, 1864, 5. Of course, the "quote" from Lincoln here is from Gulliver's recollection and cannot be taken as an exact record of his speech.

11 **"I have been told":** Herndon's recollection quoted in Jesse William Weik, *The Real Lincoln; a Portrait* (Boston: Houghton Mifflin, 1922), 240. I learned the story from Dave Richeson's wonderful book *Tales of Impossibility* (Princeton: Princeton University Press, 2019), which has everything you could possibly want to know about circle squaring, angle trisecting, and so on.

11 **"Like the geometer":** Casual translation mine. The translation of "misurar lo cerchio" as "square the circle" is convincingly justified by R. B. Herzman and G. W. Towsley in "Squaring the Circle: *Paradiso* 33 and the Poetics of Geometry," *Traditio* 49 (1994): 95–125.

11 **"Being in a Gentleman's":** John Aubrey, *'Brief Lives,' Chiefly of Contemporaries, Set down by John Aubrey, between the Years 1669 & 1696*, vol. 1, ed. Andrew Clark (Oxford: Oxford University Press, 2016), 332. https://www.gutenberg.org/files/47787/47787-h/47787-h.htm.

12 **Hobbes retorted:** F. Cajori, "Controversies in Mathematics Between Hobbes, Wallis, and Barrow," *Mathematics Teacher* 22, no. 3 (March 1929): 150.

12 **"[A]ll they know":** Review of *Geometry without Axioms*, from *Quarterly Journal of Education* XIII (1833): 105.

12–13 **A "proposition":** This insight comes from Adam Kucharski, "Euclid as Founding Father," *Nautilus*, Oct. 13, 2016. http://dev.nautil.us/issue/41/selection/euclid-as-founding-father.

13 **"One would start":** Abraham Lincoln, *The Collected Works of Abraham Lincoln*, vol. 3, eds. Roy P. Basler et al. (New Brunswick, NJ: Rutgers University Press, 1953), 375. Accessed at http://name.umdl.umich.edu/lincoln3.

13 **"but a luxury":** Thomas Jefferson, *The Essential Jefferson*, ed. Jean M. Yarbrough (Indianapolis: Hackett Publishing, 2006), 193.

13 **"I have given up":** Thomas Jefferson, *The Papers of Thomas Jefferson, Retirement Series*, vol. 4, ed. J. Jefferson Looney (Princeton: Princeton University Press, 2008), 429. Accessed at https://press.princeton.edu/books/ebook/9780691184623/the-papers-of-thomas-jefferson-retirement-series-volume-4.

14 **For Lincoln, unlike Jefferson:** On the difference between Jefferson's and Lincoln's conception of geometry and who it was for, see Drew R. McCoy, "An 'Old-Fashioned' Nationalism: Lincoln, Jefferson, and the Classical Tradition," *Journal of the Abraham Lincoln Association* 23, no. 1 (2002): 55–67.

15 **"many a youth":** The same writer, and the same review, quoted on circle-squarers earlier: *Quarterly Journal of Education*, vol. XIII (1833): 105. That anonymous reviewer was extremely quotable!

16 **"Place a cube":** William George Spencer, *Inventional Geometry: A Series of Problems, Intended to Familiarize the Pupil with Geometrical Conceptions, and to Exercise His Inventive Faculty* (New York: D. Appleton, 1877), 16. The British edition appeared in 1860.

17 **"It is this living":** James J. Sylvester, "A Plea for the Mathematician," *Nature* 1 (1870): 261–63.

17 **A survey conducted:** Kenneth E. Brown, "Why Teach Geometry?," *Mathematics Teacher* 43, no. 3 (1950): 103–6. Accessed at https://www.jstor.org/stable/27953519.

18 **"[M]any a time":** H. C. Whitney, *Lincoln the Citizen* (New York: Baker & Taylor, 1908), 177. If I'm to be honest, the example of a fallacy Whitney adduces right after this, involving a question about whether a corporation can be said to have a soul, does not really seem to me like a failure of deductive logic.

19 **"it was morally":** Whitney, *Lincoln the Citizen*, 178.

19 **"A proof is":** The Orlin material is from his Oct. 16, 2013, "Two-Column Proofs That Two-Column Proofs Are Terrible," *Math with Bad Drawings* (blog), http://mathwithbaddrawings.com/2013/10/16/two-column-proofs-that-two-column-proofs-are-terrible/.

19 **The most typical format:** The material on the Committee of Ten and the history of the two-column proof is from P. G. Herbst, "Establishing a Custom of Proving in American School Geometry: Evolution of the Two-Column Proof in the Early Twentieth Century," *Educational Studies in Mathematics* 49, no. 3 (2002): 283–312.

22 ***gradient of confidence*:** Ben Blum-Smith, "Uhm *Sayin*," *Research in Practice* (blog), http://researchinpractice.wordpress.com/2015/08/01/uhm-sayin/.

24 **he esteemed his diagram:** Bill Casselman, "On the Dissecting Table," *Plus Magazine*, Dec. 1, 2000. https://plus.maths.org/content/dissecting-table.

25 **"You have doubtless":** Henri Poincaré, *The Value of Science*, trans. G. B. Halsted (New York: The Science Press, 1907), 23.

31 **at least one study:** M. J. Nathan, et al., "Actions Speak Louder with Words: The Roles of Action and Pedagogical Language for Grounding Mathematical Proof," *Learning and Instruction* 33 (2014): 182–93.

31 **when he needed:** Jeremy Gray, *Henri Poincaré: A Scientific Biography* (Princeton: Princeton University Press, 2012), 26.

Chapter 2: How Many Holes Does a Straw Have?

33 **a 1970 paper:** David Lewis and Stephanie Lewis, "Holes," *Australasian Journal of Philosophy* 48, no. 2 (1970): 206–12.

34 **reappears in 2014:** https://forum.bodybuilding.com/showthread.php?t=162056763&page=1.

34 **a Snapchat clip:** The video has been reproduced in many online locations: for example, at metro.co.uk/2017/11/17/how-many-holes-does-a-straw-have-debate-drives-internet-insane-7088560/.

34 **shot a video:** www.youtube.com/watch?v=W0tYRVQvKbM.

34 **No—you roll:** Truth be told, among bagel makers there are those who snake and connect and those who make a dough ball and open up the middle, but definitely none who remove a hole from a fully baked no-holed pastry.

38 **"I recall above all":** Galina Weinstein, "A Biography of Henri Poincaré—2012 Centenary of the Death of Poincaré," ArXiv preprint server, July 3, 2012, 6. Accessed at https://arxiv.org/pdf/1207.0759.pdf.

38 **the loss of Alsace:** Jeremy Gray, *Henri Poincaré: A Scientific Biography* (Princeton: Princeton University Press, 2012), 18–19.

39 **In 1889 he won:** June Barrow-Green, "Oscar II's Prize Competition and the Error in Poincaré's Memoir on the Three Body Problem," *Archive for History of Exact Sciences* 48, no. 2 (1994): 107–31.

39 **precise habits:** Weinstein, "A Biography of Henri Poincaré," 20.

39 **the joke in Paris circles:** Tobias Dantzig, *Henri Poincaré: Critic of Crisis* (New York: Charles Scribner's Sons, 1954), 3.

39 **He was not only:** Gray, *Henri Poincaré*, 67.

40 **"Geometry is the art":** "La Géométrie est l'art de bien raisonner sur des figures mal faites." Henri Poincaré, "Analysis situs," *Journal de l'École Polytechnique* ser. 2, no. 1 (1895): 2.

40 **"The circles he drew":** Dantzig, *Henri Poincaré*, 3.

44 **Noether's innovation:** To be fair to Poincaré, Leopold Vietoris, who was there at the dawn of topology and died at the age of 110 in 2002, says Poincaré understood that holes formed a space, but didn't express it that way in his own work as a matter of "taste." I prefer Noether's taste, (Saunders Mac Lane, "Topology Becomes Algebraic with Vietoris and Noether," *Journal of Pure and Applied Algebra* 39 [1986]: 305–7). Vietoris himself, independently of Noether and around the same time, formalized the same notion; but in those days, math that happened in Vienna often wasn't known right away in Göttingen, and vice versa.

46 **"Nowadays this tendency":** "Diese Tendenz scheint heute selbstverstandlich; sie war es vor acht Jahren nicht; es bedurfte der Energie und des Temperamentes von Emmy Noether, um sie zum Allgemeingut der Topologen zu machen und sie in der Topologie, ihren Fragestellungen und ihren Methoden, diejenige Rolle spielen zu lassen, die sie heute spielt." Paul Alexandroff and Heinz Hopf, *Topologie I: Erster Band. Grundbegriffe der Mengentheoretischen Topologie Topologie der Komplexe· Topologische Invarianzsätze und Anschliessende Begriffsbildungen· Verschlingungen im n-Dimensionalen Euklidischen Raum Stetige Abbildungen von Polyedern* (Berlin: Springer-Verlag, 1935), ix. Thanks to Andreas Seeger for helping me translate this paragraph.

46 **"one of the many":** Biographical material about Listing is all drawn from Ernst Breitenberger, "Johann Benedikt Listing," *History of Topology*, ed. I. M. James (Amsterdam: North-Holland, 1999), 909–24.
47 **"Nobody doubts nowadays":** Poincaré, "Analysis situs," 1.

Chapter 3: Giving the Same Name to Different Things

57 **"that he may enslave A":** From private notes written prior to the Civil War. Michael Burlingame, *Abraham Lincoln: A Life* (Baltimore: Johns Hopkins University Press, 2013), 510.
59 **In 1904, the city of St. Louis:** Information about the exposition is mostly from D. R. Francis, *The Universal Exposition of 1904*, vol. 1 (St. Louis: Louisiana Purchase Exposition Company, 1913).
60 **"[T]here are symptoms":** Henri Poincaré, "The Present and the Future of Mathematical Physics," trans. J. W. Young, *Bulletin of the American Mathematical Society* 37, no. 1 (Dec. 1999): 25.
61 **"victorious and intact":** Poincaré, "The Present and the Future," 38.
63 **"crap":** In German, "Mist." Colin McLarty, "Emmy Noether's first great mathematics and the culmination of first-phase logicism, formalism, and intuitionism," *Archive for History of Exact Sciences* 65, no. 1 (2011): 113.
63 **"[S]he discovered methods":** "Professor Einstein Writes in Appreciation of a Fellow-Mathematician," *New York Times*, May 4, 1935, 12.

Chapter 4: A Fragment of the Sphinx

64 **"Mosquito Man Coming":** *St. Louis Post-Dispatch*, Sep. 17, 1904, 3.
64 **Ross spoke on the afternoon:** The time of Ross's lecture can be found in Hugo Munsterberg, *Congress of Arts and Science, Universal Exposition, St. Louis, 1904: Scientific Plan of the Congress* (Boston: Houghton, Mifflin, 1905), 68.
64 **while elsewhere:** D. R. Francis, *The Universal Exposition of 1904*, vol. 1 (St. Louis: Louisiana Purchase Exposition Company, 1913), 285.
65 **"The full mathematical":** Ronald Ross, "The Logical Basis of the Sanitary Policy of Mosquito Reduction," *Science* 22, no. 570 (1905): 689–99.
72 **A 2018 paper:** Houshmand Shirani-Mehr et al., "Disentangling Bias and Variance in Election Polls," *Journal of the American Statistical Association* 113, no. 522 (2018): 607–14. One of the authors is Columbia statistician Andrew Gelman, whose blog is the white-hot center of the internet when it comes to salty statistical commentary. See also a popular writeup of this paper by two more of its authors, David Rothschild and Sharad Goel, "When You Hear the Margin of Error Is Plus or Minus 3 Percent, Think 7 Instead," *New York Times*, Oct. 5, 2016.
72 **"Just last week":** A. Prokop, "Nate Silver's Model Gives Trump an Unusually High Chance of Winning. Could He Be Right?," *Vox*, Nov. 3, 2016. https://www.vox.com/2016/11/3/13147678/nate-silver-fivethirtyeight-trump-forecast.
74 **"After 2016":** *The New Republic*, Dec. 14, 2016.
75 **"Mr. Pearson would":** Egon S. Pearson, "Karl Pearson: An Appreciation of Some Aspects of His Life and Work," *Biometrika* 28, no. 3/4 (Dec. 1936): 206. Egon S. Pearson is Karl Pearson's son.
75 **His charisma enabled:** Except as noted, the biographical data on Pearson in these two paragraphs is from M. Eileen Magnello, "Karl Pearson and the Establishment of Mathematical Statistics," *International Statistical Review* 77, no. 1 (2009): 3–29.
75 **"Knowing Karl":** Letter of June 9, 1884, quoted in Pearson, "Karl Pearson," 207.
76 **"[I]f I only had":** Letter of Nov. 12, 1884, cited in M. Eileen Magnello, "Karl Pearson and the Origins of Modern Statistics: An Elastician Becomes a Statistician," *New Zealand Journal for the History and Philosophy of Science and Technology* 1 (2005). Accessed at http://www.rutherfordjournal.org/article010107.html.
76 **Once he flung:** M. Eileen Magnello, "Karl Pearson's Gresham Lectures: W. F. R. Weldon, Speciation and the Origins of Pearsonian Statistics," *British Journal for the History of Science* 29, no. 1 (Mar. 1996): 47–48.
76 **"I believe that":** Pearson, "Karl Pearson," 213.
77 **"I felt sadly":** Pearson, "Karl Pearson," 228.
77 **"Here, as always":** Letter of Feb. 11, 1895, quoted in Stephen M. Stigler, *The History of Statistics* (Cambridge: The Belknap Press of Harvard University Press, 1986), 337.

78 **"But I am horribly":** Letter of Mar. 6, 1895, quoted in Stigler, *History of Statistics*, 337.
78 **The mathematical statement:** "Karl Pearson and Sir Ronald Ross," *Library and Archives Service Blog*, http://blogs.lshtm.ac.uk/library/2015/03/27/karl-pearson-and-sir-ronald-ross.
78 **"The lesson of":** Karl Pearson, "The Problem of the Random Walk," *Nature* 72 (Aug. 1905), 342.
80 **He struggled:** At least, so says Bernard Bru in Murad S. Taqqu, "Bachelier and His Times: A Conversation with Bernard Bru," *Finance and Stochastics* 5, no. 1 (2001): 5, from which most of this account is drawn. Jean-Michel Courtault et al., in "Louis Bachelier on the Centenary of *Theorie de la Speculation*," *Mathematical Finance* 10, no. 3 (July 2000): 341–53, says on p. 343 that Bachelier's grades were quite good.
81 **Dreyfus was convicted:** All material on Poincaré and the Dreyfus affair is from Gray, *Henri Poincaré*, 166–69.
81 **"[O]ne might fear":** Courtault et al., "Louis Bachelier on the Centenary of *Théorie de la Spéculation*," 348.
82 **Bachelier did end up:** The story of Bachelier is drawn from Taqqu, "Bachelier and His Times," 3–32.
83 **"a fragment":** Robert Brown, "XXVII. A Brief Account of Microscopical Observations Made in the Months of June, July and August 1827, on the Particles Contained in the Pollen of Plants; and on the General Existence of Active Molecules in Organic and Inorganic Bodies," *Philosophical Magazine* 4, no. 21 (1828): 167.
83 **The Academy read:** Material on the Olympia Academy is from Maurice Solovine's introduction to Albert Einstein, *Letters to Solovine, 1906–1955* (New York: Philosophical Library/Open Road, 2011). Solovine refers to an unspecified "scientific work by Karl Pearson" as the first item read, but other sources identify this as *The Grammar of Science.*
85 **"Nekrasov sharply":** Facts and quote about Nekrasov are from E. Seneta, "The Central Limit Problem and Linear Least Squares in Pre-Revolutionary Russia: The Background," *Math Scientist* 9 (1984): 40.
85 **becoming first rector:** E. Seneta, "Statistical Regularity and Free Will: L. A. J. Quetelet and P. A. Nekrasov," *International Statistical Review/Revue Internationale de Statistique* 71, no. 2 (Aug. 2003): 325.
85 **In protest of:** G. P. Basharin, A. N. Langville, and V. A. Naumov, "The Life and Work of A. A. Markov," *Linear Algebra and Its Applications* 386 (2004): 8.
85 **"determinedly striving":** Seneta, "Statistical Regularity and Free Will," 331.
86 **"Finally, I received":** The story of the shoes is from Basharin et al., "The Life and Work of A. A. Markov," 8. The relation of KUBU, the shoe-senders, with the party is from N. Krementsov, "Big Revolution, Little Revolution: Science and Politics in Bolshevik Russia," *Social Research* 73, no. 4 (Baltimore: Johns Hopkins University Press, 2006): 1173–1204.
86 **Statistics of human behavior:** Seneta, "Statistical Regularity and Free Will," 322–23.
90 **"I, of course":** Basharin et al., "The Life and Work of A. A. Markov," 13.
92 **Peter Norvig, a director:** P. Norvig, "English Letter Frequency Counts: Mayzner Revisited, or ETAOIN SRHLDCU," 2013, available at http://norvig.com/mayzner.html. Some of the bigram and trigram frequencies are taken from Norvig's earlier "Natural Language Corpus Data," in *Beautiful Data*, eds. T. Segaran and J. Hammerbacher, eds. (Sebastopol, CA: O'Reilly, 2009).
93 **It was the engineer:** Claude E. Shannon, "A Mathematical Theory of Communication," *Bell System Technical Journal* 27, no. 3 (1948): 388.
94 **Here's some text:** All the Markov-chain-generated text here was carried out by Brian Hayes's incredibly fun "Drivel Generator," available at http://bit-player.org/wp-content/extras/drivel/drivel.html, using public baby-name data from the U.S. Social Security Administration. A more serious attempt to Markovize baby-naming practices would weight the names by how often they were used; I just used the whole list of names without any attention to which ones were popular. See Brian Hayes, "First Links in the Markov Chain," *American Scientist* 101, no. 2 (2013): 252, which covers some of the same ground as this section and has really nice pictures.

Chapter 5: "His Style Was Invincibility"

97 **"Oh, how she'd cackle":** L. Renner, "Crown Him, His Name Is Marion Tinsley," *Orlando Sentinel*, Apr. 27, 1985.
97 **Starting in 1944:** G. Belsky, "A Checkered Career," *Sports Illustrated*, Dec. 28, 1992.

98 **In 1975, he lost:** The biographical material on Tinsley's early life is mostly from Jonathan Shaeffer, *One Jump Ahead* (New York: Springer-Verlag, 1997), 127–33. Some material on Tinsley vs. Chinook is also drawn from A. Madrigal, "How Checkers Was Solved," *Atlantic* (July 19, 2017).

98 **"His style":** Renner, "Crown Him, His Name Is Marion Tinsley."

98 **"I am just free":** Schaeffer, *One Jump Ahead*, 1.

98 **But it took:** Schaeffer, *One Jump Ahead*, 194.

99 **"No one was happy":** Quoted in J. Propp, "Chinook," *American Chess Journal*, November 1997, available at http://www.chabris.com/pub/acj/extra/Propp/Propp01.html.

100 **So let's say Akbar and Jeff:** Matt Groening, *Life in Hell*, 1977–2012.

104 **the branching of the arteries:** This is figure 6 of Ronald S. Chamberlain, "Essential Functional Hepatic and Biliary Anatomy for the Surgeon," IntechOpen, Feb. 13, 2013. https://www.intechopen.com/books/hepatic-surgery/essential-functional-hepatic-and-biliary-anatomy-for-the-surgeon.

105 **here are the vices:** From the Walters Art Gallery, http://www.thedigitalwalters.org/Data/WaltersManuscripts/W72/data/W.72/sap/W72_000056_sap.jpg.

109 **every whole number:** Ahmet G. Agargün and Colin R. Fletcher, "Al-Fārisī and the Fundamental Theorem of Arithmetic," *Historia Mathematica* 21, no. 2 (1994): 162–73.

115 **Nim is first attested:** L. Rougetet, "A Prehistory of Nim," *College Mathematics Journal* 45, no. 5 (2014): 358–63.

117 **those hardy microbial spores:** W. Fajardo-Cavazos et al., "Bacillus Subtilis Spores on Artificial Meteorites Survive Hypervelocity Atmospheric Entry: Implications for Lithopanspermia," *Astrobiology* 5, no. 6 (Dec. 2005): 726–36. www.ncbi.nlm.nih.gov/pubmed/16379527.

124 **"morbidly depressing":** Jessica Wang, "Science, Security, and the Cold War: The Case of E. U. Condon," *Isis* 83, no. 2 (1992): 243.

124 **"On the novelty side":** "Fair's Ticket Sale Is 'Huge Success,' with Late Rush On," *New York Times*, May 6, 1940, 9. The material on Mr. Nimatron is followed by an announcement that Elsie, "the star bovine performer of Borden's Dairy World of Tomorrow," is starting her residency at the fair, displayed in "a special glass boudoir."

124 **"Most of its defeats":** E. U. Condon, "The Nimatron," *American Mathematical Monthly* 49, no. 5 (1942): 331.

124 **Alan Turing, who worked:** S. Barry Cooper and J. Van Leeuwen, *Alan Turing* (Amsterdam: Elsevier Science & Technology, 2013), 626.

125 **"The reader might well ask":** Cooper and Van Leeuwen, *Alan Turing.*

127 **The longest tournament game:** There seems to actually be some dispute about the longest theoretically possible chess game, but 5,898 is the figure that seems to me most commonly claimed. The 269-move match was played in 1989 in Belgrade between Ivan Nikolić and Goran Arsović. In usual chess notation, a "move" consists of two pieces moving, one for each player; so the path in the chess tree corresponding to that game would actually be 538 branches long.

128 **a poem that captures:** Robert Lowell, "For the Union Dead" (1960) from his 1964 book of the same title. You can read the poem at https://www.poetryfoundation.org/poems/57035/for-the-union-dead.

128 **Connect Four:** Proved in 1988, almost simultaneously, by James D. Allen and Victor Allis. See Allis's masters thesis: Victor Allis, "A Knowledge-based Approach of Connect-Four—The Game is Solved: White Wins." (1988, Masters thesis, Vrije Universiteit, Amsterdam).

128 **one of the first papers:** Claude E. Shannon, "XXII. Programming a Computer for Playing Chess," *London, Edinburgh, and Dublin Philosophical Magazine and Journal of Science* 41, no. 314 (1950): 256–75.

130 **keep this table handy:** Image from C. J. Mendelsohn, "Blaise de Vigenère and the Chiffre Carré," *Proceedings of the American Philosophical Society* 82, no. 2 (Mar. 22, 1940): 107.

132 **Stigler's Law, Stigler observed:** Stephen M. Stigler, "Stigler's Law of Eponymy," *Transactions of the New York Academy of Sciences* 39 (1980): 147–58.

132 **Vigenère was a well-connected:** The information about Vigenère here is all taken from Mendelsohn, "Blaise de Vigenère and the Chiffre Carré."

133 **laid out in 1553:** A. Buonafalce, "Bellaso's Reciprocal Ciphers," *Cryptologia* 30, no. 1 (2006): 40–47.

133 **"of such marvelous":** Mendelsohn, "Blaise de Vigenère and the Chiffre Carré," 120.

133 **No reliable way:** C. Flaut et al., "From Old Ciphers to Modern Communications," *Advances in Military Technology* 14, no. 1 (2019): 81.

134 **"By which you may":** William Rattle Plum, *The Military Telegraph During the Civil War in the United States: With an Exposition of Ancient and Modern Means of Communication, and of the Federal and Confederate Cipher Systems; Also a Running Account of the War Between the States*, vol. 1 (Chicago: Jansen, McClurg, 1882), 37.
136 **Until the 1990s:** Nigel Smart, "Dr Clifford Cocks CB," honorary doctorate citation, University of Bristol, Feb. 19, 2008. Accessed at http://www.bristol.ac.uk/graduation/honorary-degrees/hondeg08/cocks.html.
137 **"Teenager Bernie Weber":** From *Pryme Knumber*, a 2012 novel by author and Wisconsin gubernatorial candidate Matt Flynn. In the 2017 sequel, Bernie proves the Riemann Hypothesis and has to go on the run from Chinese intelligence. The reason it's funny is that you can't factor prime numbers—they're prime!
138 **And twenty-eight:** Brian Christian, *The Most Human Human* (New York: Doubleday, 2011), 124. Christian, along with many other sources, says there were forty games, of which twenty-one were the same, but many people on checkers forums seem to now think 28/50 is correct.
139 **"I'm basically":** Jim Propp, "Chinook," *American Chess Journal* (1997), originally published on the ACJ website. Accessed at http://www.chabris.com/pub/acj/extra/Propp/Propp01.html.
139 **"I have a better":** Quoted in *The Independent*, Aug. 17, 1992; from Schaeffer, *One Jump Ahead*, 285.
141 **"Even if I become":** "Go Master Lee Says He Quits Unable to Win Over AI Go Players," Yonhap News Agency, Nov. 27, 2019, http://en.yna.co.kr/view/AEN20191127004800315.
142 **It closed down:** "Checkers Group Founder Pleads Guilty to Money Laundering Charges," Associated Press State & Local Wire, June 30, 2005. Accessed at https://advance-lexis-com/api/document?collection=news&id=urn:contentItem:4GHN-NTJ0-009F-S3XV-00000-00&context=1516831.
142 **"I surely have":** "King Him Checkers? Child's Play. Unless You're Thinking 30 Moves Ahead. Like a Mathematician. This Mathematician," *Orlando Sentinel*, Apr. 7, 1985.
142 **"I don't look":** Martin Sandbu, "Lunch with the FT: Magnus Carlsen," *Financial Times*, Dec. 7, 2012.
142 **"Human chess":** Quoted in *Conversations with Tyler* (podcast), episode 22, May 2017.
142 **"I was amazed":** Quoted in *Conversations with Tyler* (podcast), episode 22, May 2017.

Chapter 6: The Mysterious Power of Trial and Error

151 **a 1640 letter:** Colin R. Fletcher, "A Reconstruction of the Frénicle-Fermat Correspondence of 1640," *Historia Mathematica* 18 (1991): 344–51.
151 **"if he did not":** André Weil, *Number Theory: An Approach Through History from Hammurabi to Legendre* (Boston: Birkhäuser, 1984), 56.
151 **a conjecture about:** A. J. Van Der Poorten, *Notes on Fermat's Last Theorem* (New York: Wiley, 1996), 187.
152 **"there can hardly":** Weil, *Number Theory*, 104.
152 **the tone of the letters:** Weil suggests that the two men cagily concealed their strongest theorems from each other, and perhaps even intentionally wrote down misleading statements, in order to prevent their opposite number from getting the upper hand. Weil, *Number Theory*, 63.
154 **a persistent and wrong:** Qi Han and Man-Keung Siu, "On the Myth of an Ancient Chinese Theorem About Primality," *Taiwanese Journal of Mathematics* 12, no. 4 (July 2008): 941–49.
154 **appears to originate:** J. H. Jeans, "The Converse of Fermat's Theorem," *Messenger of Mathematics* 27 (1898): 174.
155 **concealed inside the works:** The story of the mechanical Turk is widely told, very thoroughly for instance in Tom Standage, *The Turk: The Life and Times of the Famous Eighteenth-Century Chess-Playing Machine* (New York: Berkley, 2002). The best story about the Turk, told by Alan Turing in the paper "Digital Computers Applied to Games," in *Faster Than Thought*, ed. B. V. Bowden (London: Sir Isaac Pitman & Sons, 1932), is that the scheme was exposed when someone yelled "Fire" during the game, whereupon the man inside the machine beat a hasty public exit. I can't find any convincing reason to think this story is true, though, so to the endnotes it is relegated.
158 **It turns out:** All information about the gambler's ruin problem and Pascal and Fermat's correspondence about it is from A. W. F. Edwards, "Pascal's Problem: The 'Gambler's Ruin,'" *International Statistical Review/Revue Internationale de Statistique* 51, no. 1 (Apr. 1983): 73–74.

160 **Finally, late in the afternoon:** Alexandre Sokolowski, "June 24, 2010: The Day Marathon Men Isner and Mahut Completed the Longest Match in History," *Tennis Majors*, June 24, 2010. https://www.tennismajors.com/our-features/on-this-day/june-24-2010-the-day-marathon-men-isner-and-mahut-completed-the-longest-match-in-history-267343.html.

160 **"Nothing like this":** Greg Bishop, "Isner and Mahut Wimbledon Match, Still Going, Breaks Records," *New York Times*, June 23, 2010.

160 **Most sports championships:** The material on alternate World Series formats is adapted from J. Ellenberg, "Building a Better World Series," *Slate*, Oct. 29, 2004. https://slate.com/human-interest/2004/10/a-better-way-to-pick-the-best-team-in-baseball.html.

164 **Go-playing computer programs:** S. Gelly et al., "The Grand Challenge of Computer Go: Monte Carlo Tree Search and Extensions," *Communications of the ACM* 55, no. 3 (2012): 106–13.

Chapter 7: Artificial Intelligence as Mountaineering

166 **My friend Meredith Broussard:** MSNBC, *Velshi & Ruhle*, Feb. 11, 2019. Available at www.msnbc.com/velshi-ruhle/watch/trump-to-sign-an-executive-order-launching-an-ai-initiative-1440778307720.

167 **Another nice piece:** For calculus fans only: if f(x,y) is the function we're maximizing, the formula for the derivative of an implicit function tells us that the tangent to the curve f(x,y) = c (otherwise known as the line at the topo map) has slope –(df/dx)/(df/dy), while the gradient is the vector (df/dx, df/dy), which is orthogonal to this.

169 **by Frank Rosenblatt:** Frank Rosenblatt, "The perceptron: a probabilistic model for information storage and organization in the brain." *Psychological Review* 65, no. 6 (1958): 386. Rosenblatt's perceptron was a generalization of a less refined mathematical model of neural processing developed in the 1940s by Warren McCulloch and Walter Pitts.

182 **"Visualize a 3-space":** Lecture 2c of Geoffrey Hinton's notes for "Neural Networks for Machine Learning." Available at www.cs.toronto.edu/~tijmen/csc321/slides/lecture_slides_lec2.pdf.

182 **his great-grandfather:** For the familial relation between the two Hintons, see K. Onstad, "Mr. Robot," *Toronto Life*, Jan. 28, 2018.

Chapter 8: You Are Your Own Negative-First Cousin, and Other Maps

191 **The geometry of chords:** Dmitri Tymoczko, *A Geometry of Music* (New York: Oxford University Press, 2010).

192 **In 1968:** Seymour Rosenberg, Carnot Nelson, and P. S. Vivekananthan, "A Multidimensional Approach to the Structure of Personality Impressions," *Journal of Personality and Social Psychology* 9, no. 4 (1968): 283. But I read about it as a kid in Joseph Kruskal's chapter "The Meaning of Words," in *Statistics: A Guide to the Unknown*, ed. Judith Tanur (Oakland: Holden-Day, 1972), a great work of mathematical exposition whose lessons remain fresh today and that should be more widely read.

194 **You can rank:** Here I'm referring to DW-Nominate scores, developed by Keith Poole and Howard Rosenthal and available at http://voteview.com. The method of producing these scores isn't actually multidimensional scaling and doesn't strictly speaking involve the notion of "distance" between legislators; for more details, see Keith T. Poole and Howard Rosenthal, "D-Nominate After 10 Years: A Comparative Update to Congress: A Political-Economic History of Roll-Call Voting," *Legislative Studies Quarterly* 26, no. 1 (Feb. 2001): 5–29.

199 **Male for "Karen":** The source for all of these is my laptop; the word vectors produced by Word2vec are freely downloadable and you can mess around with them yourself in Python.

Chapter 9: Three Years of Sundays

200 **If we were honest:** What I say about "looking stupid" is largely inspired by a Twitter thread posted by my colleague Sami Schalk (@DrSamiSchalk) on May 8, 2019.

201 **"At the 1903 meeting":** Andrew Granville, *Number Theory Revealed: An Introduction* (Pawtucket, RI: American Mathematical Society, 2019), 194.

201 **I just factored it:** Using the ntheory.factorint command in the Python package SymPy, should you want to see for yourself how fast it is.

204 **algorithm may not:** Andrew Trask et al., "Neural Arithmetic Logic Units," *Advances in Neural Information Processing Systems* 31, NeurIPS Proceedings 2018, ed. S. Bengio et al. Accessed at https://arxiv.org/abs/1808.00508. The introduction explains how traditional

neural network architectures fail on this particular problem, and the main body of the paper suggests a possible fix.

204 **"I am convinced":** CBS, "The Thinking Machine" (1961), YouTube, July 16, 2018. David Wayne and Jerome Wiesner at 1:40 to 1:50 of the video compilation. Available at www.youtube.com/watch?time_continue=154&v=cvOTKFXpvKA&feature=emb_title.

205 **solved a long-standing:** Lisa Piccirillo, "The Conway Knot Is Not Slice," *Annals of Mathematics* 191, no. 2 (2020): 581–91. For a nontechnical account of Piccirillo's discovery, see E. Klarreich, "Graduate Student Solves Decades-Old Conway Knot Problem," *Quanta*, May 19, 2020. https://www.quantamagazine.org/graduate-student-solves-decades-old-conway-knot-problem-20200519.

205 **my own most cited results:** Jordan S. Ellenberg and Dion Gijswijt, "On Large Subsets of $F^{n}/_{q}$ with No Three-Term Arithmetic Progression," *Annals of Mathematics* (2017): 339–43.

205 **topologist named Mark Hughes:** Mark C. Hughes, "A Neural Network Approach to Predicting and Computing Knot Invariants," *Journal of Knot Theory and Its Ramifications* 29, no. 3 (2020): 2050005.

Chapter 10: What Happened Today Will Happen Tomorrow

207 **"I was really":** Ronald Ross, *Memoirs, with a Full Account of the Great Malaria Problem and Its Solution* (London: J. Murray, 1923), 491.

208 **"[S]ome members":** E. Magnello, *The Road to Medical Statistics* (Leiden, Netherlands: Brill, 2002), 111.

208 **"Sir Ronald Ross left behind":** M. E. Gibson, "Sir Ronald Ross and His Contemporaries," *Journal of the Royal Society of Medicine* 71, no. 8 (1978): 611.

208 **His lecture in St. Louis:** E. Nye and M. Gibson, *Ronald Ross: Malariologist and Polymath: A Biography* (Berlin: Springer, 1997), 117.

209 **"Regarding mathematics":** Ross, *Memoirs*, 23–24.

209 **"up to the end":** Ross, *Memoirs*, 49.

210 **"education must be":** Almost identical with "the only true education is self-education," from William Spencer's *Inventional Geometry*—could Ross have read it?

210 **"It was an aesthetic":** Ross, *Memoirs*, 50.

210 **"Nearly all the ideas":** Ross, *Memoirs*, 8.

211 **"We shall end":** This quote, and much of the rest of the material on Ross, Hudson, and the theory of happenings, owes much to Adam Kucharski's *The Rules of Contagion* (New York: Basic Books, 2020).

211 **Her first publication:** Hilda P. Hudson, "Simple Proof of Euclid II, 9 and 10," *Nature* 45 (1891): 189–90.

213 **straightedge and compass:** Hilda P. Hudson, *Ruler & Compasses* (London: Longmans, Green, 1916).

213 **"We can practice":** Hilda P. Hudson, "Mathematics and Eternity," *Mathematical Gazette* 12, no. 174 (1925): 265–70.

214 **"[T]he thoughts of":** Hudson, "Mathematics and Eternity."

215 **by some accounts considered:** Luc Brisson and Salomon Ofman, "The Khora and the Two-Triangle Universe of Plato's *Timaeus*" (preprint, 2020), arXiv:2008.11947, 6.

216 **"Now the best bond":** Plato, *Timaeus*, trans. Donald J. Zeyl (Indianapolis: Hackett Publishing, 2000), 17.

218 **In the spring 1918 wave:** R_0 values are from P. van den Driessche, "Reproduction Numbers of Infectious Disease Models," *Infectious Disease Modelling* 2, no. 3 (Aug. 2017): 288–303.

219 **an outbreak constantly:** These pictures are made by Cosma Shalizi, a wonderfully opinionated statistician and network theorist, and appear in his lecture notes on epidemics. Available at www.stat.cmu.edu/~cshalizi/dm/20/lectures/special/epidemics.html#(16).

220 **77 trillion infections:** M. I. Meltzer, I. Damon, J. W. LeDuc, and J. D. Millar, "Modelling Potential Responses to Smallpox as a Bioterrorist Weapon," *Emerging Infectious Diseases* 7, no. 6 (2001): 959–69.

220 **"Every now and again":** Mike Stobbe, "CDC's Top Modeler Courts Controversy with Disease Estimate," Associated Press, Aug. 1, 2015.

222 **He was a geometer:** Some material in this section is adapted from Jordan Ellenberg, "A Fellow of Infinite Jest," *Wall Street Journal*, Aug. 14, 2015.

223 **"the murder weapon":** István Hargittai, "John Conway—Mathematician of Symmetry and Everything Else," *Mathematical Intelligencer* 23, no. 2 (2001): 8–9.

223 He was a compulsive: R. H. Guy, "John Horton Conway: Mathematical Magus," *Two-Year College Mathematics Journal* 13, no. 5 (Nov. 1982): 290–99.

223 began to create numbers: Donald Knuth, *Surreal Numbers: How Two Ex-Students Turned on to Pure Mathematics and Found Total Happiness* (Boston: Addison-Wesley, 1974). The Knuth book introduces Conway's novel number system, but the connection of these numbers with games comes in Conway's 1976 book *On Numbers and Games.*

223 here are some: This is Figure 1 in John H. Conway, "An Enumeration of Knots and Links, and Some of Their Algebraic Properties," *Computational Problems in Abstract Algebra* (Oxford: Pergamon, 1970), 330.

224 he proved with Cameron Gordon: John H. Conway and C. McA. Gordon, "Knots and Links in Spatial Graphs," *Journal of Graph Theory* 7, no. 4 (1983): 445–53. This paper has several theorems, including the one described in the book, which was also proved by Horst Sachs.

225 As of July 2020: All the statistics in this section are from Dana Mackenzie, "Race, COVID Mortality, and Simpson's Paradox," *Causal Analysis in Theory and Practice* (blog), http://causality.cs.ucla.edu/blog/index.php/2020/07/06/race-covid-mortality-and-simpsons-paradox-by-dana-mackenzie. The numbers as I write this (September 2020) are not the same, but appear to show the same Simpson-paradoxical effect.

226 Which coin has syphilis? This section is adapted from Jordan Ellenberg, "Five People. One Test. This Is How You Get There," *New York Times*, May 7, 2020.

227 a 1941 *New York Times*: Paul de Kruif, "Venereal Disease," *New York Times*, Nov. 23, 1941, 74.

227 But back in 1942: Biographical information about Dorfman and the history of group testing is from "Economist Dies at 85," *Harvard Gazette*, July 18, 2002, and Dingzhu Du and Frank K. Hwang, *Combinatorial Group Testing and Its Applications*, vol. 12 (Singapore: World Scientific, 2000), 1–4.

227 "The Detection": R. Dorfman, "The Detection of Defective Members of Large Populations," *Annals of Mathematical Statistics* 14, no. 4 (Dec. 1943): 436–40.

228 diluting the sample: Du and Hwang, *Combinatorial Group Testing*, 3.

229 in German hospitals: Katrin Bennhold, "A German Exception? Why the Country's Coronavirus Death Rate Is Low," *New York Times*, Apr. 5, 2020.

229 a state lab: Ellenberg, "Five People. One Test. This Is How You Get There."

229 Wuhan, the city: BBC report, June 8, 2000, at www.bbc.com/news/world-asia-china-52651651.

233 "he also had the habit": James Norman Davidson, "William Ogilvy Kermack, 1898–1970," *Biographical Memoirs of Fellows of the Royal Society* 17 (1971), 413–14.

234 the SIR model did: M. Takayasu et al., "Rumor Diffusion and Convergence During the 3.11 Earthquake: A Twitter Case Study," *PLoS ONE* 10, no. 4 (2015): 1–18. M. Cinelli et al., in a 2020 preprint, "The COVID-19 Social Media Infodemic," argue that the spread of information around the outset of the COVID-19 pandemic should be analyzed similarly, and that rumors on Instagram have a measurably higher R_0 than those on Twitter.

236 In Sanskrit poetry: Material on Indian prosody is from Parmanand Singh, "The So-Called Fibonacci Numbers in Ancient and Medieval India," *Historia Mathematica* 12, no. 3 (1985): 229–44.

240 "What did the ancients": Henri Poincaré, "The Present and the Future of Mathematical Physics," trans. J. W. Young, *Bulletin of the American Mathematical Society* 37, no. 1 (1999): 26.

Chapter 11: The Terrible Law of Increase

243 Sometime in the middle: Y. Furuse, A. Suzuki, and H. Oshitani, "Origin of Measles Virus: Divergence from Rinderpest Virus Between the 11th and 12th Centuries, *Virology Journal* 7, no. 52 (2010), doi.org/10.1186/1743-422X-7-52.

243 On May 19, 1865: S. Matthews, "The Cattle Plague in Cheshire, 1865–1866," *Northern History* 38, no. 1 (2001): 107–19, doi.org/10.1179/nhi.2001.38.1.107.

243 By the end of October: A. B. Erickson, "The Cattle Plague in England, 1865–1867," *Agricultural History* 35, no. 2 (Apr. 1961): 97.

244 Farr represented: The story of Snow, Farr, and the cholera epidemic is much told; this account is drawn from N. Paneth et al., "A Rivalry of Foulness: Official and Unofficial Investigations of the London Cholera Epidemic of 1854," *American Journal of Public Health* 88, no. 10 (Oct. 1998): 1545–53.

245 "We will venture to say": British Medical Association, *British Medical Journal* 1, no. 269 (1866): 207.

245 **"a more extensive":** General Register Office, *Second Annual Report of the Registrar-General of Births, Deaths, and Marriages in England* (London: W. Clowes and Sons, 1840), 71.

245 **"suddenly rise":** *Second Annual Report of the Registrar-General*, 91.

246 **"minute insects":** *Second Annual Report of the Registrar-General*, 95.

249 **probably almost 5.4:** The Farr-style "close your eyes and pretend it's an arithmetic progression" method isn't your only option; a similar approach called Newton's method (based, as the same suggests, on calculus) yields an approximation of 5.4 in this case, just about as good as 5 4/11, and does notably better for numbers whose square root is very close to a whole number. To be the Great Square Rootio requires having several arrows in one's quiver.

250 **Around the year 600:** Information on the early history of interpolation is from E. Meijering, "A Chronology of Interpolation: From Ancient Astronomy to Modern Signal and Image Processing," *Proceedings of the IEEE* 90, no. 3 (2002): 319–42.

253 **"an admirable *danseuse*":** Charles Babbage, *Passages from the Life of a Philosopher* (London: Longman, Green, 1864), 17.

253 **"One evening":** Babbage, *Passages from the Life of a Philosopher*, 42.

254 **the same heuristic argument:** That's my best guess, at any rate. Hassett didn't volunteer the exact mechanism that produced his curve beyond labeling it "cubic fit," but fitting a cubic polynomial to the log of the observed numbers, which is what Farr did, gave me a pretty good match for the curve in the press release.

254 **"Project the U.S. line":** Justin Wolfers (@JustinWolfers), Twitter, Mar. 28, 2020, 2:30 p.m.

256 **"He quite forgets":** British Medical Association, *British Medical Journal* 1, no. 269 (1866): 206–07.

258 **the Gateway Arch:** Though it turns out the Gateway Arch, despite being called a catenary by its architect, Eero Saarinen, is actually a "flattened catenary." See R. Osserman, "How the Gateway Arch Got Its Shape," *Nexus Network Journal* 12, no. 2 (2010): 167–89.

259 **His answer:** Robert Plot and Michael Burghers, *The Natural History of Oxford-Shire: Being an Essay Towards the Natural History of England* (Printed at the Theater in Oxford, 1677): 136–39. Biodiversoty Heritage Library, https://www.biodiversitylibrary.org/item/186210#page/11/mode/1up.

259 **"Coronavirus Models":** Zeynep Tufekci, "Don't Believe the COVID-19 Models," *Atlantic*, Apr. 2, 2020. https://www.theatlantic.com/technology/archive/2020/04/coronavirus-models-arent-supposed-be-right/609271.

260 **a photo of a protester:** Photo by Jim Mone/Associated Press, http://journaltimes.com/news/national/photos-protesters-rally-against-coronavirus-restrictions-in-gatherings-across-us/collection_b0cd8847-b8f4-5fe0-b2c3-583fac7ec53a.html#48.

262 **There was an intellectual dustup:** Yarden Katz, "Noam Chomsky on Where Artificial Intelligence Went Wrong," *Atlantic*, Nov. 1, 2012, https://www.theatlantic.com/technology/archive/2012/11/noam-chomsky-on-where-artificial-intelligence-went-wrong/261637/, and the combative but very informative Peter Norvig, "On Chomsky and the Two Cultures of Statistical Learning," available at http://norvig.com/chomsky.html.

Chapter 12: The Smoke in the Leaf

264 **Conway showed:** J. Conway, "The Weird and Wonderful Chemistry of Audioactive Decay," *Eureka* 46 (Jan. 1986).

269–70 **"the first we may compare":** Karl Fink, *A Brief History of Mathematics: An Authorized Translation of Dr. Karl Fink's Geschichte der Elementarmathematik*, 2nd ed., trans. Wooster Woodruff Beman and David Eugene Smith (Chicago: Open Court Publishing, 1903), 223.

270 **"the ancients called":** H. Becker, "An Even Earlier (1717) Usage of the Expression 'Golden Section,' *Historia Mathematica* 49 (Nov. 2019): 82–83.

272 **Zu Chongzhi, a fifth-century:** Information on Zu Chongzhi and milü from L. Lay-Yong and A. Tian-Se, "Circle Measurements in Ancient China," *Historia Mathematica* 13 (1986): 325–40.

277 **"The 'divine proportion'":** From a review of *Geschichter der Elementär-Mathematik in Systematischer Darstellung, Nature* 69, no. 1792 (1904): 409–10; the review is signed just GBM, but Mathews is the obvious member of the Royal Society at the time to have written this. Thanks to Jennifer Nelson for detective work here.

277 **there's no evidence:** Mario Livio's book *The Golden Ratio* (New York: Broadway Books, 2002) is fairly definitive on the long history of claims that canonical works of art are secretly golden in nature.

278 **An influential 1978 paper:** E. I. Levin, "Dental Esthetics and the Golden Proportion," *Journal of Prosthetic Dentistry* 40, no. 3 (1978): 244–52.

278 **false teeth:** Julie J. Rehmeyer, "A Golden Sales Pitch," Math Trek, *Science News*, June 28, 2007. https://www.sciencenews.org/article/golden-sales-pitch.

278 **There was a "Diet Code":** Steven Lanzalotta, *The Diet Code* (New York: Grand Central, 2006). The actual recommendations of the book are not purely golden ratio, but also involve the number 28, which "variously represents the lunar cycle, a yogic age of spiritual unfolding, one of the Egyptian cubit measures, and a fundamental Mayan calculational coordinate." http://www.diet-code.com/f_thecode/right_proportions.htm.

278 **"BREATHTAKING":** Available all over the internet, e.g. at www.goldennumber.net/wp-content/uploads/pepsi-arnell-021109.pdf.

279 **A few years afterward:** Biographical material on Elliott is drawn from the sixty-four-page biography that opens the book *R. N. Elliott's Masterworks: The Definitive Collection*, ed. Robert R. Prechter Jr. (Gainesville, GA: New Classics Library, 1994).

279 **"Man is no less":** R. N. Elliott, *The Wave Principle* (self-published, 1938), 1.

280 **Roger Babson, who believed:** All material on Babson is from Martin Gardner, *Fads and Fallacies in the Name of Science* (Mineola, NY: Dover Publications, 1957), chapter 8. Babson's grudge against gravity seems to have stemmed from a childhood accident in which his sister drowned, as Babson recounts in his essay "Gravity—Our Enemy Number One."

281 **"As with all other":** Merrill Lynch, *A Handbook of the Basics: Market Analysis Technical Handbook* (2007), 48.

281 **warned its readers:** Paul Vigna, "How to Make Sense of This Crazy Market? Look to the Numbers," *Wall Street Journal*, Apr. 13, 2020.

285 **"in a somewhat similar":** J. J. Sylvester, "The Equation to the Secular Inequalities in the Planetary Theory," *Philosophical Magazine* 16, no. 100 (1883): 267.

286 **You can make models:** There are lots of papers on this, but a particularly influential one is the preprint by M. G. M. Gomes et al., "Individual Variation in Susceptibility or Exposure to SARS-CoV-2 Lowers the Herd Immunity Threshold," medarXiv (2020). https://doi.org/10.1101/2020.04.27.20081893.

290 **the limiting probabilities:** Robert B. Ash and Richard L. Bishop, "Monopoly as a Markov Process," *Mathematics Magazine* 45, no. 1 (1972): 26–29. A later paper carries out the same computation and gets slightly different numbers, for reasons I don't fully understand, but agrees that Illinois Avenue is the most frequently visited property. Paul R. Murrell, "The Statistics of Monopoly," *Chance* 12, no. 4 (1999): 36–40.

297 **we stipulated earlier:** I first learned about this kind of argument, not in the context of quantum physics, but from a seminar on noncommutative geometry given by Tom Nevins, a wonderful geometer and award-winning teacher at the University of Illinois who died, aged just 48, on February 1, 2020.

298 **the cochlea:** See, for instance, Robert Fettiplace, "Diverse Mechanisms of Sound Frequency Discrimination in the Vertebrate Cochlea," *Trends in Neurosciences* 43, no. 2 (2020): 88–102.

Chapter 13: A Rumple in Space

299 **they used Markov processes:** David Link, "Chains to the West: Markov's Theory of Connected Events and Its Transmission to Western Europe," *Science in Context* 19, no. 4 (2006): 561–89. The Eggenberger-Pólya paper referred to is Florian Eggenberger and George Pólya, "Über die statistik verketteter vorgänge," *ZAMM—Journal of Applied Mathematics and Mechanics/Zeitschrift für Angewandte Mathematik und Mechanik* 3, no. 4 (1923): 279–89.

302 **slang for an unclassifiable:** Jim Warren, "Feeling Flulike? It's the Epizootic," *Baltimore Sun*, Jan. 17, 1998. See also the entry for "epizootic" in the *Dictionary of American Regional English*.

302 **"At least seven eighths":** A. B. Judson, "History and Course of the Epizoötic Among Horses upon the North American Continent in 1872–73," *Public Health Papers and Reports* 1 (1873): 88–109.

302 **"vast hospital":** Sean Kheraj, "The Great Epizootic of 1872–73: Networks of Animal Disease in North American Urban Environments," *Environmental History* 23, no. 3 (2018): 495–521, doi.org/10.1093/envhis/emy010.

303 **early outbreaks:** Kheraj, "The Great Epizootic," 497.

304 **"an almost impassable":** Judson, "History and Course of the Epizoötic," 108.

305 "To put it into Euclid": See J. H. Webb, "A Straight Line Is the Shortest Distance Between Two Points," *Mathematical Gazette* 58, no. 404 (June 1974): 137–38.

306 Gerardus Mercator: Biographical details on Mercator from Mark Monmonier, *Rhumb Lines and Map Wars: A Social History of the Mercator Projection* (Chicago: University of Chicago Press, 2004), chapter 3.

309 but *it cannot do both*: My daughter/fact-checker insists that with some effort you really can bend the pizza down, so perhaps better to say the U-hold makes it *harder* for the tip of the slice to bend down. For a longer exposition on the pizza theorem, see Atish Bhatia, "How a 19th Century Math Genius Taught Us the Best Way to Hold a Pizza Slice," *Wired*, Sep. 5, 2014.

311 "I have something": S. Krantz, review of Paul Hoffman's *The Man Who Loved Only Numbers*, *College Mathematics Journal* 32, no. 3 (May 2001): 232–37.

312 And the distance between: All these distances are computed with the Collaboration Distance tool, provided by the American Mathematical Society and available at http://mathscinet.ams.org/mathscinet/freeTools.html.

314 once remarked, while visiting: Melvin Henriksen, "Reminiscences of Paul Erdös (1913–1996)," *Humanistic Mathematics Network Journal* 1, no. 15 (1997): 7.

314 "he could not find words": Henri Poincaré, *The Value of Science*, trans. George Bruce Halsted (New York: The Science Press, 1907), 138.

315 in a movie with everyone: Brandon Griggs, "Kevin Bacon on 'Six Degrees' Game: 'I Was Horrified,'" CNN, Mar. 12, 2014. https://www.cnn.com/2014/03/08/tech/web/kevin-bacon-six-degrees-sxsw/index.html.

318 "a quickening and suggestive": J. J. Sylvester, "On an Application of the New Atomic Theory to the Graphical Representation of the Invariants and Covariants of Binary Quantics, with Three Appendices [Continued]," *American Journal of Mathematics* 1, no. 2 (1878): 109.

318 "In poetry and algebra": Sylvester, "On an Application."

319 the phrase "graphic notation": Material on the origin of the term "graph" is from N. Biggs, E. Lloyd, and R. Wilson, *Graph Theory 1736–1936* (Oxford: Oxford University Press, 1999), 64–67.

319 "a giant gnome": The speaker is Sylvester's first PhD student, George Bruce Halsted, quoted on p. 137 of E. E. Slosson, *Major Prophets of To-Day* (New York: Little, Brown, 1914). Halsted seems to have inherited his advisor's taste for dispute, getting fired from a string of universities for criticizing the administration and winding up working as an electrician in the family store while continuing to publish in non-Euclidean geometry.

319 "It was a treat": Letter from F. Galton to K. Pearson, Dec. 31, 1901, in *The Life, Letters, and Labours of Francis Galton*, vol. 1, ed. K. Pearson (Cambridge: Cambridge University Press, 1924).

319 "demonstrate a thorough": D. R. McCoy, "An "Old-Fashioned" Nationalism: Lincoln, Jefferson, and the Classical Tradition," *Journal of the Abraham Lincoln Association* 23, no. 1 (Winter 2002): 60. The other two requirements were to be able to read classical authors and to translate English into Latin.

319 At Yale, in 1830: Clarence Deming, "Yale Wars of the Conic Sections," *The Independent . . . Devoted to the Consideration of Politics, Social and Economic Tendencies, History, Literature, and the Arts (1848–1921)* 56, no. 2886 (Mar. 24, 1904): 667.

320 In 1840, student rioters: Lewis Samuel Feuer, *America's First Jewish Professor: James Joseph Sylvester at the University of Virginia* (Cincinnati: American Jewish Archives, 1984), 174–76.

321 He went back to England: Biographical material on Sylvester is from Karen. H. Parshall, *James Joseph Sylvester: Jewish Mathematician in a Victorian World* (Baltimore: Johns Hopkins University Press, 2006), 66–80. The precise circumstances of Sylvester's abrupt departure from Virginia are a matter of some dispute; did Sylvester resign because of the feud with Ballard, or the sword-cane incident? Feuer, *America's First Jewish Professor*, argues for the latter.

321 He applied for the Gresham: Alexander Macfarlane, "James Joseph Sylvester (1814–1897)," *Lectures on Ten British Mathematicians of the Nineteenth Century* (New York: John Wiley & Sons, 1916), 109. https://projecteuclid.org/euclid.chmm/1428680549.

322 "undergoing in the space": James Joseph Sylvester, "Inaugural Presidential Address to the Mathematical and Physical Section of the British Association," reprinted in *The Laws of Verse: Or Principles of Versification Exemplified in Metrical Translations* (London: Longmans, Green, 1870), 113.

322 "An eloquent mathematician": James Joseph Sylvester, "Address on Commemoration Day at Johns Hopkins University," Feb. 22, 1877. Collected in *The Collected Mathematical Papers of James Joseph Sylvester: Volume III* (Cambridge: Cambridge University Press, 1909), 72–73.

323 "The early study": James Joseph Sylvester, "Mathematics and Physics," *Report of the Meeting of the British Association for the Advancement of Science* (London: J. Murray, 1870), 8.

323 despite geography: Sylvester, "Address on Commemoration Day," 81.

323 "I recently paid a visit": James Joseph Sylvester, *The Collected Mathematical Papers of James Joseph Sylvester,* vol. 4 (London: Chelsea Publishing, 1973), 280.

324 Also present at the dinner: Poincaré's remarks at the November 30, 1901, dinner of the Royal Society, and Ross's presence there, are from *The Times,* Dec. 2, 1901, p. 13, a reference I obtained from G. Cantor, "Creating the Royal Society's Sylvester Medal," *British Journal for the History of Science* 37, no. 1 (Mar. 2004): 75–92. *The Times* refers only to "Major Ross" as being present, but Ronald Ross had been elected a Fellow of the Royal Society that year, would later be its vice president, and is referred to as "Major Ross" in other contemporary documents, so I'm pretty confident it's our mosquito man.

324 He got so good: All biographical info on Jordan and the mind-reading trick is from Persi Diaconis and Ron Graham, *Magical Mathematics* (Princeton: Princeton University Press, 2015), 190–91.

326 This number is usually denoted: Florian Cajori, "History of Symbols for N Factorial," *Isis* 3, no. 3 (1921): 416.

330 One early justification: Oscar B. Sheynin, "H. Poincaré's Work on Probability," *Archive for History of Exact Sciences* 42, no. 2 (1991): 159–60.

331 seven shuffles meet: The keyword for the measure of mixed-upness used here is "total variation distance from the uniform distribution."

331 "Spontaneous Knotting of an Agitated String": Dorian M. Raymer and Douglas E. Smith, "Spontaneous Knotting of an Agitated String," *Proceedings of the National Academy of Sciences* 104, no. 42 (2007): 16432–37.

332 "Physical law would then": Poincaré, *The Value of Science,* 110–11.

332 If you choose four: I am here paraphrasing a recent theorem of Harald Helfgott, Ákos Seress, and Andrzej Zuk ("Random generators of the symmetric group: diameter, mixing time and spectral gap," *Journal of Algebra* 421 [2015]: 349–68) and not exactly accurately, either, but the right idea is conveyed, I think.

333 The Rubik's cube: Clay Dillow, "God's Number Revealed: 20 Moves Proven Enough to Solve Any Rubik's Cube Position," *Popular Science,* Aug. 10, 2010.

334 a follow-up study in 1970: C. Korte and S. Milgram, "Acquaintance Networks Between Racial Groups: Application of the Small World Method," *Journal of Personality and Social Psychology* 15, no. 2 (1970): 101–08.

334 "I'm technically only": Althea Legaspi, "Kevin Bacon Advocates for Social Distancing with 'Six Degrees' Initiative," *Rolling Stone,* Mar. 18. 2020, www.rollingstone.com/movies/movie-news/kevin-bacon-social-distancing-six-degrees-initiative-969516.

335 Facebook is a small world: Information about the Facebook graph is from Lars Backstrom et al., "Four Degrees of Separation," *Proceedings of the 4th Annual ACM Web Science Conference* (June 22–24, 2012): 33–42, and Johan Ugander et al., "The Anatomy of the Facebook Social Graph" (preprint, 2011), https://arxiv.org/abs/1111.4503.

335 And getting smaller: Described on the Facebook research blog, research.fb.com/blog/2016/02/three-and-a-half-degrees-of-separation.

336 A large-scale analysis: Ugander et al., "The Anatomy of the Facebook Social Graph." The so-called "Friendship paradox" was first described in Scott L. Feld, "Why Your Friends Have More Friends Than You Do," *American Journal of Sociology* 96, no. 6 (1991): 1464–77.

337 Duncan Watts and Steven Strogatz: Duncan J. Watts and Steven H. Strogatz, "Collective Dynamics of 'Small-World' Networks," *Nature* 393, no. 6684 (1998): 440–42.

337 Stanley Milgram is the face: See Judith S. Kleinfeld, "The Small World Problem," *Society* 39, no. 2 (2002): 61–66, for an informative depiction, using extensive research in Milgram's archives, of the gap between Milgram's scientific findings and the way he presented them in the popular press.

338 smallness of the networked world: For the history of research on small-world networks I am indebted to Duncan Watts, *Small Worlds: The Dynamics of Networks Between Order and Randomness* (Princeton: Princeton University Press, 2003), and Albert-László Barabási, Mark Newman, and Duncan Watts, *The Structure and Dynamics of Networks* (Princeton: Princeton University Press, 2006).

338 studying "chain-relations": Jacob L. Moreno and Helen H. Jennings, "Statistics of Social Configurations," *Sociometry* 1, no. 3/4 (1938): 342–74.

338 **"Chain-links" ("*Láncszemek*"):** The translation used here is by Adam Makkai and appears in Barabási, Newman, and Watts, *The Structure and Dynamics of Networks*, 21–26.

Chapter 14: How Math Broke Democracy (And Might Still Save It)

341 **Only two districts:** Districts 49 and 51. I am grateful to John Johnson of Marquette University for the data underlying this fact and the scatter plot above.

343 **"If you took Madison":** Molly Beck, "A Blue Wave Hit Statewide Races, but Did Wisconsin GOP Gerrymandering Limit Dem Legislative Inroads?," *Milwaukee Journal Sentinel*, Nov. 8, 2018.

343 **a national effort:** As documented in Dave Daley's book *Ratf**ked* (New York: Liveright, 2016), or, if you prefer the horse's mouth, Karl Rove, "The GOP Targets State Legislatures: He Who Controls Redistricting Can Control Congress," *Wall Street Journal*, Mar. 4, 2010.

344 **But the race made:** Biographical information about and quotes from Joe Handrick are from R. Keith Gaddie, *Born to Run: Origins of the Political Career* (Lanham, MD: Rowman & Littlefield, 2003), 43–55.

345 **They called it "Joe Aggressive":** Information on the "Joe Aggressive" map is from pp. 14–15 of the Nov. 21, 2016, decision in *Whitford v. Gill*, available at www.scotusblog.com/wp-content/uploads/2017/04/16-1161-op-bel-dist-ct-wisc.pdf. To be precise, "Joe Aggressive" was one of several very similar maps, all of which were in turn very similar if not exactly identical to the map enacted by Act 43; see footnotes 56 and 57 of the *Whitford v. Gill* decision.

345 **a majority of districts:** The state of Wisconsin makes ward-by-ward breakdowns of past election results publicly available; so if you see a number like this without an outside citation, it means either I or my hardworking son/data assistant dug our hands into the spreadsheets and worked it out ourselves.

346 **"unredeemable flaws":** *Baumgart v. Wendelberger*, case nos. 01-C-0121, 02-C-0366 (E.D. Wis., May 30, 2002), 6.

347 **"The maps we pass":** Matthew DeFour, "Democrats' Short-Lived 2012 Recall Victory Led to Key Evidence in Partisan Gerrymandering Case," *Wisconsin State Journal*, July 23, 2017.

349 **In New Zealand:** The material on districting systems around the world, along with occasional other material in this chapter, is adapted from J. Ellenberg, "Gerrymandering, Inference, Complexity, and Democracy," *Bulletin of the American Mathematical Society* 58, no. 1 (2021), 57–77.

350 **The town of Dunwich:** C. Lynch, "The Lost East Anglian City of Dunwich Is a Reminder of the Destruction Climate Change Can Wreak," *New Statesman*, Oct. 2, 2019.

351 **"The representatives":** E. Burke, "Speech on the Plan for Economical Reform," February 11, 1780. Reprinted in *Selected Prose of Edmund Burke*, ed. Sir Philip Magnus (London: The Falcon Press, 1948), 41–44.

351 **"a government is republican":** "Proposals to Revise the Virginia Constitution: I. Thomas Jefferson to 'Henry Tompkinson'(Samuel Kercheval), 12 July 1816," Founders Online, National Archives. https://founders.archives.gov/documents/Jefferson/03-10-02-0128-0002.

351 **Well into the twentieth century:** I. L. Smith, "Some Suggested Changes in the Constitution of Maryland," July 4, 1907, published in the *Report of the Twelfth Annual Meeting of the Maryland State Bar Association* (1907), 175.

351 **"Will some one explain":** Smith, "Some Suggested Changes," 181.

352 **"the larger, dense-populated":** Oral argument, *Reynolds v. Sims*. I don't know whether transcripts of the oral arguments are available; I transcribed this quote from the recording, and to get the full effect you really have to hear the argument presented in its full Southern dudgeon. It can be accessed at https://www.oyez.org/cases/1963/23.

356 **The civic-minded Guamanians:** A. Balsamo-Gallina and A. Hall, "Guam's Voters Tend to Predict the Presidency—but They Have No Say in the Electoral College," Public Radio International, *The World*, Nov. 8, 2016. Accessed at https://www.pri.org/stories/2016-11-08/presidential-votes-are-guam-they-wont-count.

359 **in Wisconsin and many:** Opinion of Robert Warren, attorney general of Wisconsin, 58 OAG 88 (1969).

360 **across the county line:** Information about the creation of these wards and their gerrymanderish properties is from Malia Jones, "Packing, Cracking and the Art of Gerrymandering Around Milwaukee," WisContext, June 8, 2018, www.wiscontext.org/packing-cracking-and-art-gerrymandering-around-milwaukee.

362 "There was once a time": *Baldus v. Members of the Wis. Gov't Accountability Bd.*, 843 F. Supp. 2d 955 (E.D. Wis. 2012).

363 1907 PhD dissertation: E. C. Griffith, *The Rise and Development of the Gerrymander* (Chicago: Scott, Foresman, 1907).

363 the practice dates back: Griffith, *The Rise and Development of the Gerrymander*, 26–27.

363 Madison prevailed: The Henry maybe-gerrymander is treated in Griffith, *The Rise and Development of the Gerrymander*, 31–42, and in more modern terms in T. R. Hunter, "The First Gerrymander? Patrick Henry, James Madison, James Monroe, and Virginia's 1788 Congressional Districting," *Early American Studies* 9, no. 3 (Fall 2011): 781–820.

364 "A minority rule is established": United States Department of State, *Papers Relating to the Foreign Relations of the United States* (Washington, D.C.: Government Printing Office, 1872), xxvii.

365 The proportion of "floating voters": C. D. Smidt, "Polarization and the Decline of the American Floating Voter," *American Journal of Political Science* 61, no. 2 (April 2017): 365–81.

367 Goofy's neck: Trip Gabriel, "In a Comically Drawn Pennsylvania District, the Voters Are Not Amused," *New York Times*, Jan. 26, 2018.

369 It didn't get a proof: A thorough account of the history can be found in V. Blasjo, "The Isoperimetric Problem," *American Mathematical Monthly* 112, no. 6 (June–July 2005): 526–66.

371 Those gerrymandered districts: I haven't computed compactness scores for the Act 43 Wisconsin districts, but the challenges to the map in court notably do *not* attack the Act 43 districts on compactness grounds, so I feel safe in assuming they're fine. They *look* fine, at any rate.

372 "seems to have benefited": P. Bump, "The Several Layers of Republican Power-Grabbing in Wisconsin," *Washington Post*, Dec. 4, 2018.

373 there has never been a U.S. representative: Though Justin Amash of Michigan, elected as a Republican, left the party and became a registered Libertarian while serving. He declined to run for reelection after changing parties.

374 Canada's very first: Anthony J. Gaughan, "To End Gerrymandering: The Canadian Model for Reforming the Congressional Redistricting Process in the United States," *Capital University Law Review* 41, no. 4 (2013): 1050.

374 "Therefore let us re-distribute": R. MacGregor Dawson, "The Gerrymander of 1882," *Canadian Journal of Economics and Political Science/Revue Canadienne D'Economique et De Science Politique* 1, no. 2 (1935): 197.

374 "What they were arguing: "The Full Transcript of ALEC's 'How to Survive Redistricting' Meeting," *Slate*, Oct. 2, 2019, https://slate.com/news-and-politics/2019/10/full-transcript-alec-gerrymandering-summit.html.

379 as in our Crayola examples: And what if the turnout *isn't* the same in every district? The relationship between efficiency gap and popular vote in this more general situation is worked out in Ellen Veomett, "Efficiency Gap, Voter Turnout, and the Efficiency Principle," *Election Law Journal: Rules, Politics, and Policy* 17, no. 4 (2018): 249–63.

380 an amicus brief to: Brief for the State of Wisconsin as Amicus Curiae, *Benisek v. Lamone*, 585 U.S. (2018).

384 The legal battles over gerrymandering: Some of the material about *Rucho v. Common Cause* is adapted from J. Ellenberg, "The Supreme Court's Math Problem," *Slate*, March 29, 2019, https://slate.com/news-and-politics/2019/03/scotus-gerrymandering-case-mathematicians-brief-elena-kagan.html.

384 "Now let me make it": Ovetta Wiggins, "Battles Continue in Annapolis over the Use of Bail and Redistricting," *Washington Post*, March 21, 2017.

384 Maryland's unequally sized: In the case of *Maryland Committee for Fair Representation v. Tawes*, 377 U.S. 656 (1964).

385 Ents, the sentient trees: J. R. R. Tolkien, *The Two Towers* (London: George Allen & Unwin, 1954), book 3, ch. 4.

388 the colossal quantity of ways: The number 706,152,947,468,301 was computed by Bob Harris in his 2010 preprint "Counting Nonomino Tilings and Other Things of That Ilk."

389 This particular computer: Gregory Herschlag, Robert Ravier, and Jonathan C. Mattingly, "Evaluating Partisan Gerrymandering in Wisconsin" (preprint, 2017), arXiv:1709.01596.

390 Six years later: The analyses in Herschlag et al., "Evaluating Partisan Gerrymandering in Wisconsin," only go up to the 2016 election, but Greg Herschlag was kind enough to run a similar trial for me on the 2018 governor's race.

390 getting almost 52%: More precisely, 52% is the share of the statewide vote Republicans *would* have gotten, Herschlag and company estimate, had there been a contested race in every assembly district.

390 in the 2014 election: Herschlag et al., "Evaluating Partisan Gerrymandering in Wisconsin," data summarized in figure 3, p. 3.

391 As a Wisconsin voter: Material on this page is adapted from Jordan Ellenberg, "How Computers Turned Gerrymandering into a Science," *New York Times*, Oct. 6, 2017.

392 the *ReCom* geometry: Daryl DeFord, Moon Duchin, and Justin Solomon, "Recombination: A Family of Markov Chains for Redistricting" (preprint, 2019), https://arxiv.org/abs/1911.05725.

395 One of his very first papers: John C. Urschel, "Nodal Decompositions of Graphs," *Linear Algebra and Its Applications* 539 (2018): 60–71. I interviewed John and wrote about his extremely interesting and yet also weirdly typical path into mathematics in the online magazine *Hmm Daily* ("John Urschel Goes Pro," Sep. 28, 2018). Available at https://hmmdaily.com/2018/09/28/john-urschel-goes-pro.

396 In this picture: Taken from Russ Lyons's web page at http://pages.iu.edu/~rdlyons/maze/maze-bostock.html; Russ made this using an implementation of Wilson's algorithm by Mike Bostock.

398 Manna, Dhar, and Majumdar from 1992: Subhrangshu S. Manna, Deepak Dhar, and Satya N. Majumdar, "Spanning Trees in Two Dimensions," *Physical Review* A 46, no. 8 (1992): R4471–R4474.

398 proved in 2017: Melanie Matchett Wood, "The Distribution of Sandpile Groups of Random Graphs," *Journal of the American Mathematical Society* 30, no. 4 (2017): 915–58.

398 when a sequence of spanning: Alexander E. Holroyd et al., "Chip-Firing and Rotor-Routing on Directed Graphs," *In and Out of Equilibrium 2*, eds. Vladas Sidoravicius and Maria Eulália Vares (Basel, Switzerland: Birkhäuser, 2008), 331–64.

401 "to create as many": The Hofeller testimony is quoted in the majority decision by Judge James Wynn in *Rucho v. Common Cause*, 318 F. Supp. 3d 777, 799 (M.D.N.C., 2018), 803.

401 What they were asking: Brief for Common Cause Appellees, *Rucho v. Common Cause.*

402 nine Democrats and zero Republicans: M. Duchin et al., "Locating the Representational Baseline: Republicans in Massachusetts," *Election Law Journal: Rules, Politics, and Policy* 18, no. 4 (2019): 388–401.

408 The House has passed a bill: H.R. 1, 116th Congress, "For the People Act of 2019," especially title II, subtitle E.

408 Fifty-five out of: Editorial, "11 More Wisconsin Counties Should Vote 'Yes' to End Gerrymandering," *Wisconsin State Journal*, Sep. 12, 2020.

408 its Fourteenth Amendment power: G. Michael Parsons, "The Peril and Promise of Redistricting Reform in H.R. 1," *Harvard Law Review* Blog, Feb. 2, 2021, https://blog.harvardlawreview.org/the-peril-and-promise-of-redistricting-reform-in-h-r-1/; also Peter Kallis, "The Boerne-Rucho Conundrum: Nonjusticiability, Section 5, and Partisan Gerrymandering," 15, *Harvard Law and Policy Review* (forthcoming) which argues that the decision in Rucho can be read to give the U.S. Congress the power to oversee state districting as well.

409 the certification that paid: Michael R. Wickline, "3 Ballot Petitions in State Ruled Insufficient," *Arkansas Democrat-Gazette*, July 15, 2020. For the number of signatures gathered, John Lynch, "Backers of Change in Arkansas' Vote Districting Sue in U.S. Court," *Arkansas Democrat-Gazette*, Sep. 3, 2020. The referendum did not appear on the Arkansas November ballot.

Conclusion: I Prove a Theorem and the House Expands

411 formal gardens whose perfect lines: The quote from Baker and this interpretation of French formal gardens are from Amir Alexander's book *Proof! How the World Became Geometrical* (New York: Scientific American/Farrar, Straus and Giroux, 2019). Worth noting: while Baker sees strict geometric construction as an assertion of British colonial authority, an earlier Englishman, Anthony Trollope, saw the rectilinear layouts of nineteenth-century U.S. cities as strikingly un-British, remarking on the "parallelogramic fever" of Philadelphia and upper Manhattan.

412 "It's a very puzzling": *New York Times*, Feb. 23, 1885.

413 He served on the council: Edwin Abbott Abbott, William Lindgren, and Thomas Banchoff, *Flatland: An Edition with Notes and Commentary* (Cambridge: Cambridge University Press, 2010), 262.

413 In seventeenth-century Italy: For this story, see the first part of Amir Alexander, *Infinitesimal* (New York: Scientific American/Farrar, Straus and Giroux, 2014).

414 the first Black research chemist: "Comprehensive Biography of Rita Dove," University of Virginia, http://people.virginia.edu/~rfd4b/compbio.html.

414 "My brother and I": "A Chorus of Voices: An Interview with Rita Dove," *Agni* 54 (2001), 175.

415 "you also realize": "A Chorus of Voices," 175.

416 "What I have just said": Henri Poincaré, "The Future of Mathematics" (1908), trans. F. Maitland, appearing in *Science and Method* (Mineola, NY: Dover Publications, 2003), 32.

418 Perelman himself: Luke Harding, "Grigory Perelman, the Maths Genius Who Said No to $1m," *Guardian*, Mar. 23, 2010.

419 "We are not trying": William P. Thurston, "On Proof and Progress in Mathematics," *Bulletin of the American Mathematical Society* 30, no. 2 (1994): 161–77.

419 "Basically, I'm not interested": William Grimes, "David Blackwell, Scholar of Probability, Dies at 91," *New York Times*, July 17, 2010.

419 "I still remember the concept": Donald J. Albers and Gerald L. Alexanderson, *Mathematical People: Profiles and Interviews* (Boca Raton, FL: CRC Press, 2008), 15.

420 Famous Talmudic story: Bava Metzia 59a-b. See D. Luban, "The Coiled Serpent of Argument: Reason, Authority, and Law in a Talmudic Tale," *Chicago-Kent Law Review* 79, no. 3 (2004), https://scholarship.kentlaw.iit.edu/cklawreview/vol79/iss3/33, for commentary on this story and its relevance to contemporary legal thinking. The moment where a proof bends the walls of the building is echoed in Dove's "Geometry"—coincidence?

421 "He was remorseless": William Henry Herndon and Jesse William Weik, *Herndon's Lincoln*, eds. Douglas L. Wilson and Rodney O. Davis (Champaign: University of Illinois Press, 2006), 354.

421 "I also like Eliezer": Wiesel quoted in "Wiesel: 'Art of Listening' Means Understanding Others' Views," *Daily Free Press*, Nov. 15, 2011, https://dailyfreepress.com/2011/11/15/wiesel-art-of-listening-means-understanding-others-views.

421 "[I]n those critical years": From Alexander Grothendieck, *Récolltes et Semailles*, trans. Roy Lisker, available in *Ferment Magazine* at https://www.fermentmagazine.org/rands/promenade2.html.

Index